K. Jung H. Mattenheimer
U. Burchardt (Eds.)

Urinary Enzymes

in Clinical and Experimental Medicine

With 71 Figures and 41 Tables

Springer-Verlag
Berlin Heidelberg New York
London Paris Tokyo
Hong Kong Barcelona Budapest

Docent Dr. sc. med. Klaus Jung
Department of Experimental Organ Transplantation
University Hospital Charité
Humboldt University Berlin
Landsberger Allee 49, O-1017 Berlin, FRG

Prof. (emer.) Hermann Mattenheimer M.D., D.Sc.
Department of Biochemistry
Rush-Presbyterian-St. Luke's Medical Center
1653 W. Congress Parkway, Chicago, Ill. 60612, USA

Prof. Dr. sc. med. Ulf Burchardt
Department of Medicine
District Hospital Frankfurt
Müllroser Chaussee 7, O-1200 Frankfurt (Oder), FRG

ISBN-13: 978-3-642-84315-0 e-ISBN-13: 978-3-642-84313-6
DOI: 10.1007/978-3-642-84313-6

Library of Congress Cataloging-in-Publication Data
Urinary enzymes in clinical and experimental medicine / K. Jung, H. Mattenheimer, U. Burchardt, (eds.).
Includes index.

ISBN 978-3-642-84315-0

1. Clinical enzymology. 2. Urine—Analysis. I. Jung, K. II. Mattenheimer, Hermann. III. Burchardt, U. [DNLM: 1. Urologic Diseases—diagnosis. 2. Urologic Diseases—enzymology. 3. Urologic Diseases—pathophysiology. QY 185 U765] RB48.U75 1992 616.6'07566—dc20 DNLM/DLC for Library of Congress
91-5221

Softcover reprint of the hardcover 1st edition 1992

Typesetting: Macmillan India Ltd., Bangalore-25

27/3020-543210 — Printed on acid-free paper

Preface

Interest and research in urinary enzymology were incited about three decades ago by reports that urinary enzymes are elevated in diseases of the kidney and urinary tract. Of the more than 40 hydrolases, oxidoreductases, transferases, and lyases identified in human and animal urine, only ten or so are being used as diagnostic indicators. Recognition of the quantitative distribution of enzymes in the various anatomical and functional parts of the nephron and advances in our understanding of the handling of proteins by the kidney have made it possible to associate urinary enzyme activity patterns with physiological and pathophysiological functions of the nephron.

Confidence in the diagnostic value of urinary enzymes is not unanimous among clinicians and among scientists. The main reason for the difference in opinion may well be that the variability in data exceeds the variability one is accustomed to in the diagnostic enzymology of blood plasma enzymes.

In contrast to plasma enzymes, which are protected by an "enzyme friendly" milieu, enzymes released into the urine encounter an "enzyme hostile" environment: no or little protective protein, variable pH, variable volume, variable metabolite and salt concentrations, variable concentrations of enzyme inhibitors. Through advances in methodology some of these factors can now be controlled; standardization of urine collection periods and preanalytical treatment are as important as optimization of assay methods.

This book is the first comprehensive text on urinary enzymology and gives an up-to-date account of the accomplishments in this field. The recognition that urinary enzymes are valuable markers for the nephrotoxic side effects of drugs has broadened the application of urine analysis. The contributing authors are all actively engaged in basic and/or clinical research in nephrology and renal and urinary enzymology; nonetheless, different authors do not attribute the same diagnostic value to enzyme measurements in urine.

We wish to acknowledge the helpful cooperation of Dr. Wieczorek and his editorial staff of the Springer-Verlag.

Berlin (FRG) — K. Jung
Chicago (USA) — H. Mattenheimer
Frankfurt/Oder (FRG) — U. Burchardt
in May 1991

List of Contributors

Bach, Peter H.
Polytechnic of East London, School of Science, Romford Road, London E15 4LZ, United Kingdom

Bertsch, Thomas
Forsthausstraße 2, W-6830 Schwetzingen, FRG

Bhargava, Akhil S.
Schering AG, Department of Experimental Toxicology Research Laboratories, Müllerstr. 170/178, W-1000 Berlin 65, FRG

Bleuel, Helga
Boehringer Mannheim GmbH, Sandhofer Str. 116, W-6800 Mannheim 31, FRG

Bode, Gerd
Boehringer Mannheim GmbH, Sandhofer Str. 116, W-6800 Mannheim 31, FRG

Bomhard, Ernst Michael
Bayer AG, Institute of Industrial Toxicology, Friedrich-Ebert-Str. 217-333, W-5600 Wuppertal 1, FRG

Burchardt, Ulf
Department of Internal Medicine, District Hospital Frankfurt/Oder, Müllroser Chaussee 7, O-1200 Frankfurt/Oder, FRG

Fuchs, Eberhard
German Primate Center, Kellnerweg 4, W-3400 Göttingen, FRG

Grötsch, Horst
Hoechst AG, Pharma Research, Postfach 80 03 20, W-6230 Frankfurt am Main 80, FRG

Guder, Walter G.
Institute for Clinical Chemistry, Bogenhausen Community Hospital, Englschalkinger Str. 77, W-8000 Munich 81, FRG

Günzel, Peter
Schering AG, Department of Experimental Toxicology, Research Laboratories, Müllerstr. 170/178, W-1000 Berlin 65, FRG

Hofmeister, Rainer
Schering AG, Department of Experimental Toxicology, Research Laboratories, Müllerstr. 170/178, W-1000 Berlin 65, FRG

Hropot, Max
Hoechst AG, Pharma Research, Postfach 80 03 20,
W-6230 Frankfurt am Main 80, FRG

Jung, Klaus
Department of Experimental Organ Transplantation, University Hospital Charité, Humboldt University Berlin, Landsberger Allee 49, O-1017 Berlin, FRG

Kotanko, Peter
Hospital Barmherzige Brüder, Department of Internal Medicine,
Marschallgasse 12, A-8020 Graz, Austria

Maruhn, Dieter
Bayer AG, Pharma Research Center, Section Medicine and Development, Aprather Weg, Postfach 10 17 09, W-5600 Wuppertal 1, FRG

Mattenheimer, Hermann
32 Forestway Drive, Deerfield, Illinois 60015, USA

Price, Robert G.
Biochemistry Section, Division of Biomolecular Sciences, King's College London, University of London, Campden Hill Road, London W8 7AH, United Kingdom

Rebel, Wolfgang
Boehringer Mannheim GmbH, Sandhofer Str. 116,
W-6800 Mannheim 31, FRG

Scherberich, Jürgen E.
Department of Nephrology, Center of Internal Medicine, Johann-Wolfgang-Goethe-University, Theodor-Stern-Kai 7, W-6000 Frankfurt am Main 70, FRG

Schmid, Heide
Institute of Pathology, University of Tübingen, Liebermeisterstr. 8,
W-7400 Tübingen, FRG

Simane, Zdenek J.
E. Merck
Clinical Research, Frankfurter Str. 250, W-6100 Darmstadt 1, FRG

Slabke, Peter
Schering AG, Department of Experimental Toxicology, Research Laboratories, Müllerstr. 170/178, W-1000 Berlin 65, FRG

Whiting, Paul H.
University of Aberdeen, Department of Clinical Biochemistry, Medical Buildings, Foresterhill, Aberdeen AB9 2ZD, Scotland, United Kingdom

Wilks, Martin F.
Nephrotoxicity Research Group, The Robens Institute of Health and Safety, University of Surrey, Guildford, Surrey GUZ 5XH, United Kingdom

Contents

List of Enzymes Appearing in this Text

Enzyme	EC number	Abbreviation
acetoacetyl-CoA transferase	2.8.3.5	
N-acetyl-β-D-glucosaminidase	3.2.1.30	NAG
adenosine deaminase	3.5.4.4	
adenylate cyclase	4.6.1.1	
adenylate kinase	2.7.4.3	AK
alanine aminopeptidase (microsomal)	3.4.11.2	AAP
alanine aminotransferase	2.6.1.2	ALAT
aldose reductase	1.1.1.21	
D-amino acid oxidase	1.4.3.3	
aminopeptidase A (angiotensinase A)	3.4.11.7	ATA
α-amylase	3.2.1.1	
angiotensin I converting enzyme	2.4.15.1	ACE
arylsulfatase	3.1.6.1	ASA, ARS
aspartate amininotransferase	2.6.1.1	ASAT, AST
ATPases	3.6.1.3	
catalase	1.11.1.6	
cathepsin B	3.4.22.1	
cathepsin L	3.4.22.15	
citrate synthase	4.1.3.7	
collagenase	3.4.24.3	
cysteine conjugate β-lyase =		
(glutamine transaminase K)	4.4.1.13	
deoxyribonuclease I	3.1.4.5	
dipeptidylpeptidase IV	3.4.14.5	
elastase 1	3.4.21.11	
α-fucosidase	3.2.1.51	
fructose-1,6-biphosphatase	3.1.3.11	FBP
α-galactosidase	3.2.1.22	
β-galactosidase	3.2.1.23	GAL
β-glucosidase	3.2.1.21	GLU, GLUC
β-glucuronidase	3.2.1.31	GRS
α-glucosidase	3.2.1.20	
glutamate dehydrogenase	1.4.1.2	GLDH

Enzyme	EC number	Abbreviation
glutaminase, phosphate dependent	3.5.1.2	
γ-glutamyltransferase	2.3.2.2	GGT
glutathione-S-transferase	2.5.1.18	GST
glycylproline-diaminopeptidase	3.4.14.4	GP-DAP
hexokinase	2.7.1.1	
hexosaminidase	3.2.1.30	NAG
isocitrate dehydrogenase (NAD)	1.1.1.41	ICDH
kallikrein	3.4.21.35	
lactate dehydrogenase	1.1.1.27	LDH
leucine aminopeptidase, cytosolic	3.4.11.1	
leucine aminopeptidase, microsomal	3.4.11.2	LAP
leucine arylamidase, microsomal	3.4.11.2	LAS
lipase	3.1.1.3	
lysozyme	3.2.1.17	
malate dehydrogenase	1.1.1.37	MDH
maltase, acid (= α-glucosidase)	3.2.1.20	
α-mannosidase	3.2.1.24	
α-neuraminidase	3.2.1.18	
5′-nucleotidase	3.1.3.5	
nucleotide pyrophosphatase	3.6.1.9	
3-oxoacid CoA-transferase	2.8.3.5	
phosphatase, acid	3.1.3.2	ACP
phosphatase, alkaline	3.1.3.1	AP, ALP
phosphohexose isomerase	5.3.1.9	PHI
phospholipase	3.1.1.4 & 3.1.1.32	
prostaglandin synthase	1.14.99.1	
purine nucleoside phosphorylase	2.4.2.1	
pyruvate kinase	2.7.1.40	PK
ribonuclease	3.1.4.22	RNAase
sphingomyelinase	3.1.4.12	
succinate dehydrogenase	1.3.1.6	
trehalase	3.2.1.2	

Part 1

Biochemical, Pathobiochemical, and Morphological Fundamentals of the Excretion of Urinary Enzymes

1 Origin of Enzymes in Urine

H. Mattenheimer and U. Burchardt

Although it is now undisputed that renal tissue is the main source of most enzymes in the urine, some other potential sources should be considered: blood plasma, blood cells, bacteria, epithelium of the urinary tract, and seminal fluid. A classification of diagnostically relevant urinary enzymes according to their origin and handling by the kidney is given in Table 1.

1.1 Blood Plasma

Plasma enzymes are handled by the kidney in the same way as other proteins. Contribution of plasma enzymes to physiological and pathological enzymuria may, therefore, be described in the same terms used for proteinuria.

Proteinuria is grossly classified into high molecular mass (HMM) and low molecular mass (LMM) forms: HMM proteinuria is associated with increased glomerular permeability and LMM proteinuria predominantly with impaired tubular reabsorption (see reviews [2, 14] and Chap. 3).

Glomerular filtration depends on the molecular mass of the proteins. Smaller molecules, e.g. ribonucleases of 14–44 kDa (see Sect. 8.5.2) traverse the glomerular membrane more readily than molecules exceeding about 69 kDa (albumin), which are filtered in extremely small amounts, e.g. alkaline phosphatase of 75–124 kDa (see Sect. 8.1.2). Under physiological conditions filtered proteins are reabsorbed in the proximal convoluted tubules except for trace amounts which escape reabsorption and appear in the final urine (approximately 150 mg over 24 h).

Accordingly, increased glomerular permeability results in a larger than normal contribution of HMM plasma enzymes, e.g. lactate dehydrogenase (LDH), AP, AAP, in the final urine. However, rough estimates for LDH (120 kDa) based on renal clearance values for plasma proteins with 80–160 kDa suggest that the contribution of plasma LDH to the LDH content in the final urine may be $\ll 1\%$ under physiological conditions. In renal disease with unselective HMM proteinuria, 1 g of excreted plasma protein may increase LDH in urine by not more than 10%–20% of the normal LDH activity [13]. Excessive excretion of LMM plasma enzymes, e.g. lysozyme and ribonuclease, is observed with impaired tubular reabsorption.

Urinary Enzymes
K. Jung/H. Mattenheimer/U. Burchardt (Eds.)

Table 1. Classification of diagnostically relevant urinary enzymes according to origin and handling by the kidney.

Origin	Enzyme	Remarks
Plasma		
Intact glomerular filter	Lysozyme, amylase, ribonuclease	MW < 70000, glomerular filtration, almost complete reabsorption
Increased permeability of glomerular filter	All plasma enzymes	Contribution small relative to tubular enzymes
Tubulus epithelium		
Brush border	Alanine aminopeptidase, γ-glutamyltransferase, alkaline phosphatase, α-glucosidase, trehalase, dipeptidyl-peptidase IV	Bound to brush border membrane with different strength
Lysosomes	β-Glucuronidase, *N*-acetylglucosaminidase, lysozyme, β-galactosidase, acid phosphatase	
Cytosol	Lactate dehydrogenase, leuclne aminopeptidase, ligandin, β-glucosidase	
Urinary tract and genital secretions		
	Acid phosphatase	From prostate
	dipeptidyl-peptidase IV	From sperm
Blood cells	Lactate dehydrogenase, lysozyme	From leucocytes, contribution small compared with tubular enzymes

Modified from [6]

1.2 Blood Cells

Any contribution of enzymes from blood cells depends of course on (a) activity of a given enzyme in the cells; (b) number of cells present in the urine; (c) most importantly, the degree of cell lysis in the urine. It has been estimated that a significant contribution of LDH, for example, is to be expected only in urine with grossly elevated red cell or white cell counts or both. To double the normal LDH activity in urine would require not only the presence but the complete lysis of 5×10^9 erythrocytes per liter (normal $0–5 \times 10^6$ cells per liter) or 2.5×10^8

leucocytes per liter (normal 0–20 × 10^6 cells per liter). Since there is no way of estimating how many cells are lysed, it is very difficult to correlate the actual cell counts with enzyme activities in urine. One must consider that lysis of blood cells may occur during their passage through the nephron and the urinary tract and probably during the urine collection period. Moreover, enzymes might be released and excreted into the nephron from leucocyte infiltration in renal tissue [13].

The activities of alanine aminopeptidase, β-glucuronidase, arylsulfatase A, α-glucosidase and lysozyme were investigated after incubating isolated human blood cells with human urine (incubated with frequent shaking for 5 h at 37 °C). According to the low activity or absence of these enzymes in erythrocytes, no activity increases were observed in urine. Although these enzymes occur with relatively high activities in leucocytes, a significant release into urine was found with lysozyme only, when the cell count exceeded 10^3 per μl [15].

1.3 Microorganisms

Microorganisms apparently do not contribute to enzymes in the urine, as documented for LDH [5, 18], AP [1], β-glucuronidase [3, 16], ASAT [5] and NAG [17].

1.4 Urinary Tract and Sperm

Nothing is known about the physiological release into the urine of enzymes from the epithelium of ureter and bladder; the contribution from this histologically uniform urothelium is probably insignificant when compared with the physiological enzyme release from the kidney. Pathological conditions are discussed in Chap. 12.

Spermatozoa and seminal plasma are rich in enzymes. All of the diagnostically useful enzymes were found to be present in spermatozoa and/or seminal plasma. The reader is referred to literature references in the Geigy Scientific Tables [7]. To which extent sperm enzymes might contribute to enzymes in male urine is not known.

1.5 Kidney

With the exclusion, then, of significant contributions from blood plasma, blood cells, microorganisms and the urinary tract, renal tissue is the main source of

HMM enzymes in the urine. Knowledge of the quantitative distribution of enzymes in the nephron is a prerequisite for the understanding of the origin of urinary enzymes. Quantitative activity patterns of many enzymes in the anatomical and functional units of the nephron were established through the application of Lowry's microdissection technique using lyophilized tissue sections and ultramicro-analytical methods [9]. The adaptation of these techniques to the kidney was first reported in 1958 [4]. The results obtained during the following decade have been summarized in several reviews [10–13]. Further refinement of the microdissection technique in regard to a more detailed differentiation of nephron segments, the dissection from fresh tissue and the considerable broadening of the enzyme spectrum have added decisively to the recognition of the metabolic (enzymatic) heterogeneity of the nephron [8].

Considering the origin of enzymes from the nephron, one must keep in mind the significant differences of activities in the various parts of the nephron and the relative mass of the individual structures. The proximal tubules constitute about 42% of the human kidney, and the activities of most diagnostically relevant enzymes are highest in this part (most notably the brush border enzymes AAP, AP, and GGT). Therefore, renal enzymes in urine are mainly derived from the proximal tubules, at least under physiological conditions [12].

The quantitative activity patterns are described in Chap. 2, and the mechanisms of enzyme release are discussed in Chap. 3.

References

1. Amador E, Zimmerman TS, Wacker WEC (1963) Urinary alkaline phosphatase activity. I. Elevated urinary LDH and alkaline phosphatase activities for the diagnosis of renal adenocarcinoma. JAMA 185:769–775
2. Balant L, Fabre J (1979) Clinical relevance of different electrophoretic methods for the analysis of urinary proteins. In: Dubach UC, Schmidt U (eds.) Diagnostic significance of enzymes and proteins in urine. Hans Huber, Bern, Stuttgart, Vienna, pp 216–234
3. Bank N, Bailine SH (1965) Urinary β-glucuronidase activity in patients with urinary tract infections. New Engl J Med 272:70–75
4. Bonting SL, Pollak VE, Muehrcke RC, Kark RM (1958) Quantitative histochemistry of the nephron. Science 127:1342–1343
5. Brenner BM, Gilbert VE (1963) Elevated levels of lactic dehydrogenase, oxalacetic transaminase and catalase in infected urine. Am J Med Sci 245:31–42
6. Burchardt U, Haschen RJ, Krosch H (1983) Pathobiochemische Hintergründe einer veränderten Enzymausscheidung mit dem Harn. Z Med Labor Diagn 24:187–192
7. Geigy Scientific Tables (1981) eighth edition Vol 1: Units of measurement, body fluids composition of body, nutrition Leniner C (ed). Ciba-Geigy, Basle and New Jersey pp 190–195
8. Guder WG, Ross BD (1984) Enzyme distribution along the nephron. Kidney Int 26:101–111
9. Lowry OH, Passoneau JV (1972) A flexible system of enzymatic analysis. Academic Press, New York, San Francisco, London
10. Mattenheimer H, Pollak VE (1965) Enzyme activity in the kidney: Quantitative aspects of enzyme distribution in the cortex and medulla of man, dog and rat. In: Kass E (ed) Progress in pyelonephrities. Davis, Philadelphia, pp 448–454
11. Mattenheimer H (1968) Aussagewert von Enzymbestimmungen im Urin. In: Schmidt FW (ed) Praktische Enzymologie. Hans Huber, Bern, pp 355–385

12. Mattenheimer H (1968) Enzymology of kidney tissue. In: Dubach U (ed) Enzymes in urine and kidney. Hans Huber, Bern, pp 119–145
13. Mattenheimer H (1971) Enzymes in the urine. Medical Clinics of North America 55:1493–1508
14. Mutti A (1989) Detection of renal diseases in humans; developing markers and methods. Toxicology letters 14:177–191
15. Peters JE, Thulin H, Wilhelm G (1974) Der Einfluss von Blutzellen im Harn auf die Aktivität von Harnenzymen. Z Urol 67:801–805
16. Schapiro A, Wellington P, Gonik H (1968) Urinary betaglucuronidase in urologic diseases of the kidneys. J Urol 100:146–157
17. Vigano A, Assael BM, Villa AD, Gagliardi L, Principi N, Ghezzi P, Salmona M (1983) N-Acetyl-β-D-glucosaminidase (NAG) and NAG isoenzymes in children with upper and lower urinary tract infections. Clin Chim Acta 130:297–304
18. Wacker WEC, Dorfman LE, Amador E (1964) Urinary lactic dehydrogenase activity. IV. Screening test for detection of renal disease and of arterial hypertension associated with renal disease. JAMA 188:671–676

2 Distribution of Diagnostically Relevant Enzymes Along the Nephron

H. Schmid and W.G. Guder

2.1 Introduction

Evaluation of enzymuria originating in renal tubules is considered a valuable tool in the diagnosis of impaired renal function [12, 18]. Enzymes may enter the urinary compartment by glomerular filtration, tubular or postrenal release. From the study of renal morphology, biochemistry and function, our knowledge concerning the intranephron and intracellular origin of enzymes has greatly increased over the past decades [15, 31]. Because of their relevance to the interpretation of enzymuria, we would like to review shortly the various forms of heterogeneity found in the mammalian kidney.

2.2 Heterogeneity of the Mammalian Kidney

Kriz and Bankir [20] have recently published the nomenclature of renal nephron structural elements, which is now internationally accepted (Fig. 1). The kidney tubule consists of various segments, which differ in length, cellular morphology, metabolic and transport functions as well as hormonal regulation [30, 31]. Pfaller [30] has analyzed the volume contribution of various cell types to the macroscopically discernible renal sections cortex, outer medulla, inner medulla and papilla. This intranephron heterogeneity is the main cause for the heterogeneous distribution of diagnostically relevant enzymes. Internephron heterogeneity on the other hand, is defined as the morphological, functional and biochemical difference between corresponding segments of different nephrons of the same kidney. Thus, marked differences have been described between superficial and juxtamedullary nephrons [16, 40]. Since each individual tubule regulates its own function by tubuloglomerular feedback mechanisms, internephron heterogeneity may determine the scatter of results obtained from microdissected nephron segments [42]. Under pathological conditions this type of heterogeneity may increase severalfold [34, 36–38].

Although intracellular compartmentation is a general phenomenon of mammalian cells, it seems of particular significance in tubular cells, since subcellular organelles and plasma membrane structures likewise exhibit heterogeneous

Urinary Enzymes
K. Jung/H. Mattenheimer/U. Burchardt (Eds.)

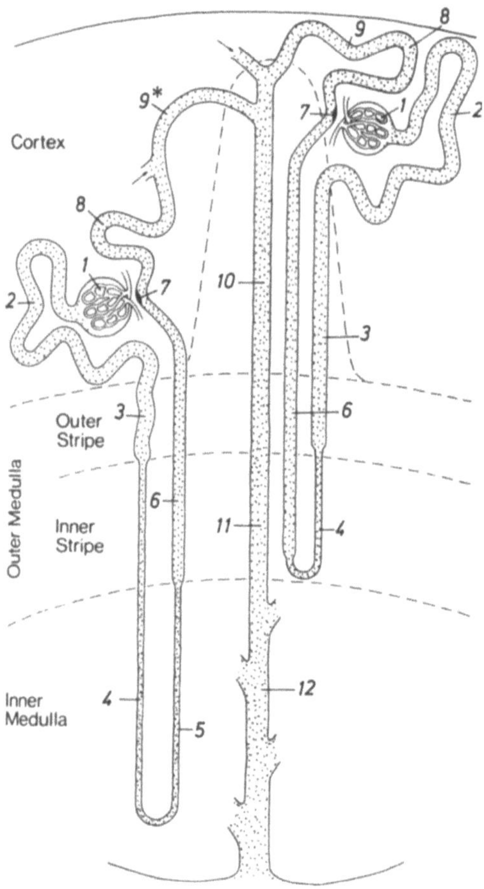

Fig. 1. Anatomically discernible segments of mammalian kidney with a superficial (short-looped) and juxtaglomerular (long-looped) nephron entering into a common collecting tubule (From Kriz and Bankir [20] with permission). The individual segments are numbered as follows *1* glomerulus (GL); *2* proximal convoluted tubule (PCT); *3* proximal straight tubule (PST); *4* thin descending limb (TDL) of Henle's loop; *5* thin ascending limb (TAL) of Henle's loop; *6* medullary (MAL) and cortical (CAL) thick ascending limb of Henle's loop; *7* macula densa (MD); *8* distal convoluted tubule (DCT); *9* connecting tubule (CNT); *9∗* connecting tubule of juxtamedullary nephron forming an arcade; *10* cortical collecting duct (CCD); *11* outer medullary collecting duct (OMCD); *12* inner medullary collecting duct (IMCD)

distribution patterns along the nephron [30]. Thus, the proximal tubule is equipped with the luminal brush border membrane. The surface per volume of cell of this reabsorptive structure exceeds that of the distal nephron segments more than tenfold [30]. Lysosomes and peroxisomes are likewise enriched in the proximal tubule, whereas mitochondrial volume and membrane surface areas are most abundant in the medullary and cortical thick ascending limb of Henle's loop (MAL, CAL) and distal convoluted tubule (DCT) segments [30]. As can be

Fig. 2a–f. Distribution pattern of marker enzymes of cellular compartments along the rat nephron. For abbreviations, see legend to Fig. 1 **a** 5′-Nucleotidase (5′ AMP), brush border [10]; **b** Na^+, K^+-ATPase, basolateral membrane [19]; **c** succinate dehydrogenase, mitochondrial inner membrane [33]; **d** acid phosphatase, lysosomes [28]; **e** D-amino acid oxidase, peroxisomes [22]; **f** lactate dehydrogenase, cytosol [5]

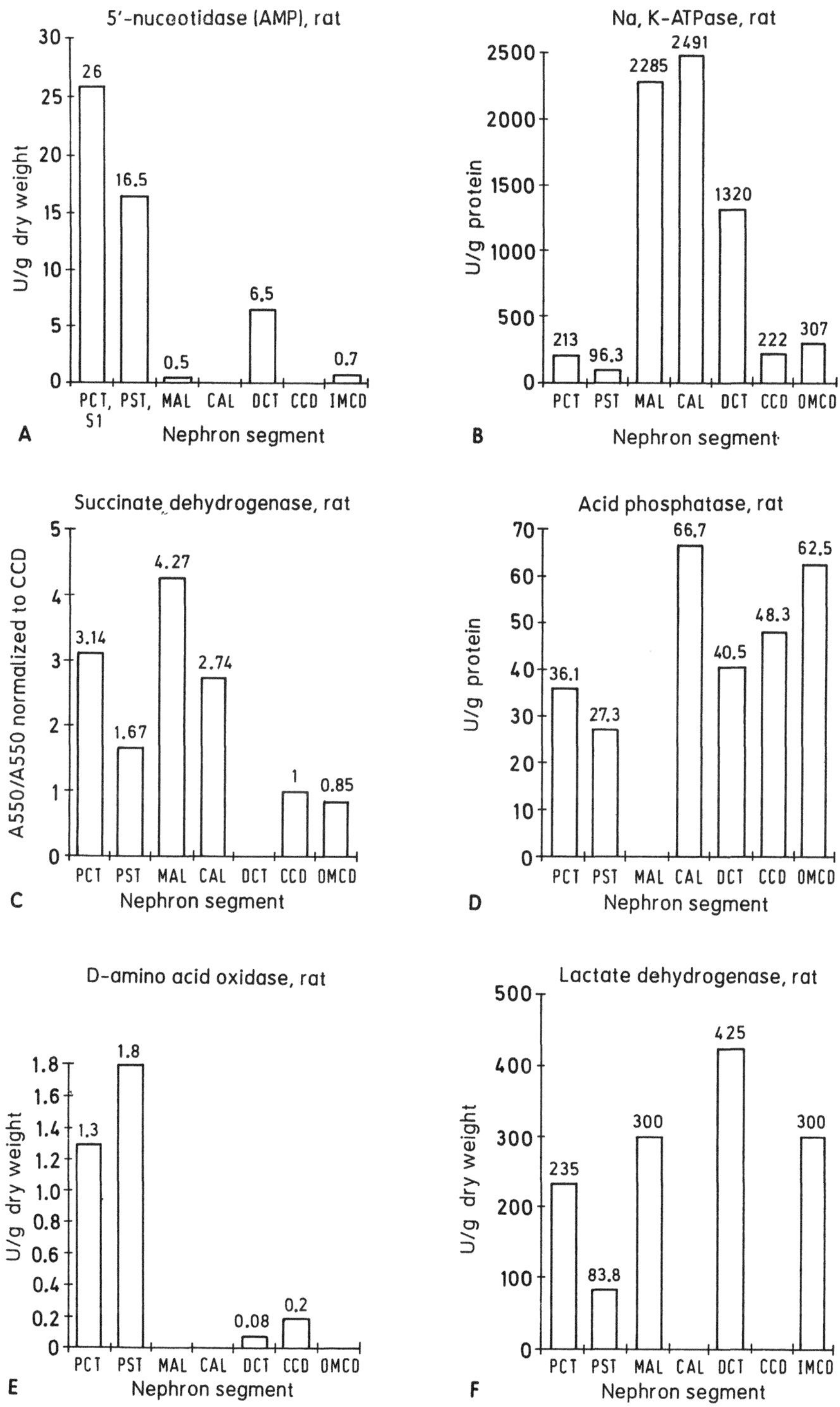
5'-nuceotidase (AMP), rat
U/g dry weight
26
16.5
0.5
6.5
0.7
PCT, S1 PST, MAL CAL DCT CCD IMCD
A
Nephron segment
Na, K-ATPase, rat
U/g protein
213
96.3
2285
2491
1320
222
307
PCT PST MAL CAL DCT CCD OMCD
B
Nephron segment
Succinate dehydrogenase, rat
A550/A550 normalized to CCD
3.14
1.67
4.27
2.74
1
0.85
PCT PST MAL CAL DCT CCD OMCD
C
Nephron segment
Acid phosphatase, rat
U/g protein
36.1
27.3
66.7
40.5
48.3
62.5
PCT PST MAL CAL DCT CCD OMCD
D
Nephron segment
D-amino acid oxidase, rat
U/g dry weight
1.3
1.8
0.08
0.2
PCT PST MAL CAL DCT CCD OMCD
E
Nephron segment
Lactate dehydrogenase, rat
U/g dry weight
235
83.8
300
425
300
PCT PST MAL CAL DCT CCD IMCD
F
Nephron segment

seen from Fig. 2, typical marker enzymes follow the distribution pattern of organelle morphology quite nicely. When various enzymes of the same compartment are compared, however, it seems that the enzymatic equipment of various subcellular compartments is heterogeneous along the nephron: mitochondria of the proximal tubule of the rat contain an inducible, phosphate-dependent glutaminase, whose activity can exceed distal tubular activity more than 20-fold in acidosis [11]. The activity of matrix 3-oxoacid COA-transferase in rabbit nephron is much higher in the CAL segment than in the proximal tubule [8], although citrate synthase activities are similar in both [26]. Likewise, lysosomal enzymes, plasma membrane ATPases and cytosolic enzymes show segment- and species-specific distribution patterns [19, 21, 31]. Species differences may in addition complicate the interpretation of results obtained in a particular experiment if the distribution of the enzyme under study has not been documented in the same species. As can be seen from Fig. 3, the activity of lysosomal *N*-acetyl-β-D-glucosaminidase is higher in the proximal straight tubule (PST) than in the proximal convoluted tubule (PCT) in human and mouse kidney, whereas the reverse was found in rat and rabbit nephron. Likewise, alkaline phosphatase activity is similar in the PCT and PST of human and rat nephron but decreases from PCT to PST in the mouse nephron [4]. It must be pointed out that in spite of 30 years' experience with the use of urinary enzymes in diagnosing renal diseases, most enzymes have not been quantitated along the human nephron yet.

2.3 Methods to Quantitate Enzymes Along the Nephron

The intranephron distribution of enzymes can be studied histochemically and by the measurement of catalytic activities in microdissected nephron segments. The microdissection procedure can be performed by two principles, which have their respective pros and cons discussed elsewhere [43]. The method developed by Burg et al. [9] and further elaborated by Morel et al. [27] uses fresh tissue slices resulting in whole nephron segments defined on their light microscopic appearance and their localization in the surrounding tissue with tubular length as the reference unit. In the method described by Lowry and Passonneau [25], microdissection is performed from lyophilized tissue sections of 10–20 μm thickness. This technique yields dehydrated tubule cross-sections, whose dry weight serves as the reference unit. Nephron segments are identified by comparison with parallel stained sections [25, 36]. In this way one is able to determine several enzyme activities from one single biopsy of human kidney, which is hardly possible with the former technique [14].

Enzymes may be quantitated by either activity measurements using histochemical stains [33], the cycling procedures of Lowry and Passonneau [6–8, 10, 11, 23, 26, 35, 36, 38–40], by direct fluorometry [3, 4, 21, 28, 29, 34, 37],

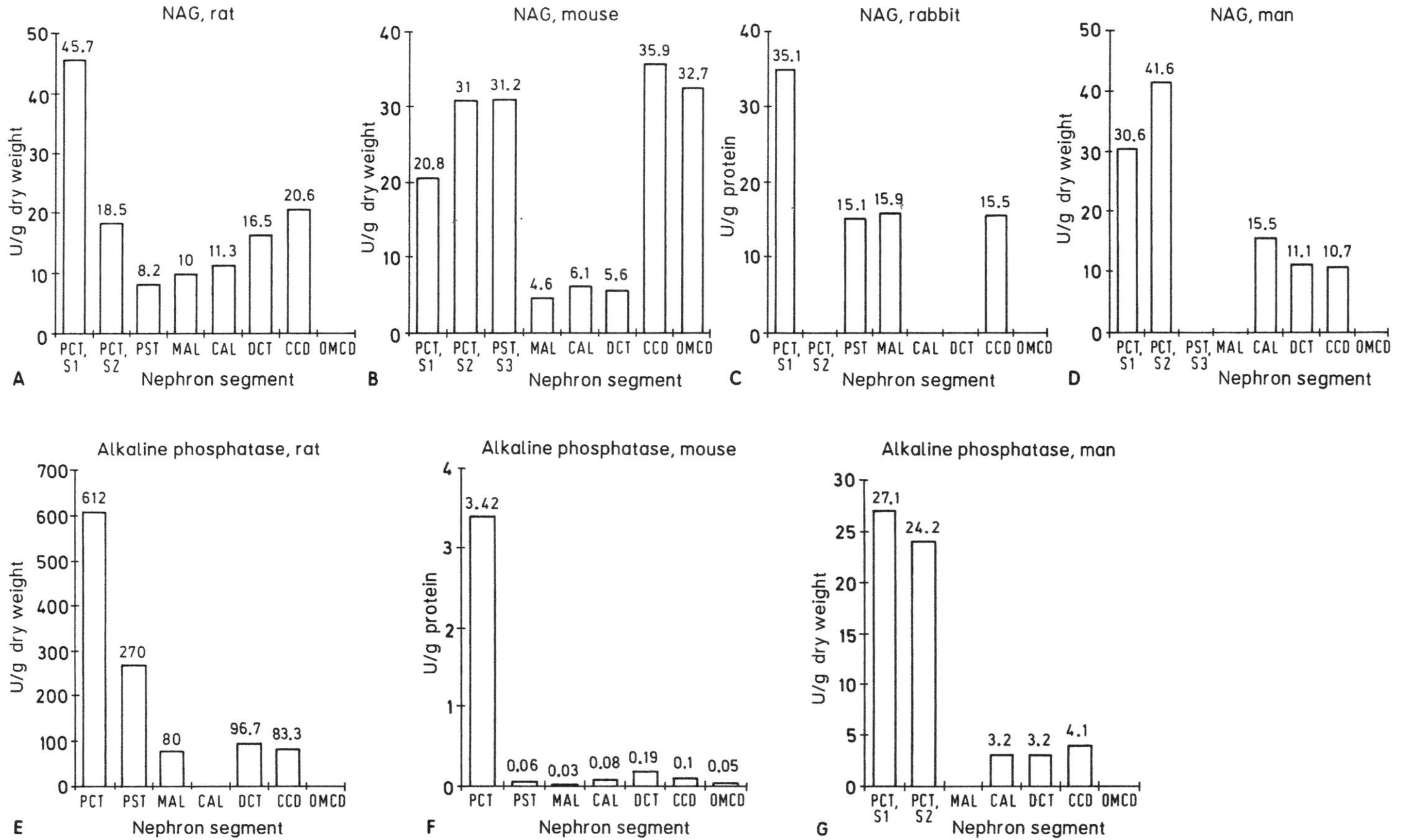

Fig. 3a–g. Species differences of the distribution pattern of *N*-acetyl-β-D-glucosaminidase activity (lysosomes) and alkaline phosphatase activity (brush-border membranes) along the nephron. For abbreviations see legend to Fig. 1. **a–d** N-acetyl-β-D-glucosaminidase activity along the rat [21], mouse (H. Schmid, unpublished), rabbit [3] and human [37] nephron. **e–g** alkaline phosphatase activity along the rat [40], mouse [4] and human [37] nephron

radiochemically [13, 17, 19, 23, 27, 42] or luminometrically [13, 14]. These methods achieve a sensitivity as fine as 1 pmol per sample, which allows the measurement of enzyme activities in 1–10 ng tissue dry weight. Recently, tubular enzymes have also been localized on the basis of immunohistochemical procedures [2] and with the Northern blot technique, which detects the mRNA of a specific tubular protein [44].

2.4 Localization of Diagnostically Relevant Enzymes Along the Nephron

Kidney enzymes which are released into the urine and considered diagnostically relevant are located predominantly in the cytosolic compartment, in the brush border membrane and in the lysosomes. The cytosolic enzymes show marked segment specificity: the gluconeogenic key enzyme fructose-1,6-bisphosphatase is confined to the proximal tubules, the glycolytic key enzyme pyruvate kinase is highly enriched in the distal tubules (Fig. 4), while lactate dehydrogenase is distributed more equally along the nephron (Fig. 2). Aldose reductase catalysing the formation of the osmolyte sorbitol from glucose, on the other hand, is exclusively located in the thin limbs and collecting tubules of the medullary and papillary regions (Fig. 4). While the distribution of these enzymes can be attributed clearly to their metabolic function, this is not yet possible for the brush border enzyme alkaline phosphatase. This enzyme (Fig. 3) appears to be equally distributed in the luminal membrane of the different nephron segments of the rat [15] in parallel with 5′-nucleotidase (5′-AMP cleaving) (Fig. 2) and thus can be considered a good marker for this membrane. It remains to be investigated whether this is also true for the different distribution patterns of these enzymes in the nephrons of mouse and man (Fig. 3). On the other hand, the different distribution of 5′-nucleotidase (Fig. 2), nucleotide pyrophosphatase in the rat and γ-glutamyltransferase in the rabbit (Fig. 4), similar to that in the rat [17], point to very specific functions of these enzymes in the brush border membrane. The enzymatic equipment of the mitochondria along the rat nephron, as along the rabbit nephron, does not seem to be completely uniform. While the distribution of succinate dehydrogenase activity (Fig. 2) correlates well with that of the mitochondrial volume and inner membranes' surface densities [33], that of the matrix enzymes citrate synthase and acetoacetyl-CoA transferase shows slight deviations [Fig. 4]. The latter enzyme is similarly distributed in the mouse and at [7] nephrons. Concerning the lysosomes, morphologically determined heterogeneity along the nephron [1] is reflected by heterogeneity of their enzymatic equipment [3, 21, 28, 29]. *N*-acetyl-β-D-glucosaminidase appears to be distributed homogeneously within the renal lysosomes of the rat [15] in contrast to acid phosphatase (Fig. 2), which is generally taken as the lysosomal marker enzyme. Again, in different species, it varies in distribution pattern (Fig. 3). In conclusion, interpretation about the renal origin of a urinary enzyme

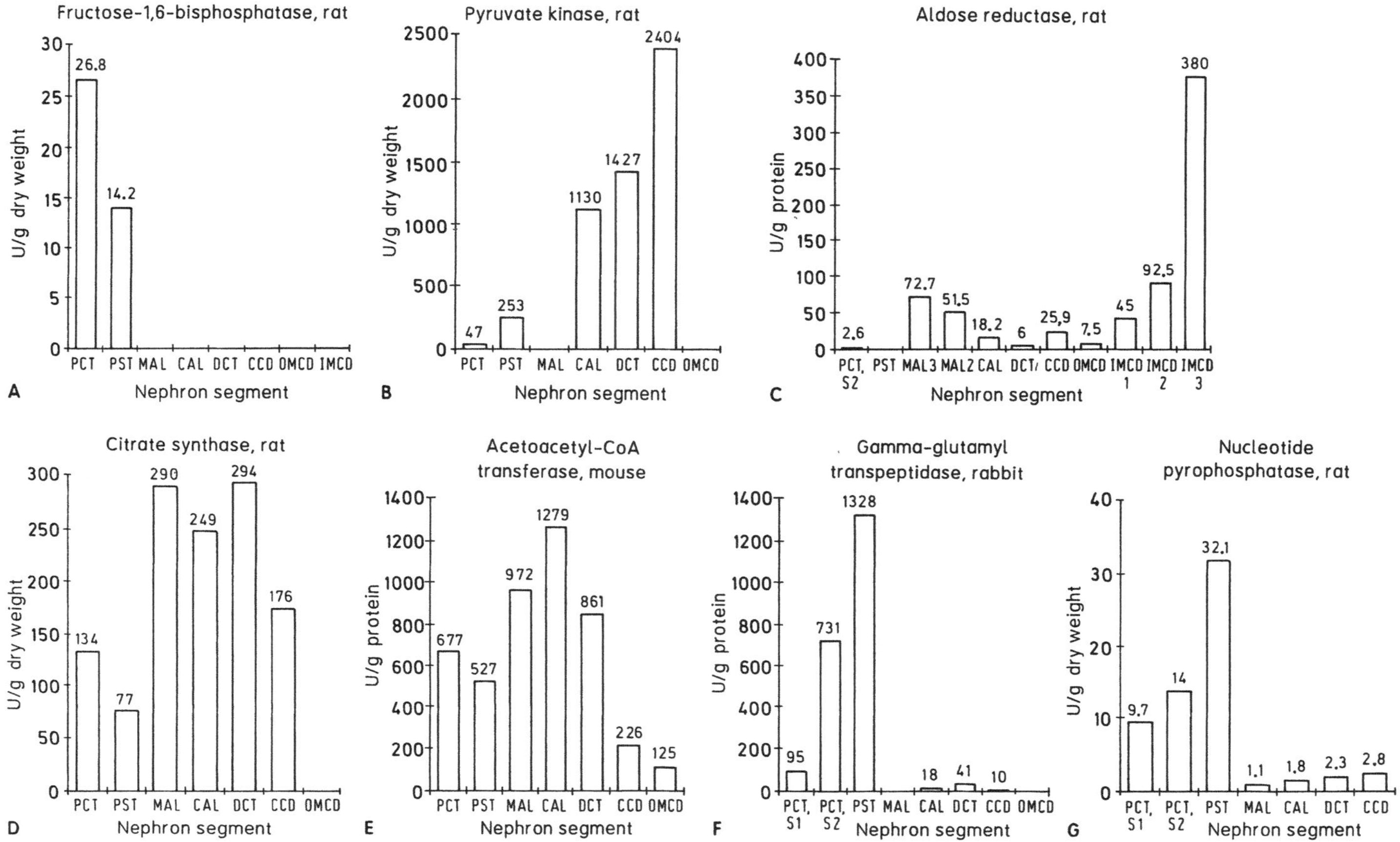

Fig. 4a–g. Heterogeneity of enzymatic equipment of intracellular compartments along the nephron. The distribution pattern of enzymes of the cytosol, mitochondrial matrix and brush border membrane are given. For comparison see Fig. 2, for abbreviations see legend to Fig. 1. **a** Fructose-1,6-bisphosphatase, rat (cytosol) [6]; **b** pyruvate kinase, rat (cytosol) [35]; **c** aldose reductase, rat (cytosol) [32]; **d** citrate synthase, rat (mitochondria) [23]; **e** acetoacetyl-CoA transferase, mouse (mitochondria) [13]; **f** gamma-glutamyltransferase, rabbit (brush border [41]; **g** nucleotide pyrophosphatase, rat (brush border) [24]

cannot be derived from studies on other marker enzymes or from studies of the same enzyme performed in different species or under other experimental conditions.

2.5 Changes in Enzyme Distribution Pattern in Disease States

The enzymatic equipment of the kidney is dynamic and undergoes adaptation to the metabolic conditions, e.g. activity increase in metabolic acidosis of phosphate-dependent glutaminase [11] and gluconeogenic enzymes [6, 16, 39]. In addition, the distribution pattern of kidney enzymes changes with disease states. Figure 5 shows that after ureter obstruction of the rabbit, pyruvate kinase activity, normally with highest activity in distal nephron segments, changes predominantly in the PST and slightly in the papillary area, while gluconeogenic and mitochondrial enzyme activities change more dramatically along the whole nephron [8].

The lysosomal enzymes *N*-acetyl-*β*-D-glucosaminidase and cathepsins B and L, which are similarly distributed in the rat nephron [21, 29] with highest activities in the proximal convoluted tubules, exhibit various changes in distribution along the proximal tubules in experimentally induced diabetes [21], overload proteinuria [29] or cyclosporine A nephrotoxicity [34] (Fig. 5). However, *N*-acetyl-*β*-D-glucosaminidase activity in the maintenance phase of human acute renal failure following kidney transplantation is increased severalfold in the glomeruli, proximal and distal tubules [37]. These changes of pattern of lysosomal enzyme activities may represent variations in the mechanism which leads to damage of cells and hence tubular enzyme release. In addition, these events appear to be time-dependent. This dependence on time after the initiating insult may be important for the interpretation of the marked decrease in alkaline phosphatase activity in a case of human acute transplant rejection with moderate and heterogeneous infiltration of the interstitial spaces by inflammatory cells (Fig. 5). In the same case, fructose-1,6-bisphosphatase activity is also slightly reduced in the PST. Increased enzymuria and insufficient enzyme resynthesis may account for these effects. On the other hand, after 14 days of post-transplantation anuria, the reduced activity of the mitochondrial succinate dehydrogenase activity, in parallel with reduced Na^+, K^+-ATPase activity,

Fig. 5a–g. Changes in enzyme distribution pattern in disease states. Enzyme activities are given in percentage of activity found in the respective control states, which are partially depicted in Figs. 2–4. For abbreviations see legend to Fig. 1. **a** *N*-acetyl-*β*-D-glucosaminidase activity in diabetes mellitus [21]; **b** cathepsin B + L in overload proteinuria [29]; **c** *N*-acetyl-*β*-D-glucosaminidase activity after cyclosporine treatment in rats [34]; **d** pyruvate kinase activity after ureter ligation in rabbits [8]; **e** alkaline phosphatase activity after acute transplant rejection in human [37]; **f** fructose-1,6-bisphosphatase activity after acute transplant rejection in human [38]; **g** succinate dehydrogenase activity 2 weeks after post transplant anuria in human [36]

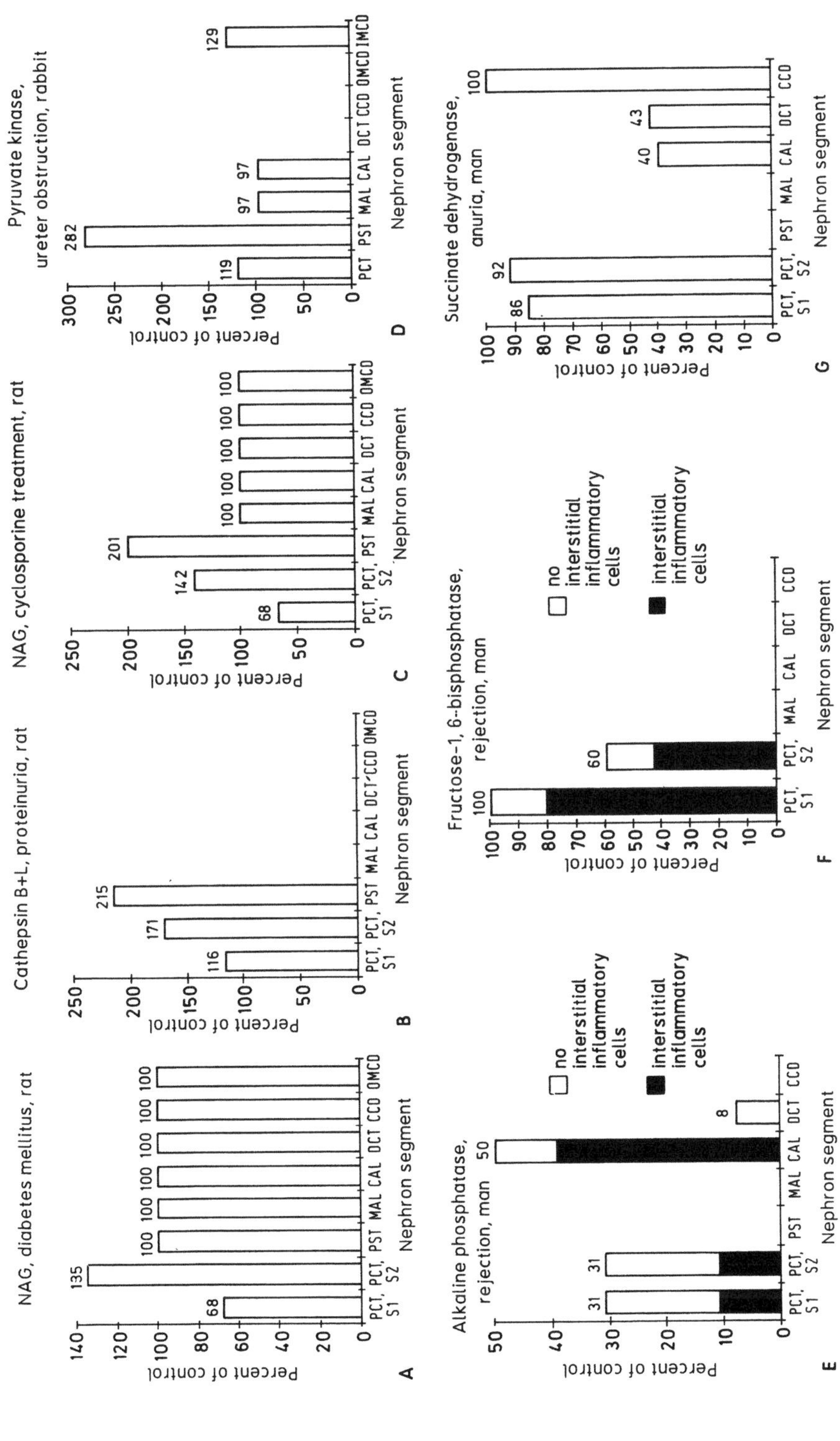
NAG, diabetes mellitus, rat
Cathepsin B+L, proteinuria, rat
NAG, cyclosporine treatment, rat
Pyruvate kinase, ureter obstruction, rabbit
Alkaline phosphatase, rejection, man
Fructose-1, 6-bisphosphatase, rejection, man
Succinate dehydrogenase, anuria, man
Percent of control
Nephron segment
PCT, S1
PCT, S2
PST
MAL
CAL
DCT
CCD
OMCD
IMCD
no interstitial inflammatory cells
interstitial inflammatory cells
A
B
C
D
E
F
G

preferentially in distal tubules [36], may characterize an advanced stage of injury.

These few examples of changes in intranephronal enzyme distribution under pathological conditions demonstrate that a variety of events at the molecular level account for functional and structural impairment of renal tubular cells. Besides tubular enzyme release, enzyme resynthesis, compensatory hypertrophy and rapid post-translational changes of activity contribute to the apparent changes in distribution pattern in disease states. This implies multiple possible mechanisms of enzyme release from tubular cells into urine, which complicates the interpretation of urinary enzyme patterns.

References

1. Andersen K-J, Haga H-J, Dobrota M (1987) Lysosomes of the renal cortex: heterogeneity and role in protein handling. Kidney Int 31:886–897
2. Barajas L, Powers K, Carretero O, Scicli AG, Inagami T (1986) Immunocytochemical localization of renin and kallikrein in the rat renal cortex. Kidney Int 29:965–970
3. Bourbouze R, Baumann F-C, Bonvalet J-P, Farman N (1984) Distribution of N-acetyl-β-D-glucosaminidase isoenzymes along the rabbit nephron. Kidney Int 25:636–642
4. Brunette MG, Chan M, Lebrun M (1981) Phosphatase activity along the nephron of mice with hypophosphatemic vitamin-D-resistant rickets. Kidney Int 20:181–187
5. Burch HB, Lowry OH, Perry SG, Fan L, Fagioli S (1974) Effect of age on pyruvate kinase and lactate dehydrogenase distribution in rat kidney. Am J Physiol 226:1227–1231
6. Burch HB, Narins RG, Chu C, Fagioli S, Choi S, McCarthy W, Lowry OH (1978) Distribution along the rat nephron of three enzymes of gluconeogenesis in acidosis and starvation. Am J Physiol 235:F246–F253
7. Burch HB, Bross TE, Brooks CA, Cole BR, Lowry OH (1984) The distribution of six enzymes of oxidative metabolism along the rat nephron. J Histochem Cytochem 32:731–736
8. Burch HB, Brooks CA, Cambon N, Bastin J, Lowry OH (1985) Effect of ureter obstruction on metabolic enzymes in different parts of the nephron. In: Dzúrik R, Lichardus B, Guder W (eds) Kidney metabolism and function. Nijhoff, Boston pp 251–255
9. Burg M, Grantham J, Abramow M, Orloff J (1966) Preparation and study of fragments of single rabbit nephrons. Am J Physiol 210:1293–1298
10. Cole BR, Hays AE, Boylan JG, Burch HB, Lowry OH (1982) Distribution of enzymes of adenylate and guanylate nucleotide metabolism in rat nephron. Am J Physiol 243:F349–F355
11. Curthoys NP, Lowry OH (1973) The distribution of glutaminase isoenzymes in the various structures of the nephron in normal, acidotic, and alkalotic rat kidney. J Biol Chem 248:162–168
12. Dubach UC, Schmidt U (eds) (1979) Diagnostic significance of enzymes and proteins in urine. Huber, Bern (Current problems in clinical biochemistry, vol 9)
13. Guder WG, Pürschel S, Wirthensohn G (1983) Renal ketone body metabolism. Distribution of 3-oxoacid CoA-transferase and 3-hydroxybutyrate dehydrogenase along the mouse nephron. Hoppe-Seyler's Z Physiol Chem 364:1727–1737
14. Guder WG, Pürschel S, Vandewalle A, Wirthensohn G (1984) Bioluminescence procedures for the measurement of NADP dependent enzyme catalytic activities in submicrogram quantities of rabbit and human nephron structures. J Clin Chem Clin Biochem 22:129–140
15. Guder WG, Ross BD (1984) Enzyme distribution along the nephron. Kidney Int 26:101–111
16. Guder WG, Schmidt U (1976) Enzymatic organization of carbohydrate metabolism along the nephron. In: Giovanetti S, Bonomini V, D'Amico G (eds) Proceeding of the 6th international congress of nephrology. Karger, Basel, pp 187–195 (Advances in nephrology, physiology, hypertension, renal diseases, renal failure, dialysis and transplantation)
17. Heinle H, Wendel A, Schmidt U (1977) The activities of the key enzymes of the γ-glutamyl cycle in microdissected segments of the rat nephron. FEBS Lett 73:220–224

18. Jung K, Burchardt U (1987) Harnenzyme im Experiment und in der Klinik. Bericht über ein Symposium der Humboldt-Universität Berlin und des Bezirkskrankenhauses Frankfurt/Oder (DDR), 22.–25.4.1987. J Clin Chem Clin Biochem 25:823–828
19. Katz AJ, Doucet A, Morel F (1979) Na-K-ATPase activity along the rabbit, rat and mouse nephron. Am J Physiol 237:F114–F120
20. Kriz W, Bankir L (1988) A standard nomenclature for structures of the kidney. Kidney Int 33:1–7
21. Le Hir M, Dubach UC, Guder WG (1980) Distribution of acid hydrolases in the nephron of normal and diabetic rats. Int J Biochem 12:41–45
22. Le Hir M. Dubach UC (1981) The activity pattern of two peroxisomal oxidases in the rat nephron. FEBS Lett 127:250–252
23. Le Hir M, Dubach UC (1982) Activities of enzymes of the tricarboxylic acid cycle in segments of the rat nephron. Pflügers Arch 395:239–243
24. Le Hir M, Dubach UC, Angielski S (1985) localization of nucleotide pyrophosphatase in the rat kidney. Histochemistry 86:207–210
25. Lowry OH, Passonneau JV (1972) A flexible system of enzymatic analysis. Academic, New York
26. Marver D, Schwartz MJ (1980) Identification of mineralocorticoid target sites in the isolated rabbit cortical nephron. Proc Natl Acad Sci USA 77:3672–3676
27. Morel F, Chabardes D, Imbert-Teboul M (1978). Methodology for enzymatic studies of isolated tubular segments: adenylate cyclase. In: Martinez-Maldonado M (ed) Methods in pharmacology, vol 4B. Plenum, New York, pp 297–323
28. Olbricht CJ, Garg LC, Cannon JK, Tisher CC (1984) Acid phosphatase activity in the mammalian nephron. Am J Physiol 247:F252–F259
29. Olbricht CJ, Cannon JK, Garg LC, Tisher CC (1986) Activities of cathepsins B and L in isolated nephron segments from proteinuric and nonproteinuric rats. Am J Physiol 250:F1055–F1062
30. Pfaller W (1982) Structure function correlation on rat kidney. Quantitative correlation of structure and function in the normal and injured rat kidney. Adv Anat Embryol Cell Biol 70:1–106
31. Ross BD, Guder WG (1982) Heterogeneity and compartmentation in the kidney. In: Sies H (ed) Metabolic compartmentation. Academic, London, pp 363–409
32. Sands JM, Terada Y, Bernard LM, Knepper MA (1989) Aldose reductase activities in microdissected rat renal tubule segments. Am J Physiol 256:F563–F569
33. Schmid H (1984) Quantitative microphotometric succinate dehydrogenase histochemistry in human nephron. Basic Appl Histochem 28:221–231
34. Schmid H, Lindmeier I, Schmitt H, Eissele R, Neuhaus G, Wehrmann M (1991) Reversible changes in N-acetyl-β-D-glucosaminidase activity along the proximal renal tubules of cyclosporine-A treated rats. In: Bach PH, Gregg NJ, Wilks MF, Delacruz L (eds) Nephrotoxicity: mechanisms, early diagnosis, and therapeutic management. Proceedings of the 4th international nephrotoxicity symposium. Marcel Dekker, New York, pp 199–204
35. Schmid H, Mall A, Scholz M, Schmidt U (1980) Unchanged glycolytic capacity in rat kidney under conditions of stimulated gluconeogenesis. Determination of phosphofructokinase and pyruvate kinase in microdissected nephron segments of fasted and acidotic animals. Hoppe-Seyler's Z Physiol Chem 361:819–827
36. Schmid H, Mall A, Bockhorn H (1985) Altered distribution pattern of Na^+-K^+-ATPase and succinate dehydrogenase activities along the nephron in human acute post-transplant renal failure. J Clin Chem Clin Biochem 23:27–34
37. Schmid H, Mall A, Bockhorn H (1986) Catalytic activities of alkaline phosphatase and N-acetyl-β-D-glucosaminidase in human cortical nephron segments: heterogeneous changes in acute renal failure and acute rejection following kidney allotransplantation. J Clin Chem Clin Biochem 24:961–970
38. Schmid H, Mall A, Bockhorn H (1987) Specific fructose-1,6-bisphosphatase activities in microdissected proximal tubules of human kidneys in function and dysfunction. In: Kovačević Z, Guder WG (eds) Molecular nephrology, biochemical aspects of kidney function. De Gruyter, Berlin, pp 339–345
39. Schmid H, Scholz M, Mall A, Schmidt U, Guder WG, Dubach UC (1978) Carbohydrate metabolism in rat kidney: heterogeneous distribution of glycolytic and gluconeogenic key enzymes. In: Guder WG, Schmidt U (eds) Biochemical nephrology. Huber, Bern, pp 282–289 (Current problems in clinical biochemistry, vol 8)
40. Schmidt U, Dubach UC (1971) Quantitative Histochemie am Nephron: Oxydoreduktasen und Na K stimulierte ATPase. Prog Histochem Cytochem 2:185–298

41. Shimada H, Endou H, Sakai F (1982) Distribution of gamma-glutamyl transpeptidase and glutaminase isoenzymes in the rabbit single nephron Jpn J Pharmacol 32:121–129
42 Vandewalle A, Wirthensohn G, Heidrich H-G, Guder WG (1981) Distribution of hexokinase and phosphoenolpyruvate carboxykinase along the rabbit nephron. Am J Physiol 240:F492–F500
43. Wirthensohn G, Guder WG (1990) Metabolism of isolated kidney tubule segments. In: Fleischer S, Fleischer B (eds) Methods in enzymology. vol 191 Academic, New York pp 325–339
44. Xiong W, Chao L, Chao J (1989) Renal kallikrein mRNA localization by in situ hybridization. Kidney Int 35:1324–1329

3 Mechanisms of Enzyme Release and Causes of Altered Enzyme Excretion

U. Burchardt and J.E. Scherberich

Several mechanisms have been recognized or are postulated in the pathogenesis of enzymuria. A synopsis of current concepts is presented in Fig. 1. Two basic principles can be deduced for changes of enzymuria: (a) abnormal composition of the primary urine, independent of the cause for any alterations, which may be pre-renal (e.g., increased level of lysozyme) or renal (e.g., increased glomerular permeability), and (b) structural and functional aberrations of nephron cells.

The following description stresses the major aspects of pathophysiological processes and was chosen for didactical reasons. One has to keep in mind, however, that changes in the composition of the primary urine have an effect on the nephron cells (e.g., lysosomotropic agents). On the other hand, processes like peroxidation that may alter glomerular permeability change the composition of the primary urine.

3.1 Altered Composition of Primary Urine

3.1.1 Low Molecular Mass Proteins (Especially Enzymes)

Proteins including enzymes with a molecular mass below 70000 daltons are practically freely filtered through the glomerular basement membrane. They are almost totally reabsorbed by the proximal tubular cells. The most important enzymes are listed in Table 1.

Renal handling of these enzymes in man is characterized by a low-threshold, high-capacity tubular reabsorption system [1, 2] (Fig. 2). Normal urine contains only small amounts of these enzymes. The extraction rate corresponds to the portion filtered, reabsorbed, stored, and catabolized in the kidney. Thus, two mechanisms lead to increased excretion of low molecular mass proteins:

1. High plasma concentrations. The serum concentration of low molecular enzymes represents a balance between primary production of the enzymes and their removal and degradation by various tissues (kidney, liver, lung). For example, in monocytic leukemia, lysozyme production within monocytes and tissue macrophages is increased [3]. The marked increase of serum lysozyme concentration results from this high synthetic activity. As the serum

Urinary Enzymes
K. Jung/H. Mattenheimer/U. Burchardt (Eds.)

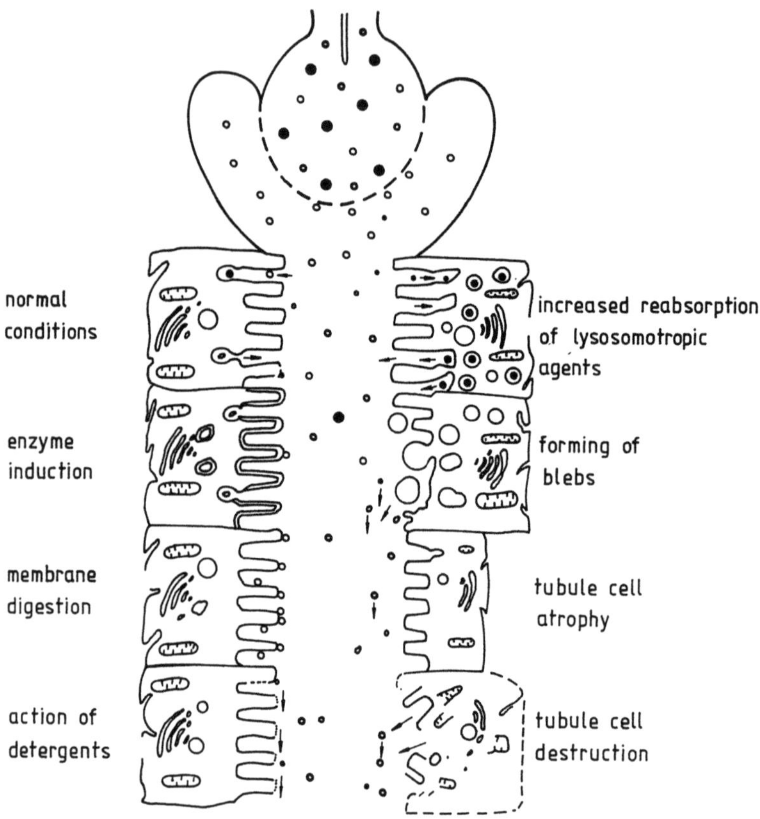

Fig. 1. Synopsis of pathobiochemical mechanisms of altered enzyme excretion. (**a**) Under *normal conditions*, only a few enzymatic activities of the urine derive from the serum, entering the urine by glomerular filtration. Enzymes exceeding a molecular mass of 70 000 are not normally excreted into the urine, at least not in significant amounts. The vast majority of urinary enzymes must derive from the tubular cells. Normal turnover rate of the tubular cells, their cellular permeability, and their metabolism account for the renal enzymes excreted in normal urine. (**b**) The amount of enzymes present in the tubular cell depend on the state of activation and/or inhibition of the gene transcription of specific enzymes. *Enzyme induction* leads to an increased output of cellular enzymes. (**c**) In connection with *membrane digestion*, especially brush border enzymes are liberated in the tubular lumen. (**d**) *Detergentlike substances* with their tension-lowering effect can increase the liberation of brush border as well as cytosolic enzymes. (**e**) According to the concept of the correlation between the reabsorption of *lysosomotropic agents* by the tubular epithelium and urinary excretion of protein and enzymes, the enzyme excretion depends on the amount of lysosomotropic substances in the primary urine. (**f**) *Bleb formation* and sequestration of proximal tubular cell material into the tubular lumen (so-called tissue proteinuria) is found under many pathological circumstances (especially connected with diminution of energy supply). (**g**) *Tubule cell atrophy* of various genesis. The protein concentration in the final urine is elevated through the diminished reabsorption of protein. The release of cell enzymes is decreased. (**h**) Due to the *tubule cell destruction*, serum and cell enzymes and structural proteins appear in large amounts in the final urine (acute renal failure)

Table 1. Low molecular mass enzymes in urine

Enzyme	EC	Molecular mass (daltons)	Origin	Isoelectric point (pH)
Lysozyme	3.2.1.17	14000	Neutrophilic granulocytes	10.5–11.0
Amylase	3.2.1.1	54000	Pancreas, parotis	6.5–7.1
Lipase	3.1.1.3	48000	Pancreas	5.8
Ribonuclease	3.1.27.5	13000	Pancreas, liver	5.0–10.0
Esterase A 1	3.1.1.1	28000	Kidney	8.0–9.0
Trypsinogen	3.4.21.4	23000	Pancreas	
Urokinase	3.4.21.31	53000	Kidney	8.7–9.6
Pepsinogen A	3.4.23.1	43000	Gastric mucosa, proximal duodenum	4.5–5.0
Pepsinogen C	3.4.23.3	40500	Chief cells in the gastric fundic mucosa	

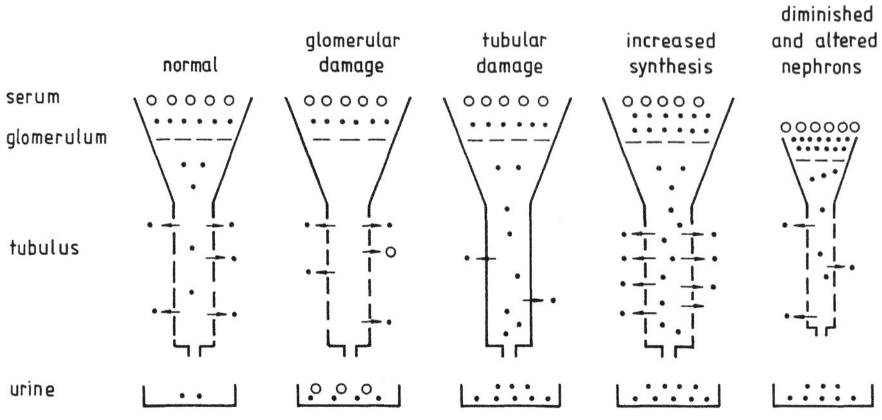

Fig. 2. Handling of high (○) and low (●) molecular mass proteins and enzymes by the nephron (adapted from Jung [9])

lysozyme level and filtered load of lysozyme increase more than 3- to 5-fold in leukemic patients, there is a marked increase in lysozyme up to 1000-fold [4]. Lysozymuria increases markedly when the serum level exceeds 25–30 mg [1, 5, 6]. All patients suffering from chronic renal failure (glomerular filtration rate below 40 ml/min per 1.73 m^2) show an increase of serum lysozyme. In addition, lysozyme is excreted from tubular cells into urine at an increased rate. In renal disease, lysozymuria and rise of serum lysozyme activity indicate severe renal damage. Increased activities of ribonucleases and amylase may be associated with acute or chronic renal failure.

2. Decreased tubular reabsorption
   A. Morphological causes. Decreased tubular reabsorption due to atrophic or damaged tubule epithelium and/or low rate of intracellular degradation increases the amount of low molecular enzymes in the final urine.
   B. Functional causes. An inhibition of small protein reabsorption suggests competition with other substances for common endocytotic transport sites at the luminal membrane of proximal tubular cells. Possibly, inhibitors play a role. Inhibitors of lysozyme reabsorption, for instance, are lysine and cytochrome *c* [7].

Experimental and clinical data suggest a different handling of individual low molecular mass enzymes by the tubular cells. Protein reabsorption, especially of low molecular mass proteins and enzymes, by the renal tubular system may be governed by highly selective uptake mechanisms, as was shown for lipase and amylase. Both enzymes are removed from the serum mainly by glomerular filtration at nearly the same rate. In contrast to amylase the reabsorption of lipase is almost complete. Differences in the renal handling of both enzymes may be due to their different affinities for hydrophilic and hydrophobic receptor sites of the epithelial surface membranes [6, 8].

Similar to this, pepsinogens A and C are almost identical low molecular mass proteins (see Table 1) with marked differences in renal handling. Some 68% of pepsinogen A and 98% of pepsinogen C extracted from the circulation were metabolized by the kidney, and fractional metabolism is closely related to the fractional reabsorption of both molecules from the glomerular filtrate. Therefore, pepsinogen is almost absent in the urine of healthy subjects [10, 11].

3.1.2 High Molecular Mass Proteins

In man, through the blood supply 50 kg of soluble protein reaches about kidneys each day. Under normal conditions only 1 g crosses the glomerular capillary wall, however, and appears in the glomerular ultrafiltrate (Fig. 2). The protein composition of the ultrafiltrate reflects a molecular mass distribution pattern which is different from that of the plasma. In contrast to plasma in which large molecular mass proteins predominate, the glomerular filtrate contains more small than large protein molecules. These are subsequently reabsorbed by the proximal tubule cells such that less than 150 mg escapes each day into normal urine. Therefore, under normal conditions, only very little urinary enzymatic activity derives from the serum [12, 13] (see also Chap. 1).

The first step in protein uptake is binding of serum components to the epithelial plasma membrane. The binding affinity of various proteins (and enzymes) for the plasma membrane receptor sites varies greatly. Most proteins (and enzymes) are reabsorbed by the luminal site of the proximal tubule epithelium (Fig. 3).

The two main cellular components which participate in protein uptake and disposal in the proximal tubule cells are the endocytotic vacuoles and the lysosomes. Some peptides and most proteins are lysosomotropic agents. Endocytotic vacuoles do not have any characteristic marker enzymes. However, lysosomes are characterized by a more or less typical pattern of acid hydrolases. The lysosomal enzymes are not capable of degrading all proteins to the same extent. For instance, lysozyme can only be partially degraded into amino acids, even in the presence of an excess of lysosomal enzymes.

Renal lysosomes are abundant in proximal tubular cells, while fewer are found in the distal nephron structures. The activity in the renal lysosomal system is clearly increased during proteinuria as indicated by the increased activity of acid hydrolases in the large lysosomes or "protein droplets". Proteinuria also has significant effects on the small lysosomes in lower parts of the nephron, suggesting a greater role of more distal segments in the handling of extra protein [14].

The increased activity of lysosomal enzymes in the large lysosomes is closely correlated with highly increased activity of urinary lysosomal enzymes in proteinuria caused by glomerulonephritis [15, 16]. The increased urinary activity of these enzymes may be due to excretion of the lysosomes and/or their contents into urine [17]. In glomerulonephritic rats, an observed change in the

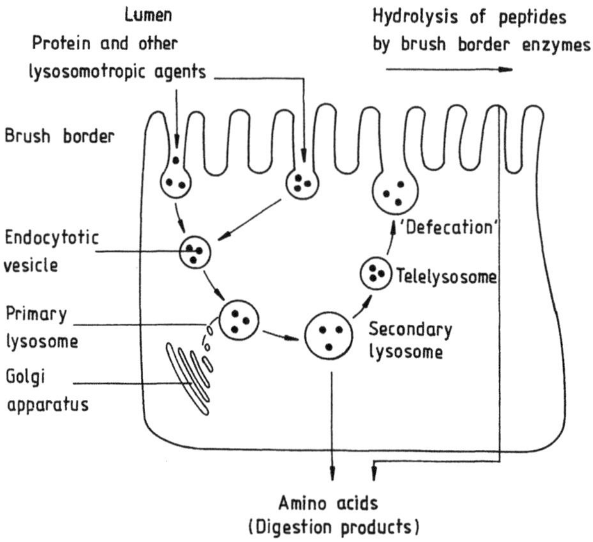

Fig. 3. Schema of the cellular mechanism of the proximal tubule for reabsorption and catabolism of protein or polypeptide molecules in connection with enzymes. Protein is taken up by endocytotic vesicles, which form phagosomes into which primary lysosomes empty their hydrolytic enzymes. Enzymatic cleavage of proteins occurs in the phagolysosomes. Liberated amino acids diffuse into the interstitium. Due to excretion of the lysosomes and/or their content into the tubule lumen ('defecation') urinary activity of lysosomal enzymes increases. The activity of urinary acid hydrolases probably reflects the activity in the renal lysosomal system. Hydrolysis of small, linear peptides (for instance small peptide hormones) occurs at the luminal site of brush border enzymes

isoenzyme pattern of urinary *N*-acetyl-β-D-glucosaminidase (NAG) [15, 16, 18, 19] may indicate that the enzymes are derived from a lysosome population in the renal cortex different from lysosomes in the healthy kidney.

3.1.3 Lysosomotropic Agents

De Duve et al. [20] introduced the term "lysosomotropic agents." This designates all substances that are taken up selectively into lysosomes, irrespective of their chemical nature or mechanism of uptake. This property characterizes many biologically active agents. Some substances (e.g., sucrose, gentamicin) are naturally lysosomotropic (see Table 2). Others can be generated by a suitable combination with a lysosomotropic carrier (e.g., an antibody).

A number of substances are known to accumulate selectively within lysosomes, sometimes in high concentrations (e.g., lysozyme 1.983 $\mu g \cdot g^{-1}$ of kidney [7]). There are distinct modes of entry into the lysosomes: endocytosis, permeation, and so-called "piggyback" endocytosis. "Piggyback" endocytosis is the transfer into cells of a substance that is either mixed or combined with a carrier.

Table 2. Lysosomotropic agents and lysosomal enzyme excretion

Lyosomotropic agent	Molecular mass (dalton)	Marker enzymes of lysosomal origin	References
Sucrose	342	NAG	25, 26
Mannitol	182	β-glucuronidase, NAG	22, 27
Dextran	40000	β-glucuronidase,	22
Contrast media	520	NAG	28, 29
		NAG, β-glucuronidase	30, 31
Gentamicin	694	NAG	21, 23, 32
		NAG, β-glucuronidase	24, 30
Proteins/peptides	varied	NAG, β-glucuronidase, acid phosphatase	17, 23
Adriamycin	560	NAG, β-glucuronidase	33

Glomerular filtered protein also appears in lysosomes [17]. Proteins have quite different resistance to renal catabolism. Lysosomes contain many hydrolytic enzymes which have been shown to digest a wide variety of proteins. "Physiological" tracers, such as homologous albumin, are more readily digested than heterologous proteins or tracers which are normally not present in the tubule, such as ferritin or peroxidase. The lysosomal enzymes do not have the same ability to degrade all proteins. A "protein storage" syndrome may therefore develop under certain conditions: Lysosomal overloading can also be induced by the administration of indigestible lysosomotropic substances.

Furthermore, it may result from an "inborn error" due to congenital deficiency of one of the intralysosomal enzymes. Under some circumstances, clearing of the lysosomes is chemically and physically impossible [21]. The only remaining measure is exocytotic discharge of the lysosomal contents into the extracellular spaces. Thereby, lysosomal enzymes enter the tubular lumen (see Fig. 3).

A well-investigated model is the action of gentamicin on tubular cell metabolism followed by hyperenzymuria [22–24]. After glomerular filtration, these highly charged molecules bind to a charge-mediated receptor site on the brush border membrane. The uptake occurs predominantly by pinocytosis at the luminal surface, and the vesicles rapidly fuse with lysosomes. The cortical concentration of aminoglycosides can reach up to 50 times the plasma concentration [21]. Gentamicin induces an impairment of lysosomal digestion by inhibition of sphingomyelinase. A lysosomal phospholipid-overloading syndrome with intracellular myeloid bodies occurs [24]. The enzymuria is temporally related to the appearance of myeloid body in urine, seen by electron microscopy study [21].

Some anticancer agents (e.g., doxorubicin, daunorubicin) are lysosomotropic detergents. They act by selectively rupturing the membranes of lysosomes, causing cell death through the action of the lysosomal enzymes that are released.

3.1.4 Detergentlike Substances

Detergents have a surface tension lowering effect. The most important detergent in urine may be bile acids, as indicated by Hay's test [34]. The daily excretion of conjugated cholic acid amounts to 2 μmol/l [35]. The occurrence of sulfate esters of cholic acid, deoxycholic acid, and chenodeoxycholic acid has been described in patients with liver disease, thereby providing evidence for more rapid renal excretion of the derivatives than for bile acids conjugated with taurine or glycine [35]. Mills et al. [36] found an inverse relationship between urine surface tension and conjugated bile salts of deoxycholate and cholate. The total daily excretion of bile acids in urine correlates with the degree of cholestasis in patients with liver disease [35]. The release of enzymes from the renal brush border membrane may be increased by as much as 15-fold in the case of liver disease with cholestasis [37]. Cholestasis is also marked by a significant rise in the serum activities of the membrane enzymes such as alkaline phosphatase, gamma-glutamyltransferase, and alanine aminopeptidase (AP, GGT, and AAP). It has been demonstrated that enzymes in cholestasis are increased only when the bile salt level in liver and serum is elevated [38]. These findings support the concept that the increased serum activities of membrane enzymes in cholestasis are caused by solubilization through bile salts [39]. Taurocholate at physiological levels liberates enzymes from the intestinal epithelium [40]. Bossmann and Haschen [41] applied the technique of segmental perfusion of the proximal jejunum of the rat to study the liberation of enzymes from the mucosa cells under the influence of bile salt. The majority of cytosolic enzymes were released to a greater extent than membrane enzymes were. The release of enzymes and protein from the mucosal surface, however, was inhibited by 200 μg cycloheximid 100 g body weight injected i.m. 1 h before the start of the experiments. Methionine, actinomycin, and pyromycin behave similarly. Cycloheximide interferes with nucleic acid and/or protein synthesis. On the other hand, lysosomal or mitochondrial enzyme levels were low or absent in the perfusion medium, a finding that excludes serious damage to the cell.

A continuing increased excretion of urinary (tubular) AAP was observed in chronic cholestasis [42]. There is evidence that during the action of detergents an increasing proportion of the enzymes originates from de novo synthesis [40].

3.2 Influences and Mechanisms Altering Tubular Cells

3.2.1 Modification of Energy Supply

Hypoxia, anoxia, and ischaemia contribute to the release of cellular enzymes in most organs. The cellular damage leading to an enzyme's release must not necessarily be irreversible [43]. The amount of enzymes lost correlates with the degree of the deterioration of the cellular energy state [44, 45]. In addition, the absolute enzyme amount released depends on the cellular content of the enzyme, its intracellular compartmentation, and its biochemical properties.

The glomerulum itself is quite resistant to ischaemic injury [46]. Experimental studies in the rat revealed that the brush border of the proximal tubule is especially susceptible to ischaemic injury [44, 47]. It is likely that shedding of membrane-bound enzymes of the proximal tubule is related to a depletion of cellular nucleotides, as has been observed during the initial phase of renal ischaemia [48, 49] (Fig. 4). An energetic insufficiency of proximal tubular cells leads to potocytosis, an extrusion (or "blebbing") of cytoplasm and cellular organelles into the tubular lumen and a concomitant membrane loss [50]. Kehrer et al. [51] studied the glomerular filtration rate and urinary lactate dehydrogenase (LDH) excretion after renal ischaemia in dogs. They found the best conformity with the extent of ischaemic stress duration by using the ratio of LDH activity to GFR. They did not find any enzyme release as a consequence of reperfusion injury. Renewed enzyme release in connection with reperfusion is a well-documented phenomenon of the heart muscle cells (the so-called oxygen paradox) [52]. Examination of urinary LDH release may be a useful diagnostic means for assessment of the extent of ischaemic injury [51].

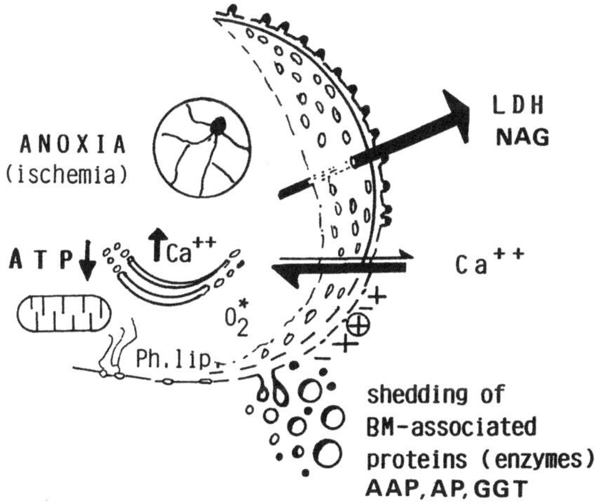

Fig. 4. Changes in cellular homeostasis during ischaemia to tubular membrane damage and blebbing (tissue protein uria): depletion of ATP, rise in cytosolic free Ca^{2+} concentration, activation of phospholipases, perturbation of the cytoskeleton, depletion of phospholipids, rise of toxic lipid breakdown products, accumulation of toxic oxygen radicals (O_2^-)

3.2.2 Modification of Membrane and Intraluminal Proteolytic Potential

In addition to the intra- and extracellular types of molecule degradation (digestion), a third type exists that may be characterized spatially as intermediate, i.e., membrane contact digestion. It occurs when substrates contact cell surfaces on the external side to which enzymes are fixed [53]. In proximal, but not distal tubules, radiolabeled angiotensin, bradykinin, glucagon and luteinizing hormone releasing hormone are hydrolyzed by brush border enzymes at the luminal membrane, followed by reabsorption of metabolites [12, 13].

Brush border enzymes are either sequestered into the external surface glycocalyx or bound in the cell membrane, with part of their molecular structure exposed to the cell exterior. Electron microscopy reveals knobby projections on the outside of brush border membranes that are removed by exposure to proteases. At the same time, enzymatic activity becomes solubilized. It is still unknown whether or not this occurs in connection with membrane digestion [12, 13].

The content of enzymatic activity in the tubular lumen is diminished as a consequence of reuptake of enzymes into the tubular cells. The endocytosis of enzymes is promoted by the presence of enzyme-specific recognition markers. A cellular pool of receptors exists, for instance, for β-glucuronidase. This enzyme cycles between intracellular and cell surface pools [19]. Haga et al. [54] also found evidence for the reabsorption of free acid hydrolases into particulated kidney vesicular material.

3.2.3 Enzyme Induction

The basic mechanism for enzyme regulation is the activation (induction) and inhibition of the gene transcription of specific enzymes. Enzyme induction may be prompted by metabolic substrates, xenobiotics, and hormonal regulation. Drugs and other xenobiotics may trigger the induction of "detoxifying" enzymes-isoenzymes of the cytochrome P-450 dependent monooxygenases and conjugation enzymes such as glutathione-S transferase (ligandin) and GGT. Hormonal regulation of enzymes via gene transcription was established for glucocorticoids, thyroid hormones, glucagon, and insulin. The probably best studied inducers are rifampicin, phenobarbital, phenytoin, chloral hydrate, glutethimide, antipyrine, and chronic alcohol ingestion [55]. Some of the enzymes used as markers of enzyme induction in the liver are also present in the tubular epithelium of the kidney. Wensing et al. [56] described an increased urinary GGT excretion both immediately and 1 week after cessation of antipyrine administration in 18 healthy control subjects. No changes were found following the administration of rifampicin. β-Glucuronidase and NAG excretion remained unchanged. Phenobarbital given over 10 days enhanced urinary excretion of GGT [55]. Similar data were also reported from patients on long-term therapy with anticonvulsant drugs [57]. Maruhn et al. [58] demonstrated in the rat that induction of monooxygenases in the kidney were not accompanied by an increase of the urinary output of GGT and/or glutathione-S-transferase.

Burchardt et al. [42, 59] noted that patients with hyperthyroidism excrete AAP at a high rate. As shown for glucocorticoids, thyroid hormones, which are involved in growth control, bind to a specific receptor molecule within the nucleus. Some hours later, the gene transcription of a set of enzymes is significantly increased compared with controls [60]. An enzyme-inducing element, i.e., a specific sequence at the DNA level may be considered responsible [61]. Insulin also rapidly increases or decreases the rate of gene transcription of specific enzymes, at least in the liver [60]. Data with respect to urinary enzymes are not available.

3.2.4 Impairment of the Cytoskeleton

There is growing evidence that the cellular damage leading to a release of enzymes with high molecular mass must not necessarily be irreversible [62, 63]. This phenomenon is also important in respect of releasing mechanisms into the serum. Certain disorders of the cytoskeletal actin filaments were recognized under electron microscopy; they are fairly common and non-specific in regard to enzyme release [49, 64]. Destruction and cross-linking of actin filaments labilize the plasma membrane, which forms protrusions in cells according to the pressure gradient and cell configuration. The so-called blebs can separate from the cell; "blebbing," "vesicle shedding," "apoptosis", and "zeiosis" are often used synonymously.

Bleb formation and sequestration of proximal tubular cell material into the tubular lumen is found in many pathological circumstances (so-called tissue proteinuria) [65] or brush border histuria [66] (see Figs. 4, 5). It was demonstrated in ischaemia [47, 64], heavy metal poisoning [67], aminoglycoside therapy [63, 68], administration of cytostatics and radiocontrast agents [49, 65]. Scherberich et al. [49] assessed the pathophysiological background of shedding of membrane bound enzymes from the proximal tubule in urine specimens of patients with renal damage. They applied immunospecific affinity chromatography, immunotitration, ultracentrifugation, electroimmunodiffusion, immunohistology, as well as negative staining technique. Compared with healthy control subjects, patients with kidney injury after drug administration revealed an increased excretion rate of vacuolar membrane fragments (50–500 nm) into urine. The brush border of the proximal tubule was identified as the main source of urinary blebs. In addition, the marker enzyme profile of urinary vacuolar blebs was similar to that of the brush border membrane from human kidney. Figure 6 shows brush border vesicles in the tubular lumina forming compact masses. These masses are able to obstruct tubular lumina and contribute to acute renal failure.

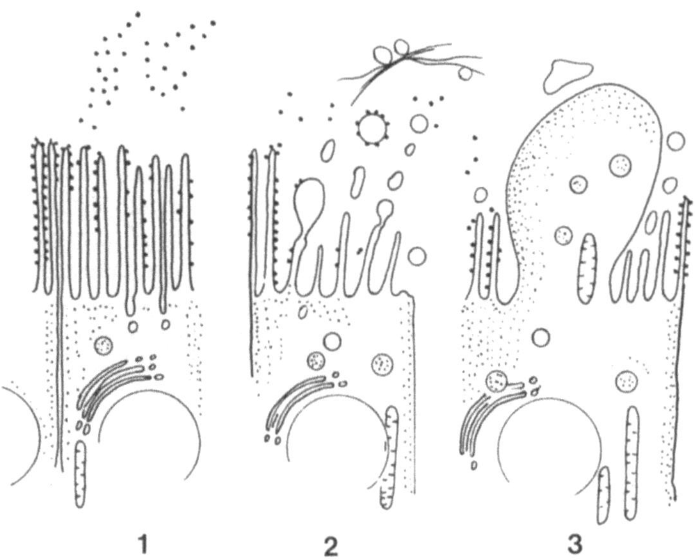

Fig. 5. Various stages of shedding of tubular proteins (and marker proteins) under pathological conditions: *1* Release of membrane surface proteins (e.g., AAP, GGT) in mild reversible tubular injury, *2* Shedding of integral brush border proteins and enzymes (e.g., AP) and vacuolar blebs. More severe damage is characterized by additional exfoliation of cytoskeletal components. *3* Severe blebbing and shedding of brush border material

During the initial phase of a tubular injury, 5–10 nmol surface glycoproteins of the brush border, among them angiotensinase A and portions of GGT, are released into urine. This might be followed by increased blebbing of macromolecular brush border fragments, indicating more severe membrane disruption.

In addition, filament components of the brush border and cytoskeleton, possibly representing actin bundles, actomyosin, fimbrin, and villin, were observed in the pellets after ultracentrifugation of urine specimens from patients receiving nephrotoxic drugs [69].

Tubule cells which have shed their brush border and part of their cytoplasm require only little energy. They do not take part in reabsorbing and metabolizing. Therefore, a greater portion of constituents of primary urine and a diminished amount of lysosomal and cytoplasmic enzymes appear in the final urine (see Fig. 1). This is the characteristic state in renal shrinkage.

3.2.5 Lipid Peroxidation

Oxygen free radicals have been implicated as inducers of tissue injury in several conditions such as inflammation, irradiation, ischaemia/reperfusion, circulatory

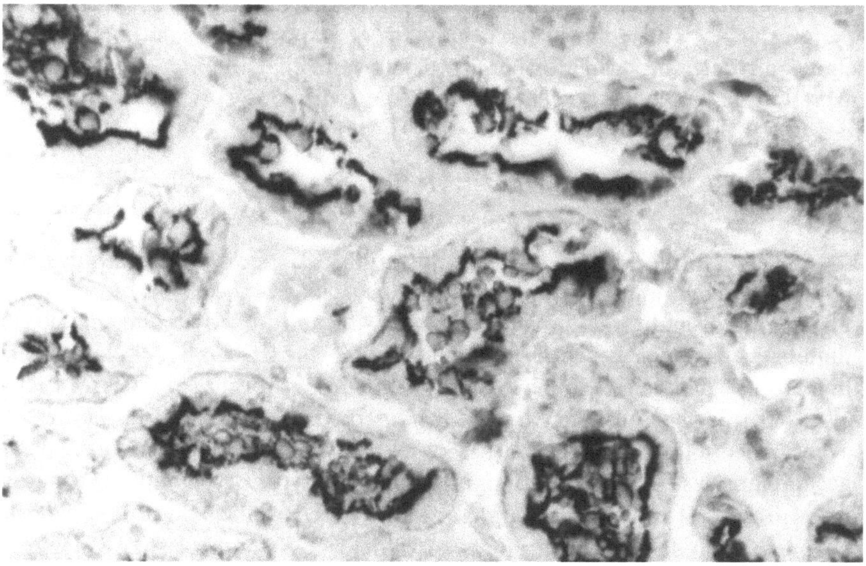

Fig. 6. Shedding of brush border enzymes under pathological conditions: paraffin-embedded kidney section of a patient with type I diabetes mellitus stained with a polyclonal antibody against AAP associated surface glycoprotein (SGP 240) hematoxylin counterstain. Intraluminal accumulation of immunoreactive material ("obstructing blebs") × 120

shock injury, and respiratory distress syndrome [14, 70–72]. They include superoxide anion, hydrogen peroxide, and the hydroxyl radical. They have been suggested to exert their cytotoxic effect by causing peroxidation of membrane phospholipids, which can result in the loss of membrane integrity [73].

The kidney's main defense mechanism against free radical induced oxidation is represented by the glutathione system and the activity of superoxide dismutase and catalase. Acute renal failure experimentally induced by either renal arterial clamping or intoxication with sodium maleate showed remarkable membrane losses in proximal and distal tubules of the rat kidney [74–76]. This was accompanied by a decrease in the activities of superoxide dismutase, catalase, and glutathione peroxidase combined with a diminished total tissue glutathione level. The occurrence of lipid peroxidation can be observed by measuring malondialdehyde levels in renal venous blood [75] and urine [77]. The pathogenic involvement of lipid peroxidation in acute renal failure is supported by data showing that scavenging of free radicals initiating lipid peroxidation leads to functional improvement [74, 77–79]. Cisplatin causes increased production of lipid peroxides in rats [79]. Gemba et al. [77] injected cisplatin into rats. Two days later, urinary NAG and GGT activities increased. The antioxidant *N–N*-diphenyl-*p*-phenylenediamine attenuated these increases in enzyme activities caused by cisplatin. The results of this study suggested that monitoring the change in urinary activities of some enzymes is the method of choice for detecting cisplatin nephrotoxicity and that the increase of enzyme excretion may involve the generation of free radicals [79].

3.2.6 Calcium Imbalance

An elevated intracellular Ca^{2+} concentration is a necessary prerequisite for initiating a membrane blebbing process. Ca^{2+} ions bind to anionic groups of phospholipids and glycolipids in the membrane and thereby reduce membrane fluidity. It has been generally understood that impaired intracellular calcium homeostasis plays a role in the loss of enzymes from the cells [52, 80] (see Fig. 4). Whether it is generally the cause of increased enzyme leakage or just an early event concomitant with the release of cytosolic enzymes or whether calcium accumulation is the ultimate common result of all types of cell injury are matters of controversy [52]. Calcium antagonists produce a blockade of slow calcium channels and reduce the intracellular calcium level. Neumayer et al. [81] found a significantly diminished AAP and GGT excretion in patients treated with radiographic contrast agents in combination with the calcium antagonist nitrendipin compared with control subjects without nitrendipin. In another study, the calcium channel blocker nifedipine minimized urinary NAG excretion after a radiographic contrast procedure [80]. This is consistent with the observation that calcium channel blockers are effective in the prevention of acute renal failure [82].

Raab and Hohenegger reported [83] that potassium depletion and kaliopenic nephropathy lead to an increase of urinary LDH and AP activity.

3.2.7 Renal Tissue Enzyme Activity, Cellular Repair, and Enzymuria

Enzymuria apparently depends on total tissue enzyme content and the capability of the nephron to recover from injury. After temporary renal ischaemia in rats, the decreased brush border enzyme activity returned to normal within 6 h, indicating effective repair mechanisms (enzyme neosynthesis and membrane protein recycling) [92]. Immunohistochemical studies on human kidney sections applying enzyme-specific antisera show variable expression of tubular marker proteins (enzymes) corresponding to the degree of cellular damage (Fig. 7–10). Compared with the glomerular marker enzyme angiotensinase A, early depletion of the tubular portion of this enzyme and also of other brush border enzymes demonstrates the predominant susceptibility and low threshold of the tubular nephron to ischaemic, inflammatory, or toxic damage (Fig. 7). In chronic progressive renal failure the total amount of tissue enzyme activity is significantly reduced (Fig. 8) [95]. However, protein handling and synthesis in remnant nephrons may be under the control of several growth promoting

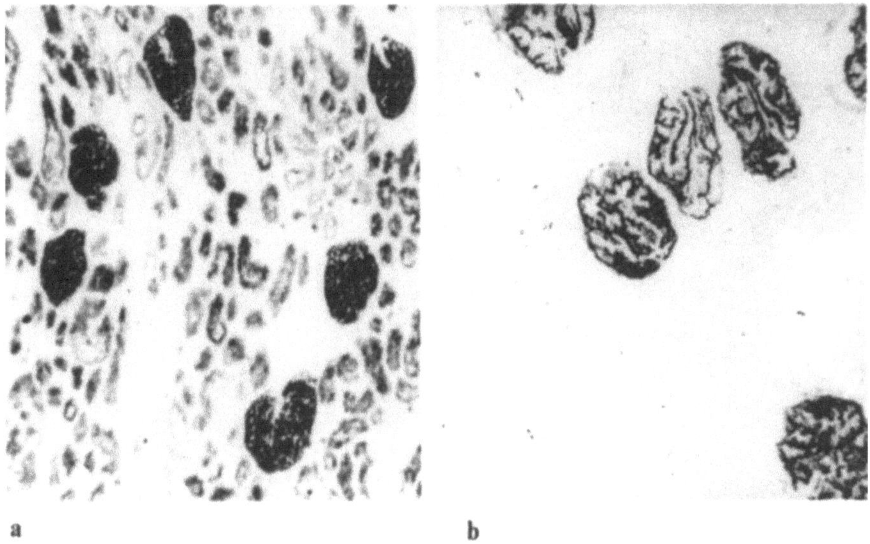

Fig. 7a, b Histochemical distribution of angiotensinase A in a normal kidney (**a**) and in a kidney with tubulointerstitial damage of a patient with renal cancer (**b**), demonstrating loss of angiotensinase A in highly vulnerable tubule segments. ×40

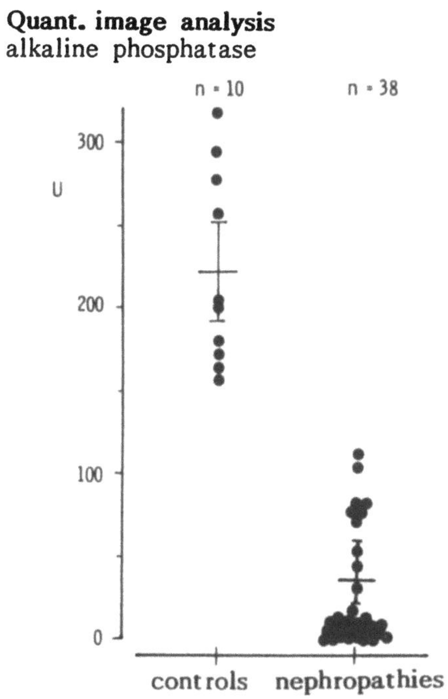

Fig. 8. Quantitative image analysis of alkaline phosphatase activity in kidney sections from normal controls and patients with various chronic nephropathies (adapted from [87–89, 94])

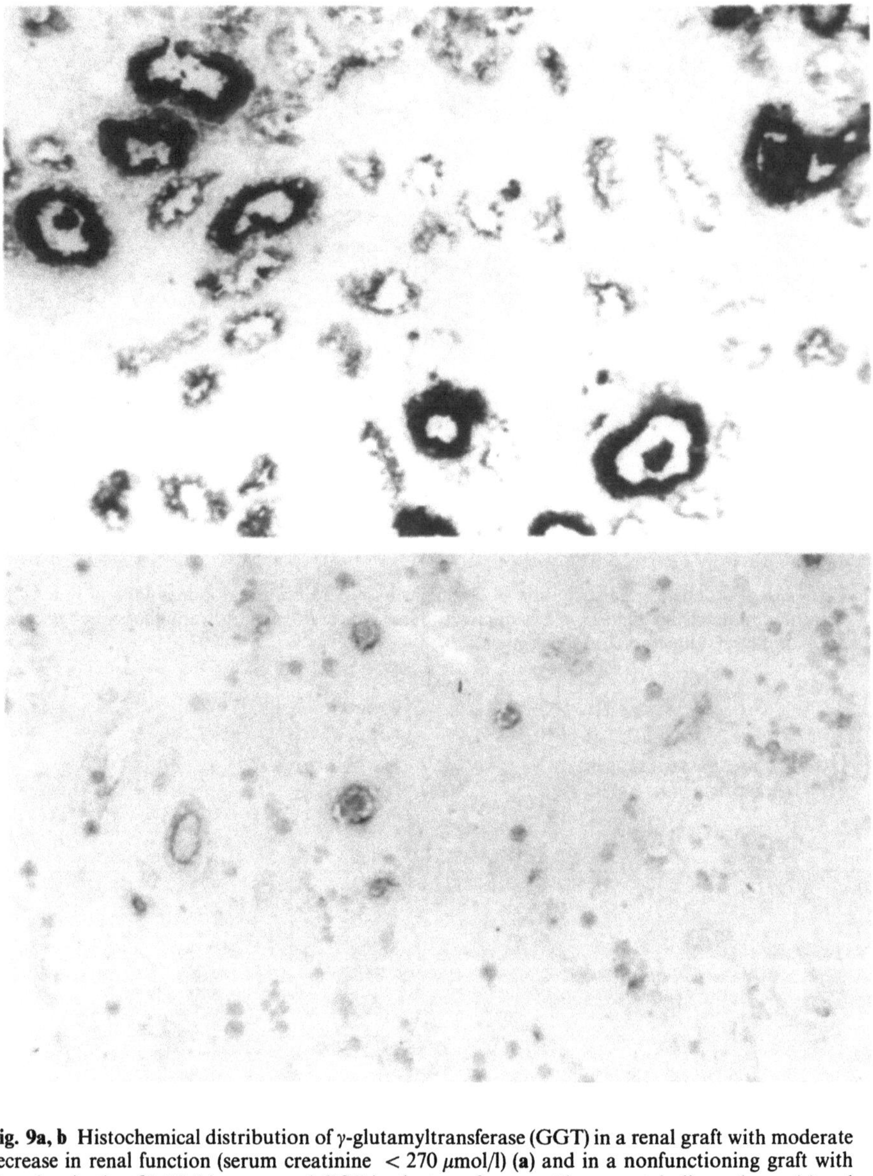

a

b

Fig. 9a, b Histochemical distribution of γ-glutamyltransferase (GGT) in a renal graft with moderate decrease in renal function (serum creatinine $< 270\ \mu mol/l$) (**a**) and in a nonfunctioning graft with chronic rejection (**b**). Highly expressed GGT in clusters of remnant hypermetabolic nephrons (**a**) and loss of tubular GGT in end-stage renal failure. × 100 (adapted from [94])

hormones and differentiation factors inducing a mitogenic response and increase of cell turnover in glomeruli and tubular epithelia (Table 3). Under these circumstances, renal epithelia show an enhanced expression of MHC class II antigens and increased enzyme activities, as was found for glomerular angioten-

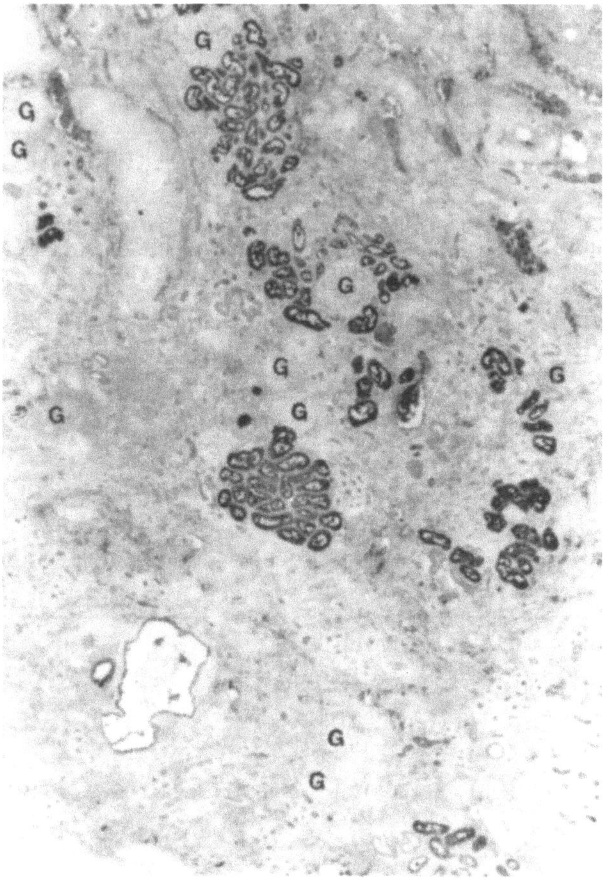

Fig. 10. Clusters of hypermetabolic tubules from remnant nephrons with high expression of alanine aminopeptidase (AAP) immunoreactivity. End-stage renal failure, kidney of a patient of chronic hemodialysis, paraffin section incubated with AAP-specific antibody, peroxidase labelling, haematoxylin counterstain. *G*, glomeruli. ×20

sinase and tubular γ-glutamyltransferase (GGT) (Fig. 9), which paralleled the compensatorily increased ammoniagenesis in remnant nephrons [84, 85, 94]. Thus, although progressive renal failure may parallel an overall decrease in marker enzyme synthesis and enzymuria, the enzyme activity per single remaining, hypermetabolic nephron may be compensatorily enhanced (Fig. 10) [93, 94].

Table 3. Growth and differentiation factors present in the kidney possibly modulating enzymuria [induction of protein (enzyme) synthesis, cell proliferation, epithelial hypertrophy and hyperplasia] (Summarized from [85, 86, 90, 91, 96, 97])

Type	Remarks
Epidermal growth factor (EGF)	Synthesis in thick ascending limb of distal tubule, luminal cell membranes; excreted in urine
Transforming growth factor β (TGFβ)	Synthesis in tubule, interstitial cells; promotion of fibronectin and procollagen (III), (IV) synthesis; responsible for interstitial fibrosis
Transforming growth factor α, lymphokines (IL-1)	Macrophages, lymphocytes of the renal interstitium
Insulin-like growth factors (IGF I, II)	Binding sites in glomeruli, proximal tubule; increased transcription of IGF-Specific mRNA after partial nephrectomy
Platelet-derived growth factor (PDGF)	Synthesis mainly in mesangial cells, endothelia, some epithelia
Fibroblast growth factor (FGF), angiotensin II (ANGII)	Interstitial fibrosis (?), angioneogenesis, tubular hypertrophy

References

1. Maunsbach AB (1976) Cellular mechanisms of tubular protein transport. Int Rev Physiol 11:145–167
2. Pruzanski W, Wilson DR (1977) Renal handling of endogenous lysozyme in man. J Lab Clin Med 90:61–67
3. Hansen NE (1974) Plasma lysozyme—a measure of neutrophil turnover. An analytical review. Series Haematol 7:7–87
4. Harrison JF, Barnes AD (1970) The urinary excretion of lysozyme. Clin Sci 38:533–540
5. Goudswaard J, Virella G (1983) Role and value of lysozyme determination for non-invasive diagnosis of kidney disease. Karg Cont Educ Ser 3:19–37
6. Maak T (1975) Renal handling of low molecular weight proteins. Am J Med 58:57–64
7. Cojocel C, Baumann K (1983) Renal handling of endogenous lysozyme in the rat. Renal Physiol 6:258–265
8. Junge W, Mályusz M, Ehrens HJ (1985) The role of the kidney in the elimination of pancreatic lipase and amylase from blood. J Clin Chem Clin Biochem 23:387–392
9. Jung K, Pergande M, Schulze BD, Precht K, May G (1989) Niedermolekulare Proteine im Serum und Urin bei nephrologischen Fragestellungen.· Biochemie—Pathobiochemie—Klinische Bedeutung. Z Klin Med 44:549–556
10. ten Kate RW, Pals G, Eriksson AW, Donker AJM, Meuwissen SGM (1989) The renal metabolism of pepsinogen A and C in man. Clin Nephrol 31:103–106
11. ten Kate RW, Pals G, Res JCJ, van Kamp GJ, Eriksson AW, Meuwissen SGM (1989) The glomerular sieving of pepsinogen A and C in man. Eur J Clin Invest 19:306–310
12. Carone FA, Peterson DR, Oparil S, Pullman T (1979) Renal tubular transport and catabolism of proteins and peptides. Kidney Int 16:271–278
13. Carone FA, Peterson DR, Flouret G (1982) Renal tubular processing of small peptide hormones. J Lab Clin Med 100:1–14
14. Pfaller W (1982) Structure function correlation on rat kidney. Adv Anat Embryol Cell Biol 70:1–106

15. Andersen KJ, Haga HJ, Dobrota M (1987) Lysosomes of the renal cortex: heterogeneity and role in protein handling. Kidney Int 31:886–897
16. Haga HJ (1989) Kidney lysosomes. Int J Biochem 21:343–345
17. Ryabov SI, Plotkin VY, Nevorotin AJ (1981) Intracellular routes of a lysosomal marker enzyme, acid phosphatase, in proximal convoluted tubule cells of human kidney in glomerulonephritis as studied by electron microscopy. Nephron 29:68–72
18. Haga HJ, Andersen KJ (1985) Isoenzymes of N-acetyl-β-D-glucosaminidase in kidney cortex and urine of normal and nephritic rats. Renal Physiol 8:348–356
19. Haga HJ, Andersen KJ, Dobrota M (1987) Latency of acid hydrolases in rat kidney cortex. Int J Biochem 19:1097–1103
20. de Duve C, Barsy T, Poole B, Trouet A, Tulkens P, van Hoof F (1974) Lysosomotropic agents. Biochem Pharmacol 23:2495–2531
21. Powell JH, Reidenberg MM (1982) In vitro response of rat and human kidney lysosomes to aminoglykosides. Biochem Pharmacol 31:3447–3453
22. Burchardt U, Krosch H, Müller G, Haschen RJ (1979) Changes of urinary enzyme excretion after drug application. In: Dubach UC, Schmidt U (eds) Diagnostic significance of enzymes and proteins in urine. Huber, Bern, pp 183–188 (Curr Probl Clin Biochem 9)
23. Burchardt U, Schinköthe G, Müller G, Neef L, Krosch H (1978) Ausscheidungskinetik von Enzymen und Protein mit dem Harn bei Applikation therapeutischer Gentamycindosen. Schweiz Med Wschr 108:1541–1545
24. Patel V, Luft FC, Yum MN, Patel B, Zeman W, Kleit SA (1975) Enzymuria in gentamicin-induced kidney damage. Antimicrob Agents Chemother 7:364–369
25. Kang SS, Price RG, Yudkin J, Worcester NA, Bruckdorfer KR (1979) The influence of dietary carbohydrate and fat on kidney calcification and the urinary excretion of N-acetyl-β-glucosaminidase. Br J Nutr 41:65–71
26. Li MK, Kavanagh JP, Prendiville V, Buxton A, Moss DG, Blacklock NJ (1986) Does sucrose damage kidneys? Br J Urol 58:353–357
27. Brouhard BH, Lagrone L, Rowe J (1985) Acute response of urinary N-acetyl-beta-D-glucosaminidase to mannitol infusion in the dog. Am J Med Sci 290:11–14
28. Hofmeister R, Bhargava A, Günzel P (1990) The use of urinary N-acetyl-β-D-glucosaminidase (NAG) for the detection of contrast—media—induced "osmotic nephrosis" in rats. Toxicol Lett 50:9–15
29. Nicot GS, Merle LJ, Charmes JP, Valette JP (1984) Transient glomerular proteinuria, enzymuria, and nephrotoxic reaction induced by radiocontrast media. J Am Med Ass 252:2432–2434
30. Schiavina G, Arrigo G, Vergnaghi D, Cavaliere G (1984) Radiocontrast media nephrotoxicity. Comparison between iopamidote and diatrizoate by measurement of urinary enzyme excretion after intravenous pyelography. Radiol Med 70:176–179
31. Goldstein EJ, Feinfeld DA, Fleischner GM, Elkin M (1976) Enzymatic evidence of renal tubular damage following renal angiography. Radiology 121:617–619
32. Langhendries JP, Battisti O, Bertrand JM (1988) Aminoglycoside nephrotoxicity and urinary excretion of N-acetyl-beta-D-glucosaminidase. Biol Neonate 53:253–259
33. Burchardt U, Schinköthe G, Anton D, Dreller E, Neef L, Ullrich D, Patsch R (1983) Enzymausscheidungsverläufe bei medikamenteninduzierten Nierenveränderungen. Dt Gesundh Wesen 38:1434–1439
34. Hay M (1986) Test for bile acids. In: Landois A, Stirling A (eds) A text book of human physiology. London
35. Back P (1973) Identification and quantitative determination of urinary bile acids excreted in cholestasis. Clin Chim Acta 44:199–207
36. Mills CO, Elias E, Martin CHB, Woo MTC, Winder AF (1988) Surface tension properties of human urine: relationship with bile salt concentration. J Clin Chem Clin Biochem 26:187–194
37. Burchardt U (1983) Harnenzymausscheidung bei Erkrankung der inneren Organe. Z Gesamte Inn Med 38: 565–570
38. Huseby NE, Vik T (1978) Membrane enzymes and bile salts. Clin Chim Acta 88:385–392
39. Bossmann B, Haschen RJ (1983) Release of enzymes from rat jejunal mucosa during in vitro incubation. J Clin Chem Clin Biochem 21:659–663
40. Bossmann B, Haschen RJ (1984) De novo synthesis of brush border membrane enzyme during intestinal perfusion with bile salt in the rat. J Clin Chem Clin Biochem 22:449–451
41. Bossmann B, Haschen RJ (1983) Release of enzymes from rat jejunal mucosa by bile salts. J Clin Chem Clin Biochem 21:1–9

42. Burchardt U, Peters JE, Neef L, Thulin H, Gründig CA, Haschen RJ (1977) Der diagnostische Wert von Enzymbestimmungen im Harn. Z Med Labor Diagn 18:190–212
43. Piper HM, Sphor R, Hütter JF, Spieckermann PG (1985) Enzyme release and glycolytic energy production. Basic Res Cardiol 80 (Suppl 1):143–148
44. Donohoe JF, Venkatachalam MA, Bernard DB, Levinsky NG (1978) Tubular leakage and obstruction after renal ischemia: structural-functional correlations. Kidney Int 13:208–222
45. Gebhard MM, Denkhaus H, Sakai K, Spieckermann PG (1977) Energy metabolism and enzyme release. J Mol Med 2:271–283
46. Bretschneider HJ, Helmchen U, Kehrer G (1988) Nierenprotektion. Klin Wochenschr 66:817–827
47. Venkatachalam MA, Jones DB, Rennke HG, Sandstrom D, Patel Y (1981) Mechanisms of proximal tubule brush border loss and regeneration following mild renal ischemia. Lab Invest 45:355–365
48. Kreisberg JI, Venkatachalam MA (1985) Membrane alterations in renal cell injury due to impaired energy metabolism. Mol Physiol 8:599–614
49. Scherberich JE (1989) Immunological and ultrastructural analysis of loss of tubular membrane-bound enzymes in patients with renal damage. Clin Chim Acta 185:271–282
50. Bohle A, Gärtner HV, Laberke HG, Krück F (1984) Die Niere. Struktur und Funktion. Schattauer, Stuttgart
51. Kehrer G, Blech M, Kallerhoff M, Bretschneider HJ (1989) Urinary LDH-release for evaluation of postischemic renal function. Klin Wochenschr 67:477–485
52. Schmidt E, Schmidt FW (1987) Enzyme release. J Clin Chem Clin Biochem 25:525–540
53. Ugulev AM (1965) Membrane (contact) digestion. Physiol Rev 45:555–596
54. Hasilik A (1980) Biosynthesis of lysosomal enzymes. Trends Biochem Sci 5:237–240
55. Heidemann HT, Ohnhaus EE (1988) Non—invasive techniques for testing enzyme induction. J Clin Chem Chin Biochem 26:499–450
56. Wensing G, Maruhn D, Ohnhaus EE (1988) The effect of antipyrine and rifampicin on the excretion of renal enzymes in human urine. Clin Nephrol 29:69–74
57. Landon MJ, Bates D, Kirkley M, Venables G (1978) Effects of anticonvulsant drugs on urinary drugs on excretion of gamma-glutamyltransferase in women. Ann Clin Biochem 15:313–315
58. Maruhn D, Schmidt U, Wehling K (1988) Induction of kidney monooxygenases, and its influence on the urinary excretion of renal enzymes. J Clin Chem Clin Biochem 26:499
59. Burchardt U, Haschen RJ, Krosch H (1979) Clinical usefulness of enzyme determination in urine. In: Dubach UC, Schmidt U (eds) Diagnostic significance of enzymes and proteins in urine. Huber, Bern, pp 106–112 (Curr Probl Clin Biochem 9)
60. Höppner W, Bünsch A, Seitz HJ (1988) Hormonal regulation of gene expression of enzymes. J Clin Chem Clin Biochem 26:506–508
61. Davis PJ, Blas SD (1981) In vitro stimulation of human red blood cell Ca^{++}-ATPase by thyroid hormone. Biochem Biophys Res Com 99:1073–1080
62. Mondorf W, Scherberich JE, Falkenberg FW, Sachse H, Gauhl C, Schoeppe W (1984) Brush border enzymes and drug nephrotoxicity. In: Solez K, Whelton A (eds) Acute renal failure. Dekker, New York
63. Price RG (1982) Urinary enzymes, nephrotoxicity and renal disease. Toxicology 23:99–134
64. De Broe ME, Wieme RJ, Logghe GN, Roels F (1977) Spontaneous shedding of plasma membrane fragments in human cells in vivo and in vitro. Clin Chim Acta 81:237–245
65. Scherberich JE, Tuengerthal S, Kollathal S, Kollath J (1983) Monitoring contrast media nephrotoxicity by specific kidney tissue proteinuria of membrane antigens. Fortschr Geb Röntgenstr Nuklearmed 118:37–42
66. Scherberich JE, Birk HW, Schoeppe W (1990) Tubule-derived membrane glycoproteins in the urine of patients (including those with AIDS) as analysed by radioimmunoblotting. J Clin Chem Clin Biochem 28:31–35
67. Ganote CE, Reimer A, Jennings RB (1975) Acute mercuric chloride toxicity. An electron microscopie and metabolic study. Lab Invest 31:633–651
68. Scherberich JE, Mondorf W, Schoeppe W (1980) Side effects of antimicrobial therapy: nephrotoxic potential of antibiotics (aminoglycosides, cephalosporins) as monitored by brush border tissue proteinuria. In: Losse H, Asscher AW, Lison AE (eds) Pyelonephritis, urinary tract infection, vol 4. Thieme, Stuttgart
69. Scherberich J, Gauhl C, Heinert G, Mondorf W, Schoeppe W (1984) Characterization and clinical significance of membrane bound proteases from human kidney cortex. In: Horl WH, Heidland A (eds) Proteases, potential role in health and disease. Plenum, New York

70. Mc Cord JM (1985) Oxygen-derived free radicals in postischemic tissue injury. New Engl J Med 312:159–163
71. Prasad K, Kalra J, Chan WP, Chaudhary AK (1989) Effect of oxygen free radicals on cardiovascular function at organ and cellular levels. Am Heart J 177:1192–1202
72. Slater TP (1984) Free radical mechanisms in tissue injury (Review). Biochem J 222:1–15
73. Freeman BA, Crapo JD (1982) Biology of disease. Free radical tissue injury. Lab Invest 47:412–426
74. Dobyan DC, Bull JM, Strebel FR, Sunderland BA, Bulger RE (1986) Protective effects of O-(β-Hydroxyethyl)-rutoside in der cis-platinum-induced acute renal failure in the rat. Lab Invest 55:557–563
75. Joannides M, Bonn G, Pfaller W (1989) Lipid peroxidation—an initial event in experimental acute renal failure. Renal Phys Biochem 12:47–55
76. Paller MS, Hoidal JR, Ferris TS (1984) Oxigen free radicals in ischemic acute renal failure in the rat. J Clin Invest 74:1156–1164
77. Gemba M, Fukuishi N, Nakano S (1988) Efect of N-N -Dipheny-*p*-phenylenediamine pretreatment on urinary enzyme excretion in cisplatin nephrotoxicity in rats. Jpn J Pharmacol 46:90–93
78. Hagen TM, Aw TY, Jones DP (1988) Glutathione uptake and protection against oxidative injury in isolated kidney cells. Kidney Int 34:74–81
79. Sugihara K, Gembe M (1986) Modification of cisplatin toxicity by antioxidants. Jpn J Pharmacol 40:353–355
80. Rodicio JL, Morales JM, Ruilope LM (1990) Calcium antagonists and the kidney. Nephrol Dial Transplant 5:81–86
81. Neumayer HH, Junge W, Küfner A, Wenning A (1989) Prevention of radiocontrast— media induced nephrotoxicity by the calcium channel blocker nitrendipine: a prospective randomised clinical trial. Nephrol Dial Transplant 4:1030–1036
82. Duggan KA, Macdonald GJ, Charlesworth JA, Pussell BA (1985) Verapamil prevents post-transplant oliguric renal failure. Clin Nephrol 24:289–291
83. Raab W, Hohenegger M (1969) Elektrolyte und renale Enzymausscheidung. Z Klin Chem Klin Biochem 7:384–389
84. Davies M, Coles GA, Thomas GJ, Martin J, Lovett DH (1990) Proteinases and the glomerulus: their role in glomerular diseases. Klin Wochenschr 68:1145–1149
85. Fine L (1986) The biology of renal hypertrophy. Kidney Int 29:619–634
86. Habenicht A (ed) (1990) Growth factors, differentiation factors and cytokines. Springer, Berlin Heidelberg New York
87. Heinert G, Wyrobnik J, Scherberich JE (1980) Quantitative computed histophotometry of membrane-integrated and lysosomal enzymes indicating inductive and alternative effects of aminoglycosides. Curr Chemother Immunother 11:252–254
88. Heinert G, Rauh W, Scherberich JE, Mondorf W, Weber W (1981) Quantitative enzymatic histophotometry of morphologic alterations caused by urologically relevant tubular kidney damage using computed image analysis device technique. Urol Int 9:178–193
89. Heinert G, Scherberich JE, Mondorf W, Weber W (1983) Quantitative enzymatic and immunologic computer-assisted histophotometry of human kidney tissue following neoplastic and other clinically significant alterations. Eur Urol 9:235–241
90. Kenny AJ, O'Hare MJ, Gusterson BA (1989) Cell-surface peptidases as modulators of growth and differentiation. Lancet II:785–787
91. Mendley SR, Toback FG (1990) Cell proliferation in the end-stage kidney. Am J Kidney Dis 16:80–84
92. Paddock JK, Taddeo AR, Paddock JP, Loewenstein LM (1985) Glycoprotein reutilisation in regenerating microvilli after renal ischaemia in rats. Kidney Int 28:498–503
93. Scherberich JE (1990) Urinary proteins of tubular origin: basic immunochemical and clinical aspects. Am J Nephrol 10 [Suppl 1]: 43–51
94. Scherberich JE, Wolf G, Albers C, Nowack, A, Stuckhardt C, Schoeppe W (1989) Glomerular and tubular membrane antigens reflecting cellular adaptation in human renal failure. Kidney Int [Suppl] 27:S38–S51
95. Scherberich JE, Wolf G, Stuckhardt C, Kugler P, Schoeppe W (1989) Characterization and clinical role of glomerular and tubular proteases from human kidney. Adv Exp Med Biol 240:275–282
96. Weinberg JM (1991) The cell biology of ischaemic renal injury. Kidney Int 39:476–500
97. Wolf G, Neilson EG (1991) Molecular mechanisms of tubulointerstitial hypertrophy and hyperplasia. Kidney Int 39:401–420

4 Enzymuria as an Indicator of Renal Pathomorphology

W. Rebel, T. Bertsch, G. Bode, and H. Bleuel

4.1 Introduction

The pathogenic mechanisms of renal injury are extremely diverse. Depending on the degree of injury, functional impairment and various pathomorphological patterns of cellular damage can develop. Disruption of the cellular integrity can lead to whole series of disorders ranging from membrane dysfunction accompanied by increased permeability to complete cellular necrosis. Considerable gains have been made in the knowledge of cellular dysfunction through ultrastructural research, biochemistry, cytochemistry and autoradiography. It has become possible to determine the biochemical data of cell organelles, establish metabolic pathways, assign cytoplasmic structures and trace the temporal course of specific metabolites within the cell. However, in the majority of cases it is still extremely difficult to establish a correlation between pathological cytoplasmic alterations and biochemically defined metabolic disorders. Usually, dysfunction cannot be concluded until structural changes have occurred. Metabolic dysfunction, as a result of cellular injury, is attributed to a pathogenic process when an appropriate morphological substrate has formed. Such disorders of cellular metabolism result in an imbalance between functional cytoplasmic structures and cellular metabolic function. Fundamentally, there are three ways in which cellular injury can affect function (metabolic performance):

1. Hyperfunction with increased performance (hypertrophy/hyperplasia)
2. Hypofunction with decreased performance (degenerative processes/atrophy)
3. Dysfunction with formation of false metabolic or differentiation products (metaplasia, tumour growth)

Not every disorder of cellular metabolism results in irreversible injury. The extremely high structural plasticity of the cell means that it is fundamentally possible for the cell to adapt to the metabolic disorder.

The various parts of the cell contain a diverse range of enzymes in differing concentrations. Detection of enzymes in a variety of body fluids can therefore be used as a diagnostic criterion for establishing changes in the pathophysiological status. Undeniably, enzyme diagnosis ought to be regarded with the highest esteem. Without enzyme detection, differential diagnosis of the most varied disease symptoms would simply not be possible.

Urinary Enzymes
K. Jung/H. Mattenheimer/U. Burchardt (Eds.)

Advances in enzymology have paved the way for our understanding of physiological and pathophysiological renal functional conditions [1]. Detection of enzymes in the urine enables recognition of changes in cellular function and the resultant morphological substrate, as a result of which it is possible to diagnose damage to renal tissue at a very early stage and therefore enable treatment to be implemented as soon as possible.

Classic clinical chemistry parameters that enable a picture to be formed of the functional status of the kidney, such as creatinine level [2, 3], blood urea nitrogen (BUN) [4, 5] and urine osmolarity, are insensitive methods which only enable the detection of increases with major functional impairment. The same problem is associated with the determination of clearance, e.g. of para-aminohippuric acid (PAH), inulin [6], etc.

Methods which enable the early detection of functional alterations of the nephron are not only beneficial instruments of medicine but are also of value for estimating the risks of novel drugs under development and for fundamental experimental biological research.

Like all tissues, the renal parenchyma contains a multitude of enzymes. According to overview tables of known enzymes [7] and an estimation by Longley [8], more than 300 enzymes should be detectable in renal tissue. In principle, tissue enzymes can be intracytoplasmic membrane-bound or bind to membranes and intracytoplasmic ultrastructures. Normal turnover or disease of the tissue causes enzymes to be released, which can then be detected by suitable methods.

Enzyme-rich, renal tubular cells undergo normal turnover [9]. The cells desquamate and are destroyed by the urine, as a result of which their enzymes can be found in the urine. This process, however, causes some of the enzymes to be deactivated and destroyed. In addition, minor quantities of intracytoplasmic enzymes [9] reach the outside by continuous membrane permeation. Mechanical or functional alterations may also cause these membrane enzymes to be supplied directly to the lumen, particularly under conditions of elevated diuresis [10–13].

The nephron, with its various functions, represents the functional anatomic unit of the kidney—the vascular system (afferent and efferent arterioles), glomerulus and tubular system. The kidney is an anatomically highly complex tissue system. It can be divided into 4 zones: *cortex* comprising the glomerulus, the proximal and convoluted tubules and some of the distal tubules; *outer strip of the outer zone of the medulla* comprising the proximal and straight tubules; *inner strip of the outer zone of the medulla* comprising the thin portion of the loop of Henle and the distal tubules; *inner medulla* comprising the thin portion of the loop of Henle and the collecting tubules.

In addition to the topographical distribution, taking into consideration both morphology and function [14, 15], species and sex-specific differences can be expounded by recording quantitative differences in the pattern of distribution of various enzymes in the different tubular segments and in part by microdissection [16, 17] and histochemical methods [18–21].

The most important tasks of the kidney are the regulation of the fluid balance, electrolyte metabolism, mineral metabolism, acid-base balance and blood pressure. In addition to these important regulatory and control functions, the kidneys produce hormones that intervene in mineral metabolism (vitamin D_3) [22, 23], blood pressure regulation (renin) [24] and erythropoietic blood formation (erythropoietin) [25].

Each part of the nephron has a specific function. The glomerulus is a specialized capillary bed, whose porous walls selectively filter the blood plasma. The subsequent tubular system then selectively reabsorbs 98%–99% of the filtrate. Actively excretable material is discharged to the urine predominantly via the proximal tubules.

4.2 Principles of Renal Pathophysiology and Morphology

4.2.1 Glomerular Diseases

Acute injury of the glomerular apparatus results in an increased permeability of the capillaries, mainly associated with changes to the basal membrane. This process results in an elevated permeability to plasma protein, erythrocytes and lymphocytes in the ultrafiltrate. With increased cellular inflammatory reaction, proliferative processes occur as well as mesangial and endothelial cellular changes. Advanced chronic and recurrent glomerular injury elicits an increased formation of connective tissue. This, on the one hand, destroys the complete glomerulus, enabling blood to flow into the tubular system unhindered. On the other hand, necrotic processes obliterate the capillaries, considerably reducing the filtration rate and through-flow.

Disease-related changes in the region of the glomerular capillary wall, in the subsequent tubular system and in the interstitium as well as its vascular system affect the process of glomerular filtration in a variety of ways, both qualitatively and quantitatively.

The relationship between glomerular injury and reactive alteration of the tubular system can be seen in toxic injury of the glomerulus (e.g. by puromycin), where an elevated urinary excretion of lysosomal enzymes such as *N*-acetyl-*β*-D-glucosaminidase (NAG) [26], *β*-glactosidase and acid phosphatase as well as the brush border enzyme alkaline phosphatase (AP) occurs. Binding of immunocomplexes to the glomerular basal membrane also results in an elevated excretion of alkaline phosphatase [27].

4.2.2 Tubular Diseases

In addition to circulatory disorders, toxic substances are primarily responsible for acute tubular damage, including necrosis of the tubular apparatus. Whereas the clinical symptoms are portrayed as disruption in the production of urine (oliguria and anuria), the morphological picture is extremely diverse: desquamation of the brush border, elevated membrane permeability, atrophy, hyperplasia, degeneration and necrosis. In addition to destructive processes, reparative processes also take place, resulting in the generation of scar tissue.

With chronic renal disease, regardless of aetiology (vascular, glomerular or inflammatory), the tubules may be destroyed or atrophy with thickening of the peritubular basal membrane and interstitial fibrosis. The consequence of these changes is an impairment of function, resulting in the disruption of both the pump mechanism and urine production as well as dysregulation of the water and electrolyte balance.

In the early stages of injury, reduction or even a complete disappearance of the brush border [28] may be observed, predominantly in the vicinity of the proximal tubular cells. This causes an increase in brush border enzymes [gamma-glutamyltransferase (GGT), alkaline phosphatase] in the urine. A decrease in the enzyme content of the tissue can also be seen [29]. With the total disappearance of the brush border, it is important to note that the initial increase of enzymes in the urine is followed by a decrease caused by the mechanism of exhaustion.

Tubular injury, e.g. by cyclosporin, can be characterized with the aid of a light microscope in terms of varying degrees of cytoplasmic vacuolic formation. In addition to sub-cellular alterations, such as increased numbers of lysosomes or the formation of myeloid bodies [30], as can also be found with aminoglycoside nephropathy [31, 32], or basal lipid droplets, it is the endoplasmic reticulum [33] that is primarily affected ultrastructurally. These injuries correlate with the rise in lysosomal enzymes, especially that of NAG in the urine [30, 34].

The elevated excretion of enzymes in the urine is used as an index of sensitivity for estimating toxic renal damage elicited by cyclosporin. The combination of cyclosporin with potential nephrotoxic substances such as gentamicin, loop diuretics or inhibitors of liver metabolism may provoke an increased excretion of NAG, which is interpreted as an increase in renal damage [30]. In contrast, combined administration of cyclosporin with stimulators of hepatic metabolism such as phenobarbital [30] or rifampicin [35] lower NAG excretion, from which a nephro-protective effect can be concluded.

In a similar manner, the combined administration of cyclosporin with angiotensin-converting enzyme (ACE) inhibitors or aldosterone antagonists in animal experiments resulted in a reduced nephrotoxicity of cyclosporin [36]. Both compounds (enalapril and spironolactone) have been shown to affect aldosterone function in different ways. As aldosterone possesses pathogenic significance in the nephrotoxicity of cyclosporin, ACE inhibitors and aldosterone antagonists are recognized as having a nephro-protective effect.

In addition to drug monitoring as well as physiological (glomerular filtration rate GFR), clinical chemistry (plasma renin concentration) and urinary (Na^+ excretion) parameters, a great deal of significance is attributed to investigations into tubular enzymuria for monitoring cyclosporin nephrotoxicity.

Injury of the proximal tubules by mercuric chloride results in an increased urinary excretion of lysosomal (β-galactosidase, NAG), cytosolic [lactate dehydrogenase (LDH), glutathione-S-transferase] and brush border enzymes (GGT, leucine arylamidase) [37].

A clear rise in GGT can be demonstrated in kidneys damaged by cisplatin, its maximum activity being demonstrable at the brush border of the proximal tubules [38]. Morphologically, there is a correlation between increased GGT excretion and the loss of microvilli from tubular cells. On administration of a single dose of cisplatin, GGT excretion rises up to day 4 and then normalizes between days 8 and 10. Ultrastructurally, this corresponds to a loss of microvilli which have once again regenerated by day 10. Biochemical and immunohistochemical tests have shown that the highest concentration of GGT is found in the pars recta of the nephron [39, 40].

Hautmann [41] has shown that LDH activity in the urine of patients with kidney tumour is significantly raised, although this is also the case in patients with other renal disorders such as inflammation, etc. Malate dehydrogenase was found to exhibit less activity in tumour tissue. Its level of activity in the urine, on the other hand, was increased. The brush border enzyme GGT was found to have significantly decreased in both renal tissue and urine.

With the various types of transplant rejection, increased activity of the cell-bound immunosystem (T cell infiltrates) [42] as well as of the humoral limb of the immune system (deposits of immunocomplexes in the kidney vessels) [43–45] can be found. Investigations of the increased enzyme excretion during this special pathophysiological situation and the resultant pathomorphological changes have been carried out [46, 47].

In general, necrosis of the tubular apparatus commences with a rise in lysosomal enzymes in the urine [48, 49]. In the case of necrotic degradation of the tubular cells, the brush border enzyme alkaline phosphatase remains present for a much longer period [50–52]. This enzyme, however, cannot be found in atrophic and regenerating cells [51–53].

Acute, experimentally induced injury of the proximal tubules by HCBD (hexachlorobutadiene) [54, 55], papillary damage by BEA (2-bromo ethylamine) [56, 57] and injury of the glomerulus by PAN (puromycin nucleoside [58, 59] have shown that alkaline phosphatase activity increases primarily with proximal tubular injury, whereas NAG activity increases when damage to the papillaries occurs. Glomerular injury was mainly characterized by proteinuria, glucosuria, biochemical alterations (rise in BUN) and changes in GFR. Alkaline phosphatase and NAG in the urine rose from day 6, with detectable histomorphological tubular damage being seen as nephrotic degradation [60].

In 1973, Ellis et al. [61] showed in canine experiments, in which damage to the kidney was elicited by $HgCl_2$, that a pathological rise in urinary enzyme activity occurred well before detectable functional disorders or tissue damage was seen.

Each structure of the nephron can be injured by the most diverse causes. The vascular system supplies the kidney with oxygen and metabolic substrates. It transports various substances to the tubular system for excretion and carries away reabsorbed substances for return to the circulation. The glomerulus, with its specialized porous capillary bed, acts as a selective filter. The subsequent tubular system selectively reabsorbs the primary filtrate, particularly in the proximal section, where active material is excreted into the urine.

The cellular response of these structures to harmful noxae can vary from scarcely visible biochemical changes to cell death (necrosis). Most noxae which damage the kidney are not exclusively nephrotoxic but elicit global cell damage instead, although the kidney tends to be the preferred target organ. This is due to the high blood flow (25% of the cardiovascular volume), the high O_2 consumption of the renal cortex, the possibility of elevated concentrations of freely filterable toxins in the tubular fluid and therefore in the renal tissues and the particular permeable properties of the proximal tubular cells.

Renal injury elicited by Hg^{2+} and CCl_4 is regarded as the prototype of acute nephrotoxic failure, with mercury complexing with proteins and enzyme structures. The general pathological principle involves the inhibition of the energy-donating or enzyme-producing processes, especially in the proximal tubular cells. Disseminated tubular necrosis with destruction of the tubular basal membrane is found pathomorphologically with these shock kidney-like disease symptoms [62]. With transient nephrotoxic renal failure, this change is uniformly restricted to the proximal tubular segment (site of enrichment) [63–70]. Normally, the tubular basal membrane remains undamaged so that later regeneration of the epithelium is possible (reversible damage). Increased excretion of enzymes following administration of mercury is known to occur [71–74].

In *acute renal failure*, caused for example by methoxyflurane, an anaesthetic agent, the fluoride impairs the tubular ion transport system, resulting in the retention of Na^+, urea and creatinine [62]. Methoxyflurane is metabolized in the liver to oxalic acid, dichloroacetic acid and difluoromethoxyacetic acid. These compounds can elicit chronic renal damage with interstitial fibrosis, which in turn can lead to persistent dysfunction [75, pp 433–438, 76].

An example of *irreversible and persistent renal insufficiency* is ethylene glycol intoxication [62, 75, pp 433–438]. Non-toxic ethylene glycol is metabolized via glycoaldehyde to oxalic acid which is then enriched and reabsorbed in the tubules and deposited as calcium oxalate. It forms crystals (inflammatory process) and causes necrotic obliteration of the tissues and consequently interstitial fibrosis.

The physical stimulus of these crystalline deposits and the resultant inflammatory processes can then elicit enzymuria. It has been shown that urinary excretion of glycosidases and acidic phosphatase, as a lysosomal enzyme, was elevated after oxalates had crystallized out [77].

Man is subjected daily to nephrotoxic substances in the form of various environmental toxins, medicines, chemicals, etc. In medicine, there is a whole series of drugs which posses a dose-dependent nephrotoxic potential, so that only their targeted use can be justified. For example, aminoglycosides are dose-dependently nephrotoxic. These antibiotic agents are bactericidal at high doses, inhibiting protein synthesis in the bacterial cell. The aminoglycoside penetrates the cell, where it is absorbed by the lysosomes and transformed to cations in their acid medium [78–80] so that it can no longer exit via the lysosomal membrane. Consequently, the enriched aminoglycoside H^+ complexes demolish the lysosomal membrane, causing the release of proteases and the subsequent destruction of the cell [81, 82]. This mechanism of intra-lysosomal aminoglycoside enrichment only occurs in the proximal tubular cells.

Using urinary NAG excretion as the parameter of nephrotoxicity, both daily and seasonal differences in sensitivity have been found with the administration of aminoglycosides. NAG excretion was greatest in the afternoon (at 2 p.m.) [83]. Circadian changes of the tubular epithelium and transitional epithelium can be demonstrated autoradiographically and caryometrically. The most voluminous cell nuclei were found around 10 a.m. and 6 p.m. in the epithelia of the primary and middle segments. The shift in the nucleus-plasma ratio, as a sign of active synthesis or increased regeneration, indicates that maximum tubular regeneration occurs late morning (10 a.m.) or late afternoon (6 p.m.) [84–86].

With regard to aminoglycoside nephrotoxicity, it can be concluded that rats in the status of highest physiological cell turnover are less susceptible to nephrotoxic irritants in the mornings, whereas in the early afternoon they are clearly more sensitive. In addition to the excretion of lysosomal enzymes, excretional characteristics of brush border enzymes such as GGT [87] and AAP [88, 89] as well as cytosolic enzymes such as LDH [90, 91] were also investigated following the administration of aminoglycosides.

Lithium nephropathy is portrayed as a disorder in urinary concentration, caused by the inhibition of adenylate cyclase. Under normal conditions, antidiuretic hormone activates adenylate cyclase, contributing to the increase in the intracellular cAMP concentration. In turn, cAMP elicits an increased permeability of the collecting tubular cells for water via various mechanisms (e.g. elevated enzyme activity). Lithium-inhibited adenylate cyclase produces inadequate water absorption (disorder of the urinary concentration) [92]. Depending on the duration of action, lithium elicits injury to various tissue structures of the kidney. Using the pattern of excretion, this can be discriminated by enzymuria [93].

The heavy metal *cadmium*, on the one hand, complexes with and denatures protein structures and is therefore harmful to the cell. On the other hand, it binds to proteins predominantly in the renal cortex and liver (half-life > 10 years). With chronic intoxication, it causes tubular proteinuria and a rise in Na^+ absorption by the proximal tubules. Elevated vesiculation of the proximal tubular cells with mitochrondrial and lysosomal alterations can be found ultrastructurally. The tubular cell membrane itself remains intact. This agrees with investigations conducted by Gatta [94] who did not find any increased excretion of membrane GGT with an intact tubular cell membrane.

Upon chronic administration of lithium or cadmium, hyalinized glomerula, ballonning of the tubular epithelia and formation of vacuoles in the cytoplasm can be seen under the light microscope [75 pp. 528–531]. Tubular atrophy with consecutive interstitial fibrosis may occur (irreversible injury).

An almost identical pathomorphological picture is presented by chronic lead nephropathy [75 pp. 528–531]. However, in this case the proximal tubular system is preferentially affected.

Contrast agent nephropathy, which usually only elicits sub-clinical, short-term, reversible damage of the tubules, can as a result of direct tubular injury (enzymuria, tubular proteinuria, tubular cell necrosis) elicit renal ischaemia through vasoconstriction and intratubular obstruction because of the precipitating out of the contrast agent.

Analgesic nephropathy [75 pp. 502–516] can be induced by vasogenic and direct tubular toxic components. The vasogenic event can, on the one hand, be explained as the result of direct toxic injury to the vasa recta, leading to capillary sclerosis and consequently to restricted perfusion. On the other hand, inhibition of renal prostaglandin synthesis with resultant vasoconstriction can occur [95]. This similarly causes reduced perfusion. O_2 deficiency of already oxygen-deficient renal papillae causes papillary necrosis and secondarily medullary fibrosis. Phenacetin should be mentioned as an example. This is metabolized to paracetamol, filtered glomerularly and then reabsorbed by the tubules passively. The paracetamol metabolite binds to tubular cellular proteins, particularly when these cells are deficient in glutathione. The causal role of phenacetin metabolites in the pathogenesis of analgesic nephropathy has been stressed by Raaflaub and Dubach [96]. Increased absorption results in cell damage, primarily seen as atrophy of the tubular cells. Salicylates also cause analgesic nephropathy through not yet fully explained mechanisms [97, 98].

Enzymuria was detected in varying degrees following the administration of different analgesic agents. After a dose of phenacetin, increased excretion of alkaline phosphatase (AP), leucine aminopeptidase (LAP) and LDH occurred in the urine [99, 100]. After one of ASA, there was elevated urinary excretion of LDH [101], and after giving salicylates, NAG increased in the urine [102].

With *allergic nephropathy* [62], necrotic vasculitis, as indirect toxic nephropathy, represents a special form. The formation of an antigen-antibody complex is regarded as being the inducer. This is deposited in the vascular wall (media) and subsequently elicits necrotic arteriitis by various mechanisms. Detection of immunocomplexes in the vascular wall provides evidence of immunogenic pathogenesis. The antigens may be of endogenous (serum proteins, tumour antigens) or exogenous (hepatitis B, streptococcae and medications) origin. Fibrinoid media necrosis, cell wall infiltrates and tiny aneurysmatic growths of the vascular wall can be found, resulting in fibrotic changes and necrotic shrinkage of the vessels.

With *immunocomplex glomerulonephritis* (GN) [62] (allergic-toxic), a causal relationship can be seen between the glomerular inflammatory process and the allergictoxic noxae under suspicion. Although not yet clarified, this must be regarded as probable on the basis of clinical and immunohistological indicators. Functional and morphological signs of allergic-toxic glomerulonephritis have occurred with penicillamine (membranous GN, anti-GBM-GN) [103, 104], trimethadione, auric salts [106] (membranous GN) [105], sulphonamides, phenylbutazone, probenezide, organic and inorganic mercury compounds [107] as well as organic bismuth compounds and nitrogen. Immunohistological evidence has been found recently (e.g. penicillamines) of finely granulated deposits of immunoglobulins and complements on the subepithelial side of the glomerular basal membrane. Auric salts also alter the surface of the tubular epithelia, causing an induction of the humoral immune response. Immunocomplexes are formed, which are also deposited on the subepithelial side of the glomerular basal membrane (perimembranous glomerulonephritis). Auric salt thereby results in the excretion of various enzymes (AAP, LDH) [108] into the urine. Combined therapy with auric salts and nonsteroidal antiphlogistics also causes increased urinary excretion of lysosomal enzymes (e.g. NAG) [109].

Acute interstitial nephritis [62, 75 pp. 494–502, 110, 111] is regarded as the expression of a drug-induced hypersensitivity reaction. The pathogenesis of methicillin-induced interstitial nephritis is best explained by its haptenic properties. With this form of nephritis, complexes of immunoglobulin plus hapten are found along the glomerular [112] and tubular basal membranes [112–114]. In addition, with acute interstitial nephritis, a delayed-type immune response with no immunocomplex deposition is most frequently seen. Monocytic, predominantly T-cell infiltrates, oedema of the interstitium and tubular cell necrosis are found histologically. These types of injury are generally reversible. However, as a result of rapid progression to interstitial fibrosis, persistent renal insufficiency ranging to end-stage kidney failure is possible.

Renal injury elicited by *non-steroidal antiphlogistics* (NSA) is based on complex pathomechanisms (functional, immunological, direct toxic, glomerular nephritic) [62]. Functionally, there is an inhibition of prostaglandin synthesis and consequently vasoconstriction, resulting in a reduced blood flow through the tissue. Immunologically, on the basis of a cell-mediated immune response, acute interstitial nephritis is induced [111]. On a direct toxic basis, accumulation of cytotoxic NSA molecules causes renal tubular cell injury. However, NSA-induced glomerular nephritis with and without the nephrotic syndrome is rare. Pathomechanically, the toxin activates *T* cells which release lymphokines, finally resulting in the development of glomerular nephritis. In addition, release of vascular permeability factors and glomerular polyanions of toxic substances is under discussion [115].

4.3 Enzymuria

Presentation of various nephropathies shows diverse pathomechanisms with an extremely wide range of cellular and tissue lesions. Even the slightest biochemical cellular alterations, such as injury of the tissue structure, can cause the most varied types of enzymes to be released and transported into the neighbouring off-flow system.

It is, however, difficult to assign appropriate enzyme excretion patterns to the various pathomechanisms which cause damage to the renal tissue. For example, it is known that the normally very low urinary enzyme activity rises with advent of tubular cell injury. Enzyme distribution in the enzyme-rich renal tissue can vary considerably [18, 19], with enzymes being excreted depending on the injury of the appropriate compartment. The most important enzymes for kidney diagnosis are predominantly found in cytoplasmic structures or are of cytosolic, membranous or brush border origin (Fig. 1).

ENZYMURIA

- **BRUSH BORDER ENZYMES**
  - **physiological:**
    - peripheral enzymes (AAP)
simple avulsion, increased diuresis
("wash-out effect")
    - deep-seated enzymes (AP)
cell turnover in the kidney
  - **slight injury:**
GGT raised
AAP: very sensitive
at periphery
  - **severe injury:**
deep-seated AP: raised
peripheral enzymes: raised
GGT, AAP
- **LYOSOMAL ENZYMES**
  - **physiological:** reverse pinocytosis (NAG)
→ membrane turnover +
lysosomal turnover
  - **slight injury:** increased lysosomal cell elements
raised reverse pinocytosis (NAG)
→ functional enzymuria
  - **severe injury:** cell destruction:
disintegration of the cell and lysosomal
membrane with consecutive release of lysosomal
enzymes → lesional enzymuria
- **CYTOSOLIC ENZYMES**
  - **physiological:** cell turnover
  - **slight injury:** disorders of membrane permeability
  - **severe injury:** cell destruction

Fig. 1. Localization and possible excretion of various enzymes according to degree of injury. *AP* alkaline phosphatase; *AAP*, alanine aminopeptidase; *NAG*, *N*-acetyl-β-D-glucosamidase; *GGT*, gamma-glutamyltransferase

4.3.1 Brush Border Enzymes

The brush border enzyme AAP plays a role in the amino acid metabolism of the kidney. This enzyme is localized on the surface of the cytoplasmic membrane as a "superficial cytoplasmic component" [116]. As a result of luminal flow within the tubules, it is probable that enzymes are released by simple avulsion. They can then be detected in terms of the physiological or basal urinary excretion rate. This process of avulsion does not damage the integrity of the tubular cells. Due to its peripheral location, it is understandable that even minor damage will lead to urinary excretion of this enzyme [117]. In addition to NAG, Price et al. therefore regard AAP as the most important enzyme for detecting renal injury [118].

GGT, an important enzyme for the metabolization of the cytoprotective substance glutathione, [119, 120] like AAP also resides at the luminal side of the renal tubular cells, being bound to the core of a multi-enzyme complex [121]. This explains the increase in enzymuria with even minor alterations of the brush border of the tubular cells.

The function and significance of AP lies in its control of transmembranous transport processes [122]. Even today, its structure and function still remain largely unexplained [123, 124]. The enzyme hydrolyzes various pathophysiological and synthetic phosphate esters [125–127]. Its low substrate specificity,

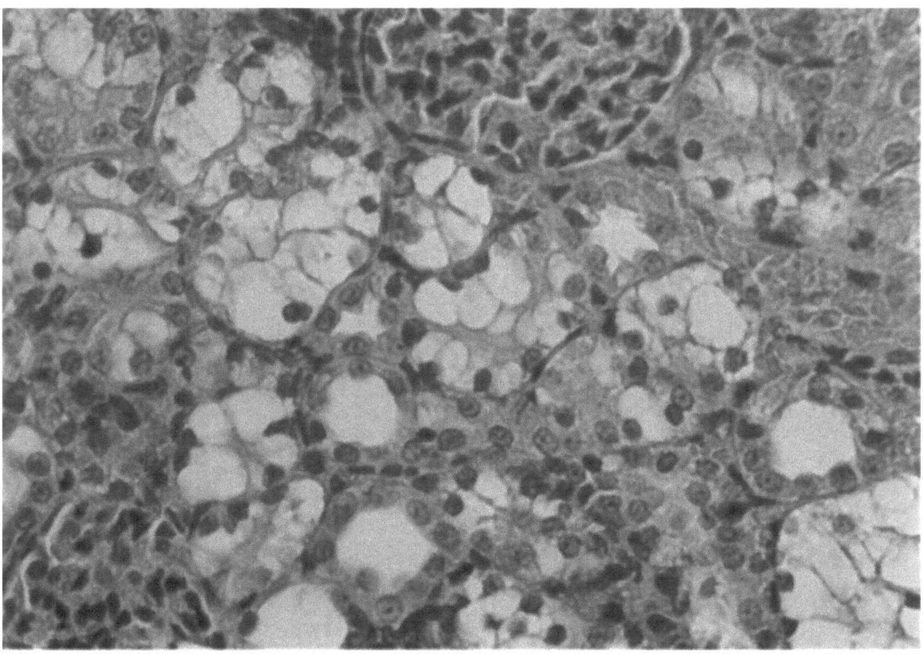

Fig. 2. Hydropically swollen tubular cells

different activators and inhibitors and its wide pH optimum (9.0–9.6) all indicate that a group of phosphatases is involved [128].

AP is located deep within the tubular membrane (intrinsic membrane enzyme) [129] and appears only to be excreted in the urine when major tubular injury occurs, i.e. when apical cellular segments split off by potocytosis. Normally detectable urinary excretion can be explained in terms of physiological cell moulting by which programmed cell death causes rupture of the cell and thus the release of even deep-seated membrane enzymes. In this context, it should be noted that in cases of extremely severe renal injury in which the nephrons are completely deactivated, the excretion of enzymes such as AAP can decrease.

4.3.2 Lysosomal Enzymes

The lysosomal enzyme NAG can also be detected in the urine along with the lysosomally localized enzymes β-glucuronidase and acid phosphatase. The

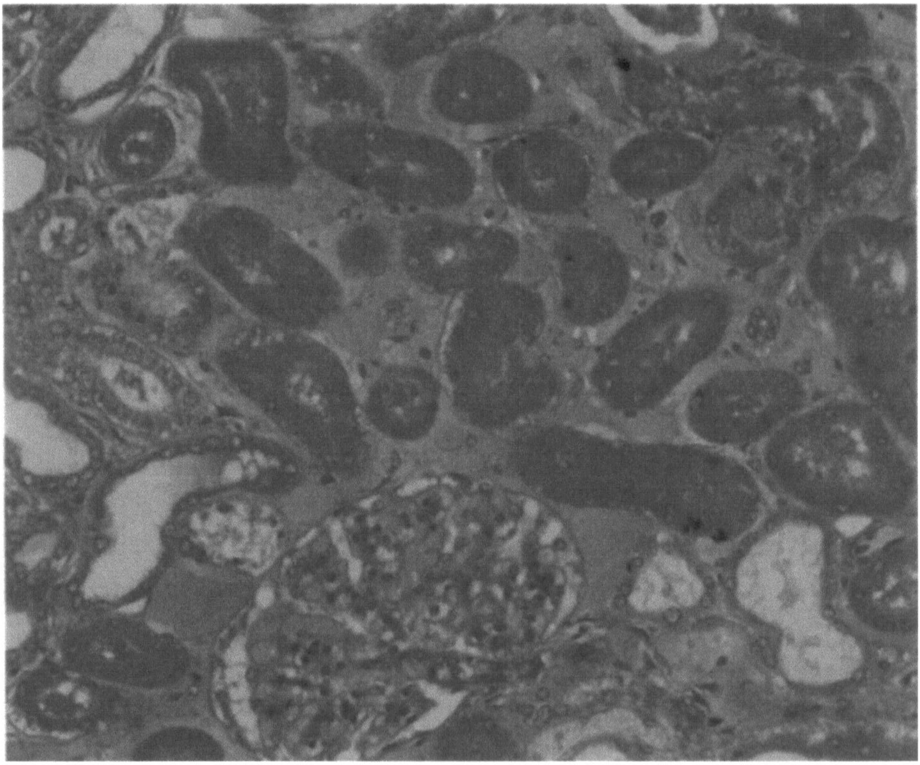

Fig. 3. Interstitial oedema

excretion characteristics of NAG were observed in conjunction with the administration of aminoglycoside antibiotics [130], for which lysosomes are the primary target organelle [131, 132]. It was found that the nephrotoxic potential of aminoglycosides can only be seen in conjunction with the NAG isoenzyme pattern [133].

Isoenzyme A is free in the lysosome, not localized on the membrane and found physiologically in the urine [134]. It is excreted by exocytosis (reverse pinocytosis), by the process of physiological membrane conversion and lysosomal turnover. Luminal release of this enzyme occurs even with minor functional alterations. These may be elicited by functional overburdening in the form of sublethal cell damage, e.g. enzyme induction. Such injuries, which are also found upon administration of cyclosporin, result in an intracellular increase in lysosomes.

This phenomenon can be explained in terms of the increased splitting-off of lysosomal fragments from the functional hypertrophic endoplasmic reticulum, which in turn cause increased exocytosis of isoenzyme A and elevated functional enzymuria [135]. The integrity of the cyto- and lysosomal membranes is not injured by this process.

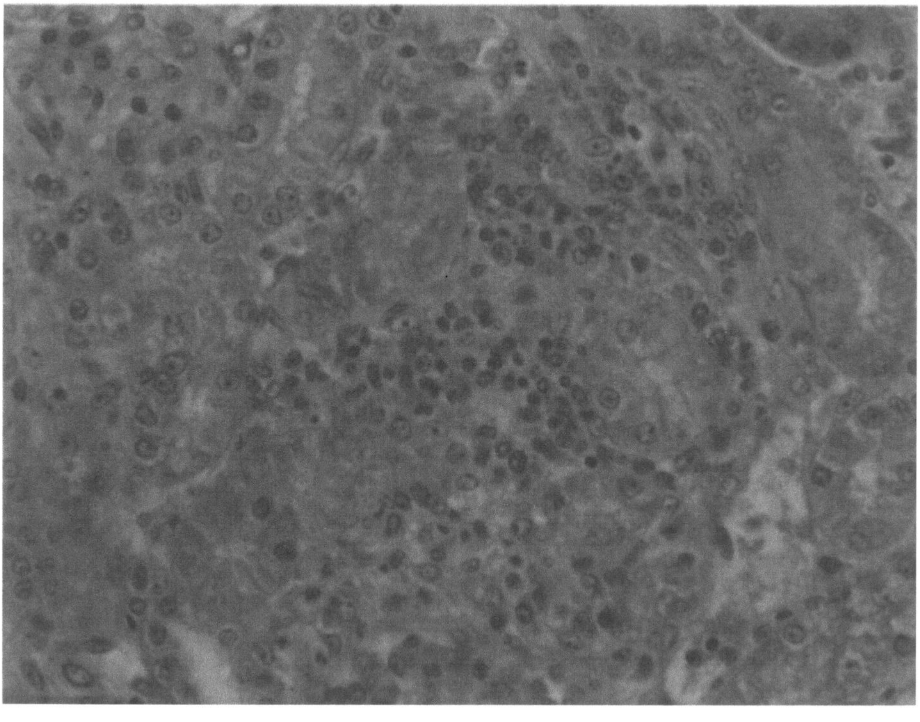

Fig. 4. Tubulo-interstitial nephropathy, cellular reaction

Isoenzyme B of NAG is not free in the lysosome but fixed to the lysosomal membrane [134]. Administration of aminoglycoside promotes the synthesis of this isoenzyme. In contrast to isoenzyme A, isoenzyme B is only excreted to a higher extent in the urine with the advent of severe nephrotoxic injury. The occurrence of isoenzyme B in the urine is regarded as lesional [135], i.e. nephrotoxic enzymuria. The cytomembranes (e.g. lysosomal membrane) rupture during the releasing process of isoenzyme B.

4.3.3 Cytosolic Enzymes

In addition to glutathione transferase, LDH is a cytosolic enzyme of the renal epithelia. Various isoenzymes of LDH can be isolated from the urine [136–138]. Of all the human organs, the kidney possesses the highest amount of LDH activity [139].

LDH is excreted in the urine, following even minute changes in the membrane architecture and disorders of membrane permeability of the tubular epithelia. This excretion usually occurs before any elevated excretion of brush border enzymes (e.g. AP) [140]. Urinary LDH excretion is accordingly an extremely sensitive parameter that can be determined even with very slight

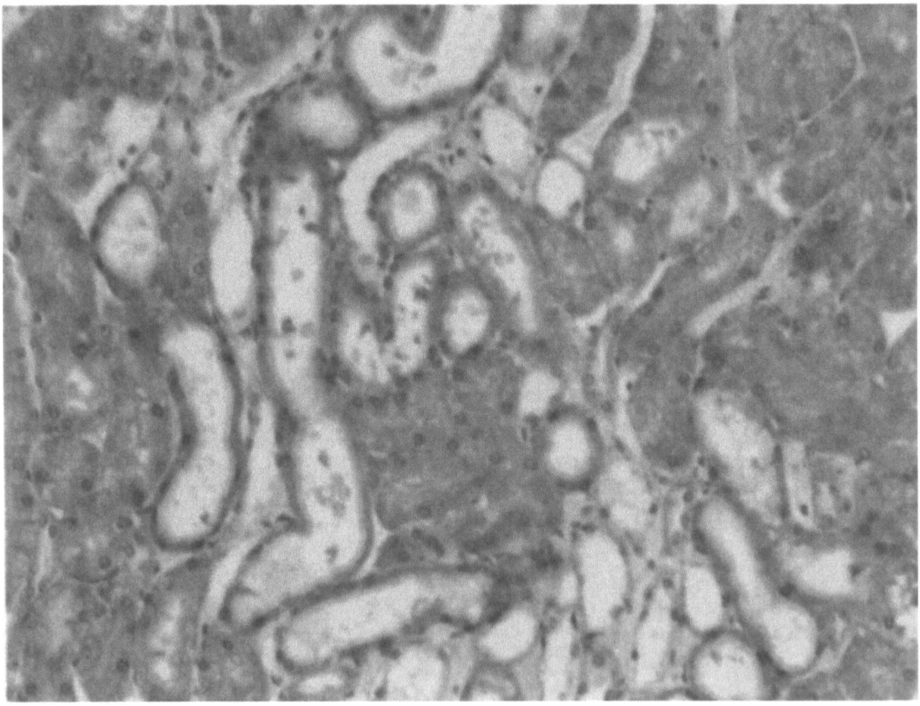

Fig. 5. Tubulus dilatation, focal flattening of the tubules

disorders of membrane permeability, before any changes to the tissue structure are visible microscopically.

4.4 Experimental Studies

In studies conducted in dogs exposed to various potential nephrotoxic substances, we have been able to show a correlation between urinary enzyme excretion and morphologically manifest renal injury. The following light microscopic and electron microscopic figures depict the kidneys of the experimental animals.

Initial histomorphologically demonstrable changes were found in the tubular system. These can partially be explained as pharmocologically functional overburdening with clear-cut diuretic action.

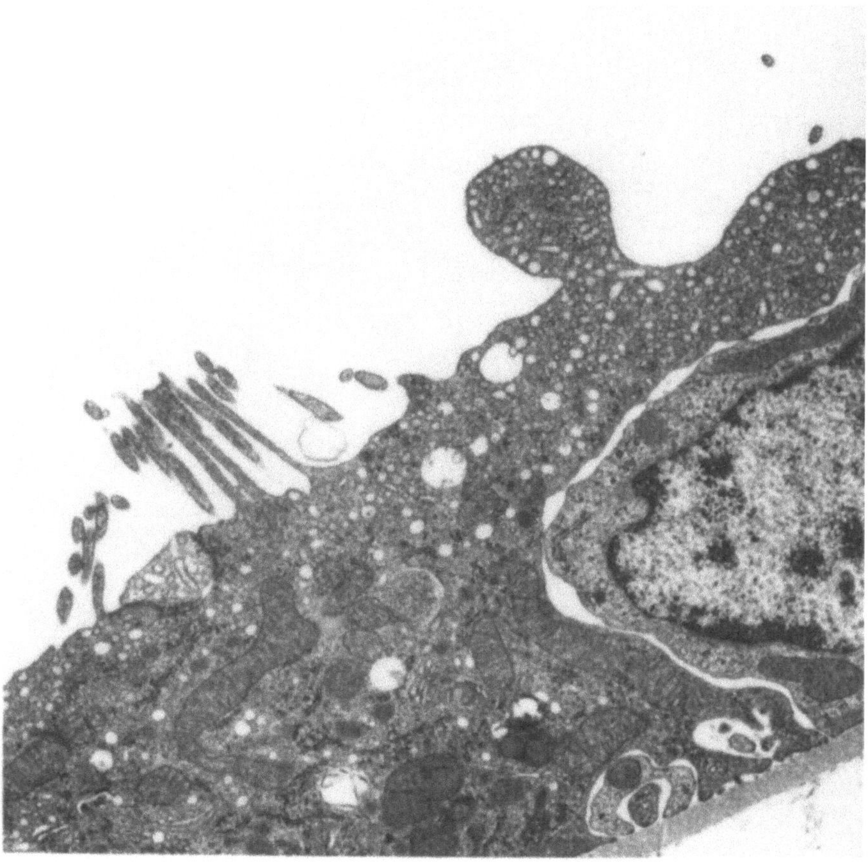

Fig. 6. Loss of apical cellular components (potocytosis) in proximal tubular cell, Elmi, × 7000 dog

The renal tubular epithelia are affected to varying degrees, distended hydropically (Fig. 2). In addition to distended epithelia, there are flattened, sometimes endothelial-like tubular epithelia encompassing the complete circumference or parts thereof. Necrotic injury of individual tubular epithelia cannot be ruled out.

The interstitium is loosened oedematously. This is due to an increased permeability of the tubular wall with leakage of the tubular fluid into the peritubular interstitium (Fig. 3). This type of oedema, which is very rarely marked, is an inflammatory stimulus for the mesenchyma; inflammatory infiltrates (lymphocytes and plasma cells) can be seen to have spread in the form of tubulo-interstitial nephropathy (Fig. 4).

The most important morphological substrate for the action of foreign substances remains the renal tubular epithelia. Injury here can be seen as a measurable distension of the epithelia or a focal flattening of the tubules (Fig. 5).

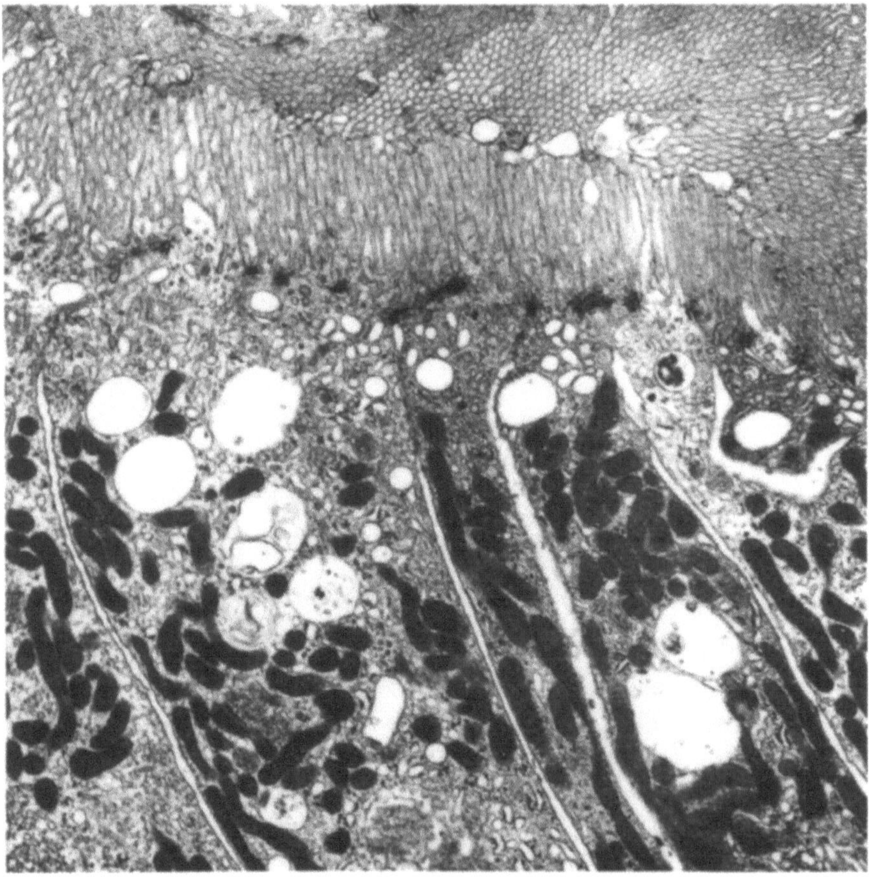

Fig. 7. Normal aspect of the brush border of the proximal tubular cell, Elmi, × 7000, dog

The flattened tubular epithelia were originally distended but lost their apical cellular components by potocytosis to the lumina of the renal tubules [75, pp. 412, 141].

The electron micrograph (Fig. 6) shows a clear loss of the brush border and potocytosis of an already atrophied tubular cell. In contrast, Fig. 7 depicts a normal tubular cell brush border.

The minor (sublethal) injuries attributable to functional overburdening or an inadequate energy supply can be seen as marked distension of the tubules and mitochondria that are frequently associated with the occurrence of small to medium-sized lysosomal cellular elements (vacuolic formation).

These two figures (Figs. 8, 9) present electron microscopically detectable mitochondrial changes (Fig. 8) as megamitochondria and in Fig. 9 as lysosomal vacuolic formation with almost complete loss of the brush border (potocytosis).

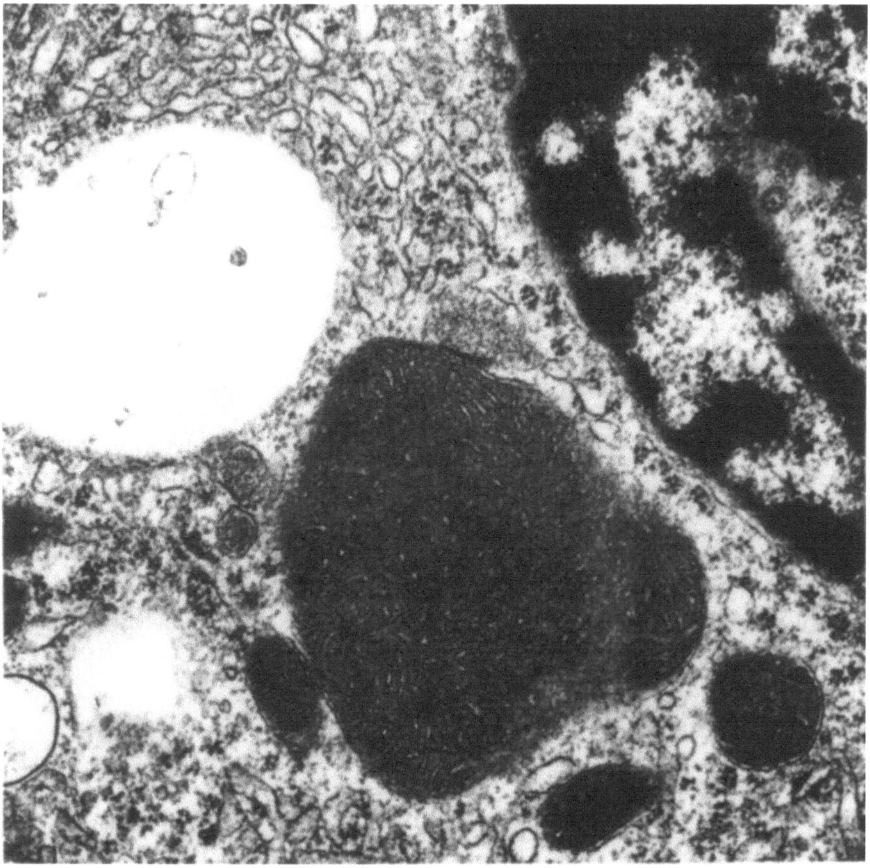

Fig. 8. Enlarged mitochondria (megamitochondria) within a swollen tubular cell, Elmi × 20 000 dog

In the presence of considerable lesions of the energy-donating enzyme groups, disintegration of the luminal brush border can be seen along with almost complete rejection, causing the epithelial cells to balloon (Figs. 6, 9). Even greater injury to individual cells results in avulsion of the brush border and larger cytoplasmic components along with their corresponding cell organelles (Fig. 6). Even with this degree of injury, a large number of these tubular cells appear to be able to survive with an extreme reduction of their cytoplasmic and organelle components, in particular by radically reducing their surface area in both the region of the brush border and basal labyrinths [75 pp. 418, 142–145].

Hydropic distension of the tubular epithelia and shedding of the apical cytoplasmic border can cause obstruction of the nephron, which can then lead to oliguria or anuria with acute renal failure [146, 147].

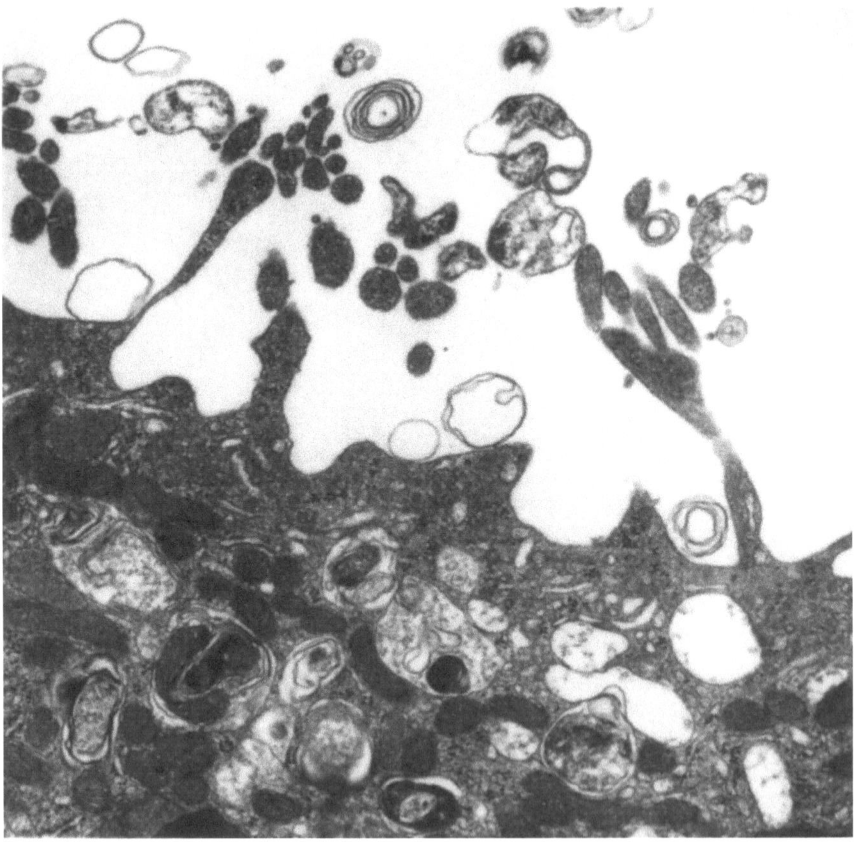

Fig. 9. Tubular cell with lysosomal vacuolic formation and disintegration of the brush border, Elmi ×12000 dog

From an energetic viewpoint, ischaemically injured tubular cells appear to be able to adjust to an energetically feasible survival level by reducing their cellular compartments, in particular their active surface area (brush border, basal labyrinth) [148–150]. It has been shown that tubular cells are able to rebuild their extracellular and intracellular sodium levels very soon after ischaemic injury. Morphologically, this reparative/regenerative process can be seen as flattened tubular epithelia with basophilic cytoplasm, larger nuclei and mitotic figures [75, pp. 558, 151, 152]. Concentrated groupings of membrane figures (fingerprints) can be seen in Fig. 10. These fingerprints can be found in regenerating cells, indicating an ultrastructural correlation to regeneration [153].

The various injuries cause distension of the tubules (tubular dilatation) (Fig. 4) resulting in partial or complete atrophy of the tubular cells with

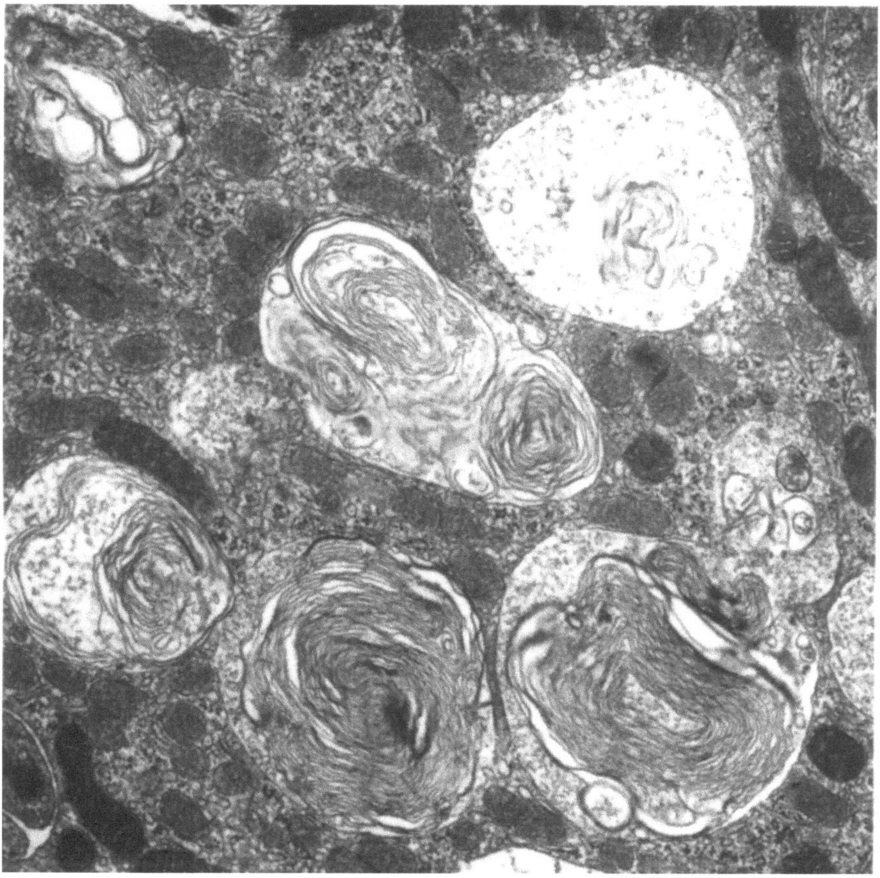

Fig. 10. Formation of concentric membrane figures (fingerprints) as a sign of tubular cell regeneration, Elmi, × 12 000 dog

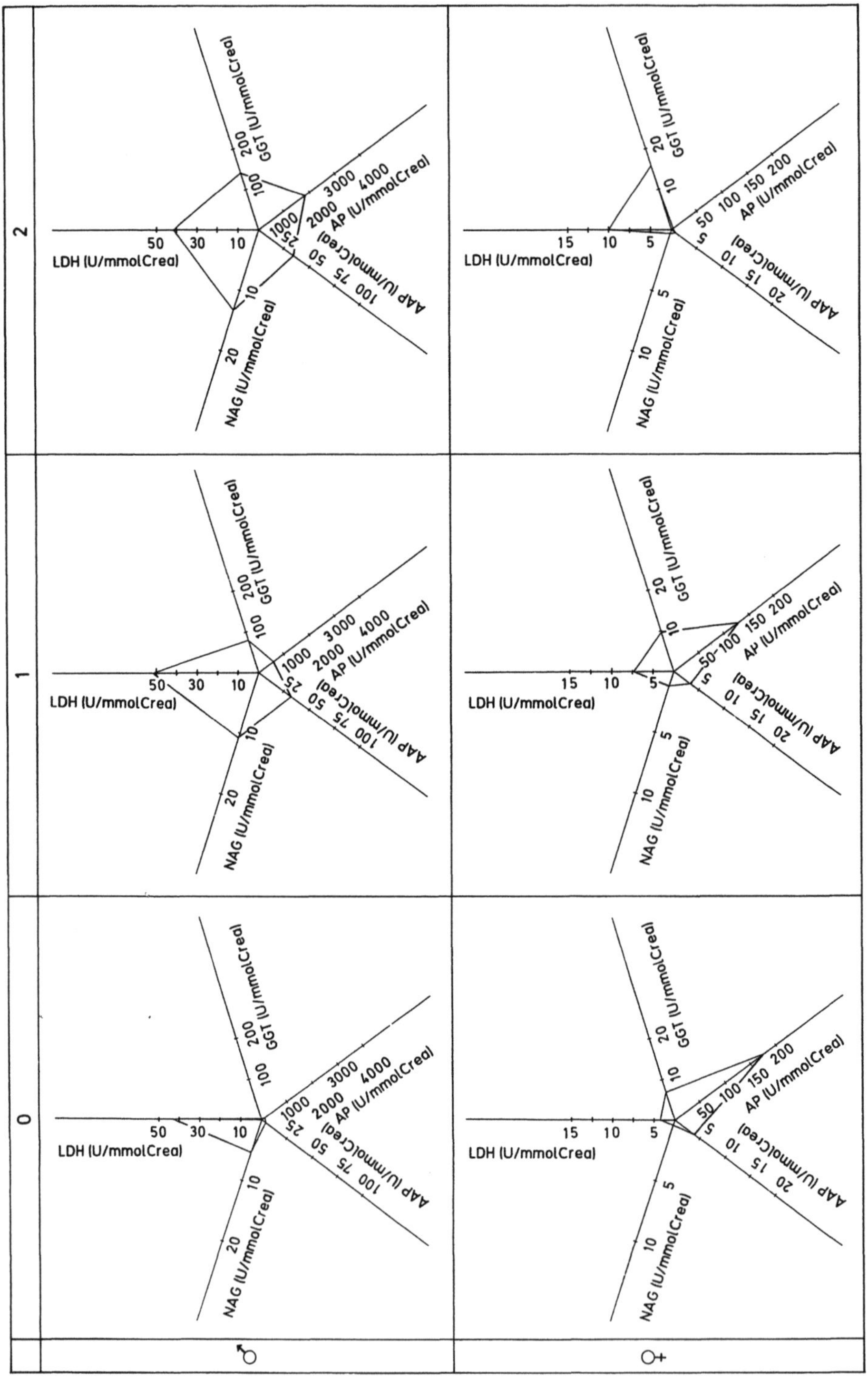
0
1
2
♂
♀
LDH (U/mmolCrea)
GGT (U/mmolCrea)
AP (U/mmolCrea)
AAP (U/mmolCrea)
NAG (U/mmolCrea)

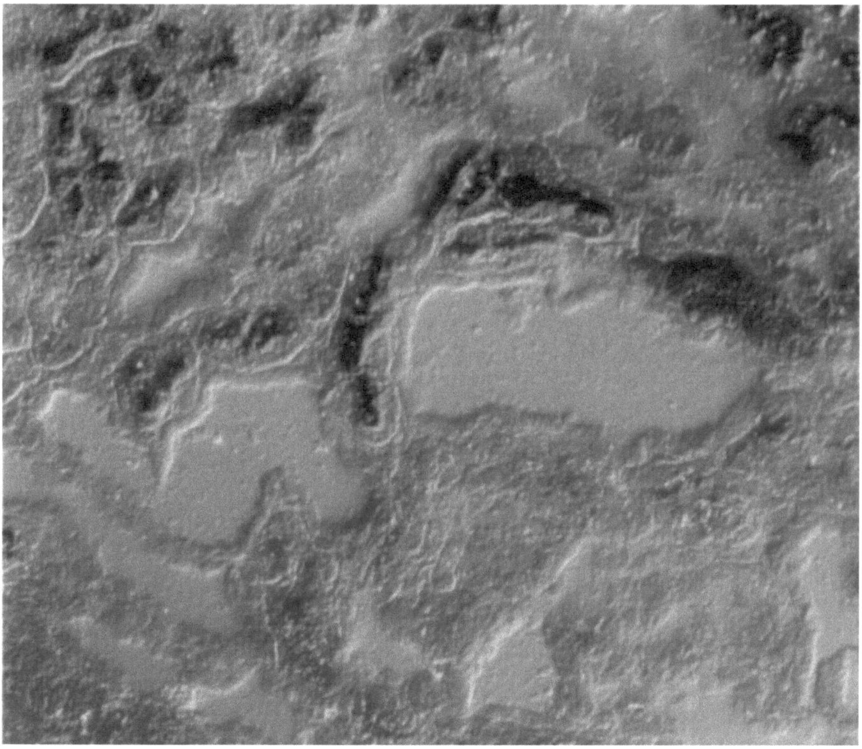

Fig. 12. Enzymehistochemical stain of GGT (Nomarsky interference filter): the tubular dilatation combined with atrophic epithelium shows an enzyme-negative reaction as an expression of the loss of brush border.

denudation of the basal membrane. The tubular walls become increasingly permeable, resulting in leakage of tubular fluid (backleak) into the interstitium [146] with a corresponding development of interstitial oedema (Fig. 3). A significant correlation has been shown between the formation of interstitial oedema and tubular dilatation [154, 155].

Localized ischaemia can then occur as a result of the longer route of diffusion between the tubules and the capillary caused by interstitial oedema and/or the elevated intratubular mechanical pressure with compression of the neighbouring blood vessels. This elicits tissue reactions that cause interstitial inflammatory processes (tubulo-interstitial nephropathy).

◄

Fig. 11. Average excretion of various enzymes in the urine by male and female dogs in different stages of kidney damage. Group 0, no light microscopic findings; group 1, minimal tubular dilatation and/or inflammatory cell reactions; group 2, moderate to severe tubular dilatation with the onset of cellular infiltration; *Crea*, creatinine

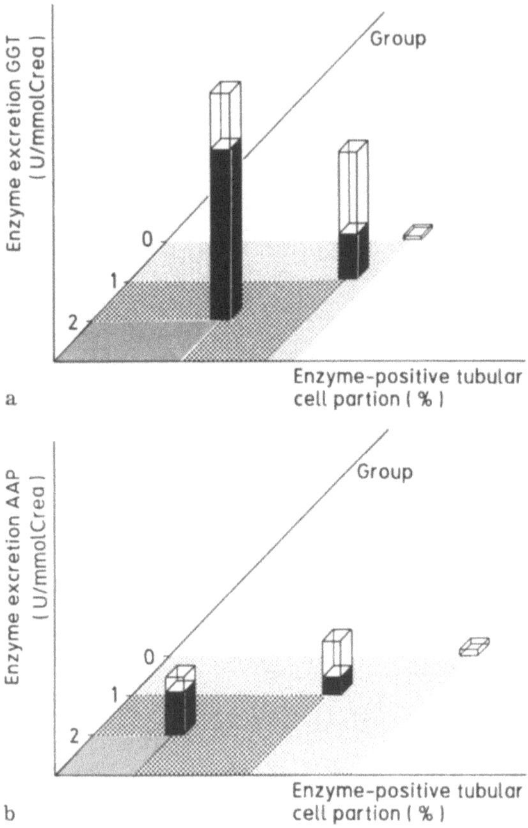

Fig. 13a, b. *Abscissa* gives the average contents of enzymepositive tubular cell portion, the *ordinate* shows the average enzyme excretion in urine. The *third line* represents the different stages of cell damage. The column visualizes the average enzyme excretion of all animals. The parts (%) demonstrate the portion of animals which have a certain share of basophilic foci in their kidney tissue. A tissue damage dependent urine excretion of the enzymes GGT (**a**) and AAP (**b**) could be demonstrated. *Crea*, creatinine

Our study shows that there is a clear correlation between the degree of tubular injury (tubular dilatation and/or infiltration) and the level of enzyme excretion (Fig. 13). Sex-specific differences can be seen. The male dogs show significantly more clear-cut enzymuria than the females. Sex differences in physiological enzyme excretion in various animal species are known for NAG, AAP and GGT (Fig. 11) [156, 157].

It can be shown in male dogs that levels of GGT, NAG, AAP and AP increase, depending on the extent of injury, whereas that of LDH shows no differences with individual injury patterns. The female animals react with a relatively low urinary excretion of enzymes; a rise in GGT and LDH may, however, be seen depending on the level of injury.

Figure 12 demonstrates dilated tubules with atrophic tubular epithelium. This epithelium shows no enzyme-positive reaction.

Fig. 14a–c. Excretion profiles: **a** animal without histomorphological damage; **b** animal with histomorphological findings and concomitant urine enzyme excretion (= responder); **c** animal with histomorphological findings and without significant urine enzyme excretion (= non-responder); *Crea*, Creatinine

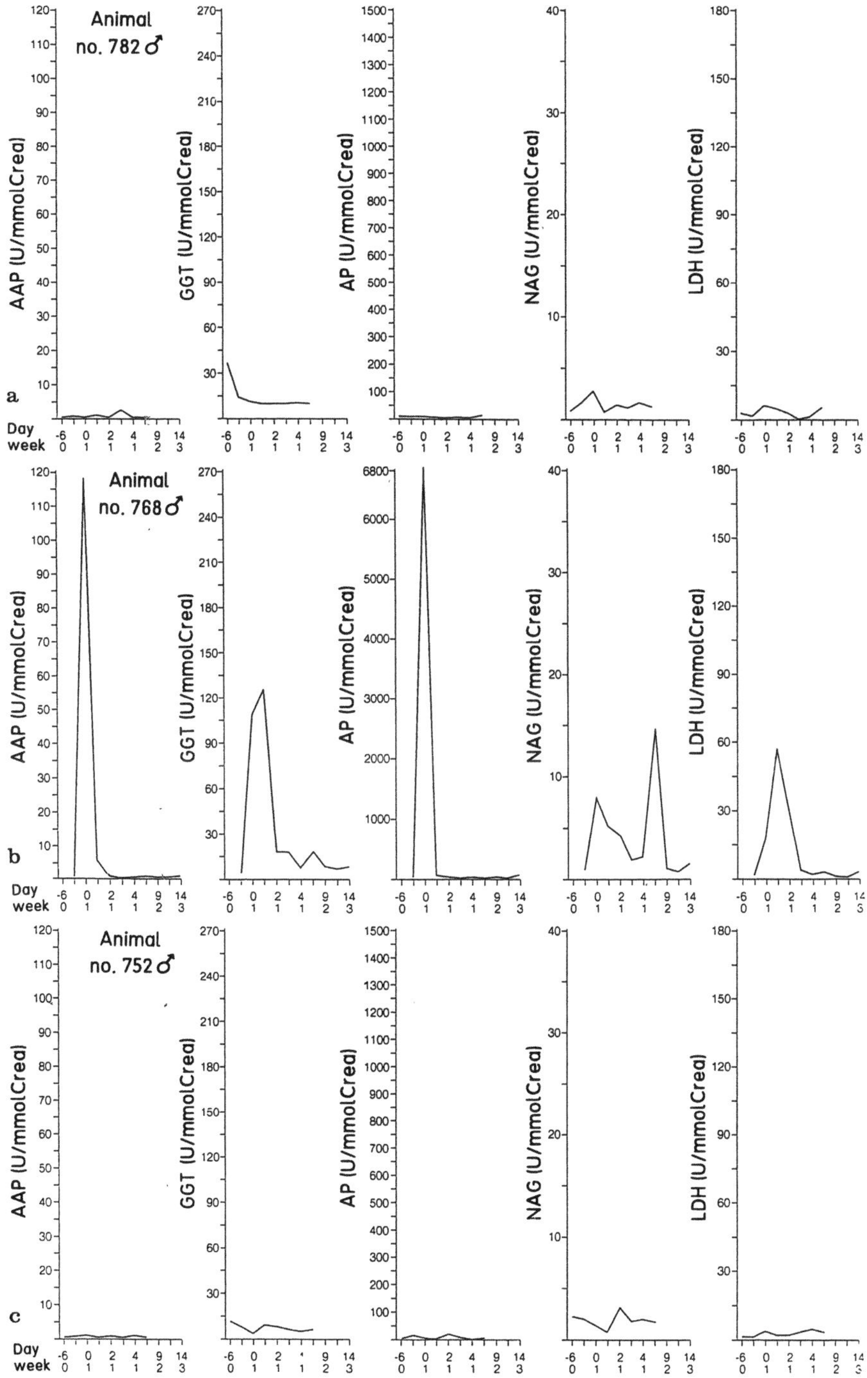

Animal no. 782 ♂
Animal no. 768 ♂
Animal no. 752 ♂
a
b
c
AAP (U/mmolCrea)
GGT (U/mmolCrea)
AP (U/mmolCrea)
NAG (U/mmolCrea)
LDH (U/mmolCrea)
Day
week

Morphometrical evaluation of the enzymes GGT (Fig. 13a) and AAP (Fig. 13b) presented histologically in the tissue of male dogs shows a correlation between the severity of tissue injury and the excretion of enzymes in the urine. A similar relationship between urinary enzyme excretion and concomitant enzyme loss from the renal parenchyma has been observed by Wachsmuth [29].

Basophilic foci of the tubules [152], which are regarded as signs of regeneratory repair, correlate with the degree of injury and enzymuria. It can be concluded that repair is most pronounced in animals with manifest tissue injury in which loss of released enzymes has not yet been regenerated.

The values of individual animals presented in Fig. 14 depict enzyme excretion characteristics for varying degrees of tubular injury. It can be seen that AAP and GGT levels either did not increase or only increased very slightly in the group in which no injury was visible under the light microscope. However, upon administration of potential nephrotoxic substances, rises in AAP, NAG and LDH activity could be seen at various times (between 8 h and 6 days), with urinary excretion of enzymes normalizing in the subsequent observation period.

It can be seen in the injured group that the majority of the animals present increased enzymuria in accordance with their tissue injury, although individual animals had to be classified as non-responders despite tissue degradation. This phenomenon of non-response has been described by Mondorf et al. for AAP excretion [158]. There is no clear-cut explanation for this phenomenon. It is, however, suspected that individual factors are partially responsible in addition to early enzyme denaturation by the xenobiotic.

4.5 Conclusions

Enzymuria offers a good possibility of being able to recognize renal tubular alterations and lesions without having to result to invasive methods. It enables the course of injury to be followed-up and evaluated. It does, however, require more than just single enzymes to be included in the evaluation. Conclusions as to tubular injury can only be drawn on considering the excretion profile of several enzymes. With regard to the degree of tubular injury, the enzymes NAG and GGT have proven to be very sensitive, followed by AAP and AP.

Understandably, the cytosolic enzyme LDH does not correlate with microscopically found changes, as it is excreted at a very early stage independently of any subsequently manifest injury and is also excreted with even ths most minor alterations. The level of urinary excretion of other enzymes correlates well with the severity of renal injury. Sex-specific differences should be taken into account here. In addition, non-response can occur. The lack of enzymuria ought to be attributable to individual factors in addition to xenobiotic-specific properties.

References

1. Mattenheimer H (1968) Enzymology of kidney tissue. In: Aebi H, Mattenheimer H, Schmidt FW (eds) Enzymes in urine and kidney. Huber, Bern, pp 119–147
2. Luft FC, Patel V, Yum MN, Patel B, Kleit SA (1975) Experimental aminoglycoside nephrotoxicity. J Lab Clin Med 86:213
3. Diezi J, Biollaz F (1979) Renal function tests in experimental toxicity studies. Pharmacol Ther 5:135
4. Rankin TL, Luft FC, Yum MN, Sloau RS, Dieswiddie CB, Isaacs LL (1979) Antimicrob Agents Chemother 16:491
5. Roxe DM (1975) Toxic nephropathy due to drugs. Ration Drug Therapy 9:1–5
6. Plummer DT, Norazar S, Obatomi DK, Haslan FD (1985/1986) Assessment of renal injury by urinary enzymes. Uremia Invest 9:97–102
7. Dixon H, Webb EC (1964) Enzymes, 2nd edn. Academic, New York
8. Longley J (1969) Histochemistry of the kidney The Kidney, morphology, biochemistry physiology Rouilur C, Muller AF (eds) In: The kidney, 3rd edn. Academic, New York
9. Raab WP (1968) Enzymes and isoenzymes in urine. In: Aebi H, Mattenheimer H, Schmidt FW (eds) Enzymes in urine and kidney. Huber, Bern, p 17
10. Jung K, Schulze G, Reinholdt C (1986) Different diuresis-dependent excretions of urinary enzymes. Clin Chem 32:529–532
11. Burchardt U, Hempel A, Höpfner JB, Hempel RD (1976) Alaninaminopeptidase- und Kreatininausscheidung unter verschiedenen Diuresebedingungen beim Menschen. Z Gesamte Inn Med 31:550–552
12. Hohenegger M, Raab W (1970) Das Verhalten der Enzymaktivität im Harn bei Veränderung der Wasser- und Ionenausscheidung. Clin Chim Acta 28:61–65
13. Jösch W, Dubach UC (1967) Einfluß der Diurese auf die Enzymurie. Clin Chim Acta 15:325–330
14. Kriz W, Bankir L (1988) Renal fluid and electrolyte physiology. Am J Physiol 254:23
15. Kriz W, Bankir L (1988) A standard nomenclature for structures of the kidney. Kidney Int 33:1–7
16. Mattenheimer H, Pollak VE, De Bruin H (1964) Quantitative histochemistry of the nephron. IX. Distribution and activity of various enzymes in the kidney of the dog. Enzymol Biol Clin 4:107
17. Bonting SL, Pollak VE, Muehrcke RC, Kark RM (1960) Quantitative histochemistry of the nephron. II. Alkaline phosphatase activity in man and other species. J Clin Invest 39: 1372
18. Jacobson NO, Jörgensen F (1973) Further enzyme histochemical observations on the segmentation of the proximal tubule in the kidney of the male rat. Histochemie 34:11
19. Jacobson NO, Jörgensen F (1975) Ultrastructural observations on the pars descendens of the proximale tubule in the kidney of the female rat. Histochemistry 43:29
20. Mühlenfeld WE (1969) Über die Entwicklung und Chemodifferenzierung der Rattenniere unter besonderer Berücksichtigung der Geschlechtsunterschiede. Histochemie 18:97
21. von Deimling O, Baumann G, Noltenius H (1965) Hormonabhängige Enzymverteilung in Geweben. V. Wirkung von Kastration und Sexualhormonen auf 5 Enzyme der Mäuseniere. Histochemie 5:1
22. Valentin H (1973) Renal function: mechanisms preserving fluid and solute balance in health. Little, Brown, Boston
23. Brenner RM, Reclar FC, Jr (eds) (1986) The kidney, vols 1, 2, 3rd edn. Saunders, Philadelphia
24. Tigerstedt R, Bergmann PG (1898) Niere und Kreislauf. Skand Arch Physiol 8:223
25. Naets J (1969) Erythropoietin In: Roullier C, Muller AF (eds) The kidney, vol 2. Academic New York p. 363
26 Price RG, Ellis BG (1976) Urinary enzyme excretion in aminonucleoside nephrosis in rats. Chem Biol Interact 13:353
27. Harding S, Munro AJ (1978) Furosemide and renal enzyme excretion Br Med J 2:1431
28. Gordon JA, Gottone VH, Schoolwerth AC (1986) Glutamyl transpeptidase excretion in cisplatin induced acute renal failure. Am J Kidney Dis 8:18–25
29. Wachsmuth ED (1981) Quantification of nephrotoxicity in rabbits by automated morphometry of alkaline phosphatase stained kidney sections. Histochemistry 71:235–248

30. Whiting PH, Thomson AW, Simpson JG (1986) Cyclosporine and renal enzyme excretion. Clin Nephrol 25:100–104
31. Humes HD, Weinberg JM, Krauss TC (1982) Clinical and pathophysiologic aspects of aminoglycoside nephrotoxicity. Am J Kidney Dis 11:5–29
32. Koloyanides GJ (1984) Aminoglycoside-induced functional and biochemical defects in the renal cortex. Fundam Appl Toxicol 4:930–943
33. Sacchi G, Benetti A, Falchetti M, Grigolato P, Cristinelli L, Strada A, Maiorca R (1987) Ultrastructural renal findings in allografted kidneys of patients treated with cycloporin A. Appl Pathol 5:101–107
34. Mc Auley FT, Simpson JG, Thomson AW, Whiting PH (1986) The predictive value of enzymuria in cyclosporin A induced renal toxicity in the rat. Toxicol Lett 32:163–169
35. Hopps V, Galione A, Biondi F, Vaccaro F, Sorrentino, MC Vetri B, Leone F (1988) Rifampicin reduces nephrotoxicity of cyclosporin A in rats; studies of renal enzyme excretion. Transplant Proc 20 [Suppl]:557–560
36. Mc Auley FF, Whiting PH, Thomson AW, Simpson JG (1987) The influence of enalapril or spironolactone on experimental cyclosporine nephrotoxicity. Biochem Pharmacol 36:699–703
37. Maruhn D, Bomhard E, Paar D (1984) Urine enzyme patterns in the rat after damage to different parts of the nephron, selected. (Topics in clinical enzymology, DeGruyter, Berlin, pp 429–434 vol 2)
38. Gordon JA, Gattone VH, Schoolwerth AC (1986) Gamma-glutamyl tanspeptidase excretion in cisplatin induced acute renal failure. Am J Kidney Dis 8:18–24
39. Albert Z, Orlowski M, Szewczuk A (1961) Histochemical demonstration of gamma-glutamyl-transpeptidase. Nature 191:767
40. Heinle H, Wendel A, Schmidt U (1977) The activities of the key enzymes of the gamma-glutamyl cycle in microdissected segments of the rat nephron. FEBS Lett 73:220
41. Hautmann R (1979) Diagnosis of renal disorders: comparison of urinary enzyme patterns with corresponding tissue patterns. Curr Probl Clin Biochem 9:58–70
42. Platt JL, Lebien TW, Michael AF (1982) Interstitial mononuclear cell populations in renal graft rejections. J Exp Med 155:17–30
43. Busch GJ, Reynolds ES, Galvanek EG, Braun WE, Dammin GJ (1971) Human renal allografts: the role of vascular injury in early graft failure. Medicine (Baltimore) 50:29–83
44. Magil A, Rubin J, Ladewig L, Johnson M, Goldstein MB, Bear RA (1980) Renal biopsy in acute allograft rejection. Significance of moderate vascular lesions in longterm graft survival. Nephron 26:180–183
45. Mc Kenzie JFC, Whittingham S (1968) Deposits of immunoglobulin and fibrin in human allografted kidneys. Lancet ii:1313–1316
46. Jung K, Diego J, Strobelt V, Scholz D, Schreiber G (1986) Diagnostic significance of some urinary enzymes for detecting acute rejection crises in renal transplant recepients: alanine aminopeptidase, alkaline phosphatase, gamma-glutamyltransferase, N-acetyl-beta-D-glucosaminidase and lysozyme. Clin Chem 32:1807–1811
47. Asfar SK, Whiting PH, Simpson JG, Catto GR, Engeset J (1984) Early diagnosis of rejection. Correlation of enzymuria and histological findings after renal transplantation in the rat. Transplantation 37:548–551
48. Ellis BG, Price RG, Topham JC (1973) The effect of papillary damage by ethyleneimine on kidney function and some urinary enzymes in the dog. Chem Biol Interact 7:131
49. Ellis BG, Price RG (1975) Urinary enzyme excretion during renal papillary necrosis induced in rats with ethyleneimine. Chem Biol Interact 11:473
50. Hepler OE, Gurley H, Simmonds JP (1945) Experimental nephropathies. II. Renal Phosphatase after poisoning with mercury bichloride uranyl nitrate and potassium dichromate. Arch Pathol (Chic) 39:133
51. Wachstein M (1947) Influence of experimental kidney damage on histochemically demonstrable lipase activity in the rat comparison with alkaline phosphatase activity. J Exp Med 84:503
52. Wachstein M (1947) Nephrotoxic action of DL-Sevine in the rat I. The localization of the renal damage, the phosphatase activity and the influence of age, sex, time and dose. Arch Pathol (Chic) 43:503
53. Breedis C, Flory M, Furth J (1943) Alkaline phosphatase level in urine in relation to renal injury. Arch Path (Chic) 36:402

54. Lock EA, Ismael J (1979) The acut toxic effects of hexachlorobutadiene on the rat kidney. Arch Toxicol 43:47–57
55. Ismael J, Pratt J, Lock EA (1982) Necrosis of the pars recta (S3-segment) of the rat kidney produced by hexachloro 1:3 butadiene. J Pathol 138:99–113
56. Murray G, Wyllie RG, Hill GS, Ramschen PW, Heptinstall RH (1972) Experimental papillary necrosis of the kidney I. Morphologic and functional data. Am J Pathol 67:285–302
57. Hill GS, Wyllie RS, Müller M, Heptinstall RH (1972) Experimental papillary necrosis of the kidney II. Electron microscopic and histochemical studies. Am J Pathol 68:213–234
58. Feldmann JD, Fisher ER (1959) Renal lesions of aminonucleoside nephrosis as revealed by electron microscopy. Lab Invest 8: 371–378
59. Marsh JB, Sparks CE (1979) Hepatic secretion of lipoproteins in the rat and the effect of experimental nephrosis. J Clin Invest 64:1229–1237
60. Stonard MD, Gore CW, Oliver GJ, Smith IK (1987) Urinary enzymes and protein patterns as indicators of injury to different regions of the kidney. Fundam Appl Toxicol 9:339–351
61. Ellis BG, Price RG, Topham JC (1973) Urinary enzymes and the detection of kidney damage in the dog. Biochem Soc Trans 1:955–977
62. Edel H, Altmeyer B (1987) Toxisch und allergisch bedingte Nephropathien. In: Siegenthaler W (ed) Klinische Pathophysiologie, vol 6 rev edn. Thieme, Stuttgart, pp 975–979
63. Berlin M, Ulberg S (1963) Accumulation and retention of mercury in the mouse. II. An autoradiographic comparison of methylmercuric dicyanamide with inorganic mercury. AMA Arch Environ Health 6:610
64. Swensson A, Ulfvarson M (1968) Distribution and excetion of mercury compounds in rats after a single injection Acta Pharmacol Toxicol 26:273
65. Aberg B, Ekman L, Falk R, Greitz U, Persson G, Sniks JO (1969) Metabolism of methyl mercury (203Hg) compounds in man: excretion and distribution. Arch Environ Health 19:478
66. Ulfvarson U (1969) The effect of the size of the dose on the distribution and excretion of mercury in rats after single intravenous injections of various mercury compounds. Toxicol Appl Pharmacol 15:517
67. Antonovych TT (1984) Drug-induced nephropathies. Pathol Annu 19:165–196
68. Gritzka TL, Trump BF (1968) Renal tubular lesions caused by mercuric chloride: electron microscopic observations: degeneration of the pars recta. Am J Pathol 52:1225
69. Rodin AE, Crowson CN (1962) Mercury nephrotoxicity in the rat I. Factors influencing the localisation of tubular lesions. Am J Pathol 41:297
70. Robinson M, Hesketh A (1976) Effect of mercuric chloride on the structure and function of the kidney of sheep. J Comp Pathol 86:307
71. Ellis BG, Price RG, Topham JC (1973) The effect of tubular damage by mercuric chloride on kidney function and some urinary enzymes in the dog. Chem Biol Interact 7:101
72. Tandon SK, Magos L, Cabrol JRP (1980) Protection against mercuric chloride by nephrotoxic agents which do not induce thionein. Toxicol Appl Pharmacol 52:227
73. Braun JP, Rico AG, Bernard P, Burgot-Sacaze V, Eghbali B, Godfrain JG (1978) La gamma-glutamyl transferase urinaine ex toxicologie renale cher le rat bases de sou utilisation interest loss de nephlite figure mercullie. Toxicology 11:73
74. Kirschbaum BB (1979) Alanine aminopeptidase excretion after mercuric chloride renal failure. Biochem Med 21:220–225
75. Bohle, A (1984) Die Niere: Struktur und Funktion. Schattauer, Stuttgart, pp. 412, 418, 433–438, 494–516, 528–531, 558
76. Williams HE (1978) Oxalic acid and the hyperoxaluric syndroms. Kidney Int 13:410–417
77. Kirschbaum BB, Bosman HB (1974) Chem Biol Interact 8:207
78. Baylis C, Rennke HR, Bremer BM (1977) Mechanisms of the defect in glomerular ultrafiltration associated with gentamycin administration. Kidney Int 12:344
79. Houghton DÇ, Campbell-Boswel MV, Bennett WM, Porter WA, Brooks RE (1978) Myeloid bodies in renal tubles of human relationship to gentamycin therapy. Clin Nephrol 10:140
80. Silverblatt FJ (1979) Autoradiography of gentamycin uptake by the rat proximal tuble cell. Kidney Int 15:335
81. Adelmann RD, Conzelman G, Spangler W, Ishizaki G (1979) Comparative nephrotoxicity of gentamicin and netilmicin. Functional and morphological correlations with urinary enzyme activities. Curr Probl Clin Biochem 9:166–180
82. Price RG (1979) Urinary N-acetyl-beta-D-glucosaminidase as an indicator of renal disease. Curr Probl Biochem 9:150–161

83. Dorian C, Catroux P, Cambar J (1987) Mise en evidence des variations saisonnieres des rythmes circadiens de la chrononephrotoxicite de l'amikacine chez le rat. Pathol Biol (Paris) 35:731–734
84. Heine WD, Stöcker E, Stöcker HD (1970) Tageszeitliche Rhythmen der Zellproliferation der Niere. Naturwissenschaften 57:94
85. Blumenfeld C (1938) Periodic and rhythmic mitotic activity in the kidney of the albino rat. Anat Rec 72:435
86. Bucher O, Kolb I, Juhasz P (1961) L'influence du rythme diurne sur l'etat fonctionel des reins et du foie. Med Exp 5:42
87. Kolb R, Markhof CM, Zekert F (1976) Verhalten von Harnenzymen nach Vitamin A, Gentamycin und Klemmung der Nierenarterien. Wien Klin Wochenschr 88:463–467
88. Mondorf AW (1979) Investigations on the potential nephrotoxicity of cefazedone and gentamycin and of their combination in comparison with the combination of cefazoline and cefalothin with gentamycin. Arzneimittel forsch 29:449
89. Mondorf AW, Hendus J, Beier J, Scherberich JE, Schoeppe W (1978) Tubular toxicity induced by aminoglycosides in human kidneys. In: Fillastre JP (ed) Nephrotoxicity: interaction of drugs and membrane systems. Mitochondria-lysosomes, Masson, New York p 172
90. Sack K, Freisleben H, Züllich B, Beck H, Schulz E (1978) Die Differenzierung der potentiellen Nephrotoxizität verschiedener Aminoglycoside im Tierexperiment. Adv Clin Pharmacol 15:69
91. Raab W, Moerth w (1976) Renal effects of gentamicin and cephaloridine. Arzneimittel forsch 26:377
92. Dousa T, Hechter O (1970) The effect of NaCl and LiCl on vasopressin-sensitive adenylatcyclase. Life Sci 9:765
93. Emanuelli G, Anfossi G, Calcamuggi G, Marcarino C, Ottone G, Dughera L (1985) Urinary enzyme excretion in acute and subacute lithium administration. Enzyme 34:177–185
94. Gatta A, Bazzerla G, Amodio P, Menon F, Angeli P, Schiaffnio E, Schmid C (1989) Detection of the early steps of cadmium nephropathia—comparison of light- and electron-microscopical patterns with the urinary enzymes excretion. An experimental study. Nephron 51:20–24
95. Caterson RJ, Duggin GG, Horvath JS, Mohandas J, Tiller DJ (1978) Aspirin, proteintransacetylation and inhibition of prostaglandin synthetase in the kidney. Br J Pharmacol 207:584–593
96. Raaflaub J, Dubach UC (1972) Zur Frage der Pathogenese der chronisch-interstitiellen Nephritis nach protrahiertem Schmerzmittelabusus. Klin Wochenschr 50:489–497
97. Kincaid-Smith P, Nanra RS (1978) Renal medullary necrosis. In: Hamburger J, Crosnier J, Grünfeld JP (eds) Nephrology Flammarion, Paris pp 583–594
98. Nanra RS (1976) Pathology, aetiology and pathogenesis of analgesis nephropathy. Aust N Z J Med 6 [Suppl 1]: 33–37
99. Raab W, Moerth C (1973) Renal enzyme excretion following administration of analgetic and antirheumatic drugs Helv Med Acta 37:73
100. Raab W, Krammer P, Moerth C (1976) An experimental study of acute and chronic effects of phenacetin and the rat kidney, using clinical chemical and biochemical methods. Enzyme 21:76
101. Laethwood PD, Plummer DT (1969) The excretion of lactate dehydrogenase in human urine after ingestion aspirin. Biochem J 114:197
102. Proctor RA, Kunin CM (1978) Salicylate-induced enzymuria Am J Med 65:987
103. Seelig HP, Seelig R, Fischer A, Maurer R, Sater A, Schnitzlein W (1977) Immunkomplex-glomerulonephritis nach peroraler Penicillamin-applikation bei Ratten. Res Exp Med 170:35
104. Donker AJ, Venuto RC, Vladutiu AO, Brentjens JR, Andres GA (1984) Effects of prolonged administration of D-penicillamin or captopril in various strains of rats. Clin Immunol Immunpathol 30:142–155
105. Nagi AH, Alexander F, Barabas AZ (1971) Gold nephropathy in rats. Light and electron microscopic studies. Exp Mol Pathol 15:354
106. Yarom R, Stein H, Peters PD (1975) Nephrotic effect of parental and intraarticular gold. Ultrastructure and electron microprobe examinations of clinical and experimental material. Arch Pathol 99:37
107. Albini B, Glurich J, Andres GA (1982) Mercuric chloride induced immunologically mediated diseases in experimental animals. In: Porter GA (ed) Nephrotoxic mechanisms of drugs and environmental toxins. Plenum, New York. Chap 37
108. Trnavasky K, Rovensky J (1973) The influence of gold salts on the urinary enzyme level. Pharmacology 9:300

109. Redondo FL, Bergon E, Tinture T, Miravalles EM, Pascual T (1987) Urinary enzymes activities in patients treated with gold and other antirheumatic drugs. Clin Biochem 20:343–347
110. Blantz RC (1980) The glomerulus passive filter or regulatory organ. Klin Wochenschr 58:957
111. Druet P, Jacquot C, Baran D, Kleinknecht D, Fillastre JP, Mery JP (1987) Immunologically mediated nephritis induced by toxins and drugs. In: Bach PH, Lock EA (eds) Nephrotoxicity in the experimental and clinical situation, pt 2. Nijhoff, Dordrecht, p 749
112. Colvin RB, Burton NE, Hyslop NE (1974) Penicillin associated interstitial nephritis. Ann Intern Med 81:404
113. Border WA, Lehmann DH, Egan JD, Sass HJ, Glade JE, Wilson CB (1974) Antitubular basement antibodies in methicillin associated interstitial nephritis. N Engl J Med 291:381
114. Lehmann DH, Wilson CB, Dixon FJ (1975) Extraglomerular immunglobulin deposits in human nephritis. Am J Med 58:765
115. Stachura I, Jayakumar S, Bourke E (1983) T- and B-lymphocyte subsets in fenoprofen nephropathy. Am J Med 75:9–16
116. Scherberich JE, Falkenberg FW, Mondorf AW, Müller H, Pfleiderer G (1976) Biochemical and immunological studies on isolated brush border membranes of human kidney cortex and their membrane surface proteins. Clin Chim Acta 55:179–197
117. Pfleiderer G, Baier M, Mondorf AW, Stefanescu T, Scherberich JE, Müller H (1980) Change in alkaline phosphatase isoenzyme pattern in urine as possible marker for renal disease. Kidney Int 17:242–249
118. Price RG (1982) Urinary enzymes, nephrotoxicity and renal disease. Toxicology 23:99–134
119. Rankin BB, McIntyre TM, Curthoys NP (1983) Role of the kidney in the interorgan metabolism of glutathione. In: Larson A et al (eds) Functions of glutathione: biochemical, physiological, toxicological and clinical aspects. Raven, New York
120. Meister A (1983) Metabolism and transport of glutathione and other gamma-glutamyl compounds. In: Larson A et al. (eds) Functions of glutathione: biochemical, physiological, toxicological and clinical aspects. Raven, New York
121. Scherberich JE, Gauhl C, Mondorf W (1978) Biochemical, immunological and ultrastructural studies on brush-border membranes of human kidney. Curr Prob Clin Biochem 8:85–95
122. Bargmann W (1978) Niere und ableitende Harnwege, Springer, Berlin Heidelberg New York, p 334 vol 7 pt 5 Handbuch der mikroskopischen Anatomie des Menschen.
123. Morton RK (1961) Acid and alkaline phosphatases. In: Long CNH (ed) Biochemists' handbook. Van Norstrand, Princeton, pp 249–251
124. Thomson SW (1966) Selected histochemical and histopathological methods. Thomas, Springfield
125. Gmori G (1949) Histochemical specificity of phosphatases. Proc Soc Exp Biol Med 70:7–11 1949
126. Newman W, Feigin I, Wolf A, Kabat EA (1950) Histochemical studies on tissue enzymes. IV. Distribution of some enzyme systems which liberate phosphate at pH 9.2 as determined with various substrates and inhibitors; demonstration of three groups of enzymes. Am J Pathol 26:257–305
127. Feigin I, Wolf A (1955) The phosphatase of the nervous system. J Neuropathol Exp Neurol 14:11–22
128. Emmel VM (1950) Effects of HCl on alkaline phosphatae in kidney and intestine-histochemical and quantitative study. Proc Soc Exp Biol Med 75:114–117
129. John-Kenny A, Both AG (1976) Organisation of the kidney proximal-tubule plasma membrane. Biochem Soc Trans 4:1011–1017
130. Luft FC, Patel V (1978) Lysosomal acid hydrolases as urinary markers of aminoglycoside nephrotoxicity in the rat. In: Fillastre JP (ed) Nephrotoxicity: interaction of drugs and membrane systems. Mitochondria—lysosomes. Masson, New York, p 127
131. Wellwood JM, Lovell D, Thompson AE, Tighe JR (1976) Renal damage caused by gentamycin: A study of the effects on renal morphology and urinary enzyme excretion J Pathol 118:171
132. Morin JP, Fresel J, Fillastre JP, Vaillant R (1978) Aminoglycoside action on rat kidney lysosomes 'in vitro' and 'in vivo'. In: Fillastre JP (ed) Nephrotoxicity: interaction of drugs and membrane systems. Mitochondria — lysosomes. Masson, New York, p 253
133. Gibey R, Dupond JL, Peltier H, Iehel-Robert M, Henry JC (1986) Un marquer percoce et specifique de la nephrotoxicite des aminoglycosides: l'isoenzyme B de la N-acetyl-beta-glucosaminidase (NAG) urinaire. Pathol Biol (Paris) 34:342–345
134. Dupond JL, Gibey R, Iehl-Robert M, Mallet H, Leconte des Floris R, Henry JC (1984)

Nephrotoxicite des aminosides et cephalosporines. Interet de l'etude des profiles iso-enzymatiques de la N-acetyl-beta-D-glucdosaminidase (NAG). Rev Med Interne 5:321–327
135. Gibey R, Dupond JL, Mallet H, Guerber F, Iehl-Robert M, Flausse-Parrot F, Becker M (1986) Les isoenzymes de la N-acetyl-beta-D-glucosominidase (NAG): une nouvelle methode d'evaluation de la nephrotoxicite des aminosides. Notre experience avec la dibekacine en geriatrie. Nephrologie 7:51–55
136. Dubach UC (1966) On the origin of lactic dehydrogenase isoenzymes in the urine. Helv Med Acta 33:139–150
137. Gerlach V (1966) Einige pathophysiologische Grundlagen der Enzymdiagnostik. Enzymologia 30:29–48
138. Wajima T, Takahashi H (1965) Isoenzymes of amylase, alkaline phosphatase, lactic dehydrogenase and leucine aminopeptidase in urine. Chem Abstr 63:13789b
139. Wroblewski F, Laude JS (1955) Lactic dehydrogenase activity in blood. Proc Soc Exp Biol Med 90:210–213
140. Plummer DT, Ngaha EO, Wright PJ, Leathwood PD, Blake ME (1979) The sensitivity of urinary enzymes measurements for detecting renal injury. In: Dubach UC, Schmidt U (eds) Diagnostic significance of enzymes and proteines in urine. Huber, Bern, pp 71–84
141. Scherberich JE (1989) Immunological and ultrastructural analysis of loss of tubular membrane bound enzymes in patients with renal damage. Clin Chim Acta 185:271–282
142. Andrews PM, Porter KR (1974) A scanning electron microscopic study of the nephron. Am J Anat 140:81–116
143. Kinne R, Schmitz JE, Kinne-Safran E (1973) The localisation of the Na^+-K^+-ATPase in the cells of rat kidney cortex: a study on isolated plasma membranes. Pflügers Arch 329:191–206
144. Maunsbach AB (1973) Ultrastructure of the proximal tubule. In: Orloff J Berliner RW, Geiger SR (eds) Renal physiology. Waverly, Baltimore, p 31 Handbook of physiology, sect 8
145. Spinelli F, Wirz H, Brucher C, Pehling G (1972) Fine structure of the kidney revealed by scanning electron-microscopy. Ciba-Geigy, Basel
146. Donohoe JF, Venkatachalam MA, Bernard DB, Levinsky NG (1978) Tubular leakage and obstruction after renal ischaemia: structural functional correlations. Kidney Int 13:208–222
147. Zager RA (1983) Glomerular filtration rate and brush border debris excretion after mercuric chloride and ischaemic acute renal failure: mannitol versus furosemide diuresis. Nephron 33:196–201
148. Beck F, Mason J, Bauer R, Dröge A, Thurau K (1978) Electronmicrobe analysis of intracellular electrolyte concentrations of rat kidneys following ischaemia. Pflügers Arch 377 [Suppl R 12] (abstr 42)
149. Dröge A, Mason J, Beck F, Bauer R, Thurau K (1978) Electronmicrobe analysis of intracellular electrolyte concentrations of rat kidneys following ischaemia. Pflügers, Arch. 373: Supp. R. 27: Abstr. 79
150. Mason J (1978) Die Nephronfunktion in der Frühphase des experimentellen akuten Nierenversagens. Habilitationsschrift, University of Munich
151. Braun L (1964) Zur Regeneration der Tubulusepithelien beim akuten Nierenversagen. Z Urol 57:789–796
152. Zollinger HU, Mihatsch MJ (1978) Renal pathology in biopsy. Light, electron and immunofluorescent microscopy and clinical aspects. Springer, Berlin Heidelberg New York, p 120
153. Thomas C (1986) Histopathologie: Lehrbuch und Atlas für die Kurse der allgemeinen und speziellen Pathologie, 10th edn. Schattauer, Stuttgart, p 17
154. Solez K, Morel-Maroger L, Sraer JD (1979) The morphology of "acute tubular necrosis" in man. Analyses of 57 renal biopsies and a comparison with the glycerol model. Medicine (Baltimore) 58:362–376
155. Trebbin H, Jahnecke J, Lange T. Bohle A (1962) Beitrag zur Frage nach den Beziehungen zwischen der Weite der Harnkanälchenlichtungen und der Breite des Interstitiums beim akuten Nierenversagen. Klin Wochenschr 40:575–580
156. Higashiyama N, Nishiyama S, Itoh T, Nakamura M (1983) Effects of castration on urinary N-acetyl-beta-D-glucosaminidase levels in male beagles. Renal Physiol 6:226–231
157. Grötsch H, Hropot M, Klaus E, Malerczyk V, Mattenheimer H (1985) Enzymuria of the rat: biorhythms and sex differences. J Clin Chem Clin Biochem 23:343–347
158. Mondorf AW, Scherberich JE, Schoeppe W (1982) Besitzen Cephalosporine eine nephrotoxische Potenz? FAC Fortschritte der antimikrobiellen und antineoplastischen Chemotherapie, 257–265

5 Physiological Factors Influencing Enzyme Excretion

U. Burchardt and K. Jung

Numerous physiological factors influence the excretion of urinary enzymes. They cause the wide variability of reference ranges of urinary enzyme excretions and have to be considered if the results of urinary enzyme measurements are to be interpreted. Influencing factors change urinary enzyme excretion independent of the assay. Invariable influencing factors are age, sex, genetic factors and race. Variable ones are body weight, muscle mass, nutrition, physical activity, biorhythms and climate.

5.1 Urinary Flow

Urinary enzyme excretion depends on urine flow rate [32]. Forced diuresis after water drinking causes an increase of time-related enzyme excretion whereas the volume activity decreases because of the dilution effect [11, 47, 73]; e.g. after a high intake of fluid the hourly excretion of the brush border enzymes alanine aminopeptidase, alkaline phosphatase and gamma-glutamyltransferase (AAP, AP and GGT) and the lysosomal *N*-acetyl-β-D-glucosaminidase (NAG) increased with the increasing urine flow rate [35] (Fig. 1). Diuresis-dependent excretions of all brush border enzymes can be exclusively considered to be a consequence of changed excretion of soluble forms (Fig. 2), but the excretion rates differ between various enzymes (Fig. 1).

The increased urinary excretion of enzymes during high urine output has not yet been explained. Jösch and Dubach [32], while investigating the excretion of lactate dehydrogenase (LDH), leucine aminopeptidase and AP in the states of diuresis and antidiuresis, discussed an altered permeability of tubular cells for enzymes and a changed diuresis-dependent membrane turnover. In addition, an increased urine flow rate should mean that more nephrons are functioning. Thus, more ultrafiltrate would flow through the tubules, which could release more enzymes. The lack of a coordinated excretion of enzymes from different cellular compartments [35] suggests that the enzymes do not come from exfoliated cells. This so-called distortion of enzyme pattern has been well established for serum enzymes in liver diseases [68]. One reason for this finding indicated in Fig. 1 could be the different localization of the corresponding

Urinary Enzymes
K. Jung/H. Mattenheimer/U. Burchardt (Eds.)

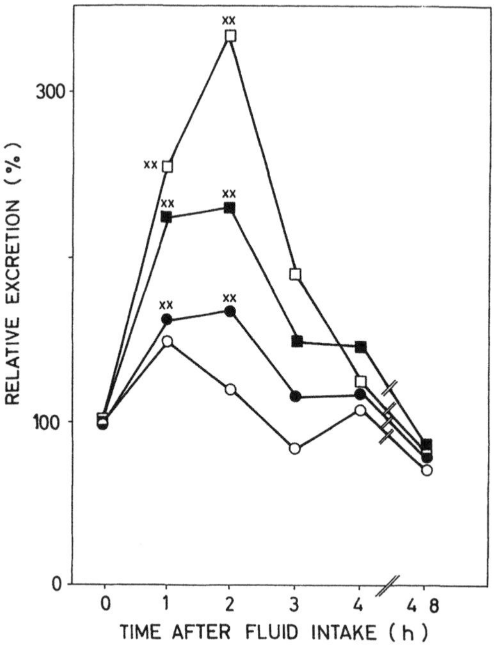

Fig. 1. Diuresis-dependent excretion rate per hour (U/h) of alanine aminopeptidase (■), alkaline phosphatase (□), gamma-glutamyltransferase (●) and *N*-acetyl-*β*-D-glucosaminidase (○) after fluid intake (according to [35]). The excretion rates during 1 h (6 to 7 a.m.) before the intake of unsweetened tea (22 ml/kg body weight) were considered as baseline values (100%). On 2 days before the drinking experiment the excretion rates were measured in six subjects at the same times under habitual conditions (normal intake of fluid). Because there was no difference of excretion rates between the first hour and the following hours, a diurnal influence on the enzyme excretion could be excluded. Therefore, the hourly excretion rates determined before the start of the fluid intake were used as baseline values. Arithmetic means of changes are given. Significance level in comparison with the starting values at *x*, $P < 0.05$; *xx*, $P < 0.01$ and *xxx*, $P < 0.001$

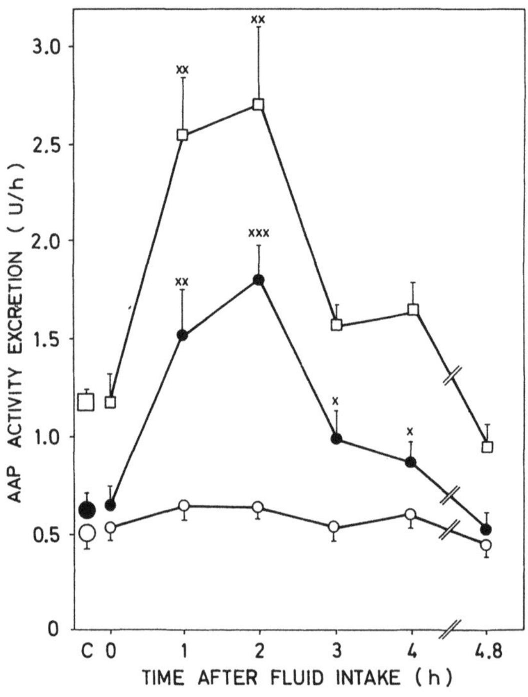

Fig. 2. Diuresis-dependent excretion rate per hour of the total alanine aminopeptidase and its multiple forms (according to [34]). The values given are arithmetic means ± SEM of six healthy men after the intake of tea equivalent to 22 ml/kg of body weight. The values indicated with *C* (as sign for controls) are the mean values of all hourly excretion rates determined between 6 a.m. and 3 p.m. on 2 days before the drinking experiment. Urinary flow rates are the same as in Fig. 1. Significant changes of excretions were calculated in comparison with the hourly excretion rate before the drinking experiment and are indicated as in Fig. 1. ○, particulate form; ●, soluble form; □, total activity

enzymes in the tubular cell: NAG is found in the lysosomes, whereas the three other enzymes are components of the brush border. Because AP shows a higher excretion rate than the more superficially located AAP and GGT, however, it must be assumed that not only the different site but also such factors as linkage, shape and size of the enzymes are responsible for the different rates. An active "package" secretion of lysosomal enzymes has been suggested [62, 63]. Since the relationships of volume activity and creatinine concentration in urine to the urine minute volumes are hyperbolic (Fig. 3), the best way to adjust the enzyme excretion to urine flow rate is to relate the enzyme activity to urinary creatinine [11]. Depending on the particular enzyme measured, the diuresis effect on enzyme excretion can be partially or completely compensated for using this reference basis (Fig. 4).

5.2 Circadian and Infradian Rhythms

Circadian rhythms have been documented for numerous renal functions and have also been described for many enzymes [44, 53, 62]. In general, maximal excretion occurs between 3 a.m. and 9 a.m. Low output values were registered during the afternoon and early night hours. These differences are valid both for

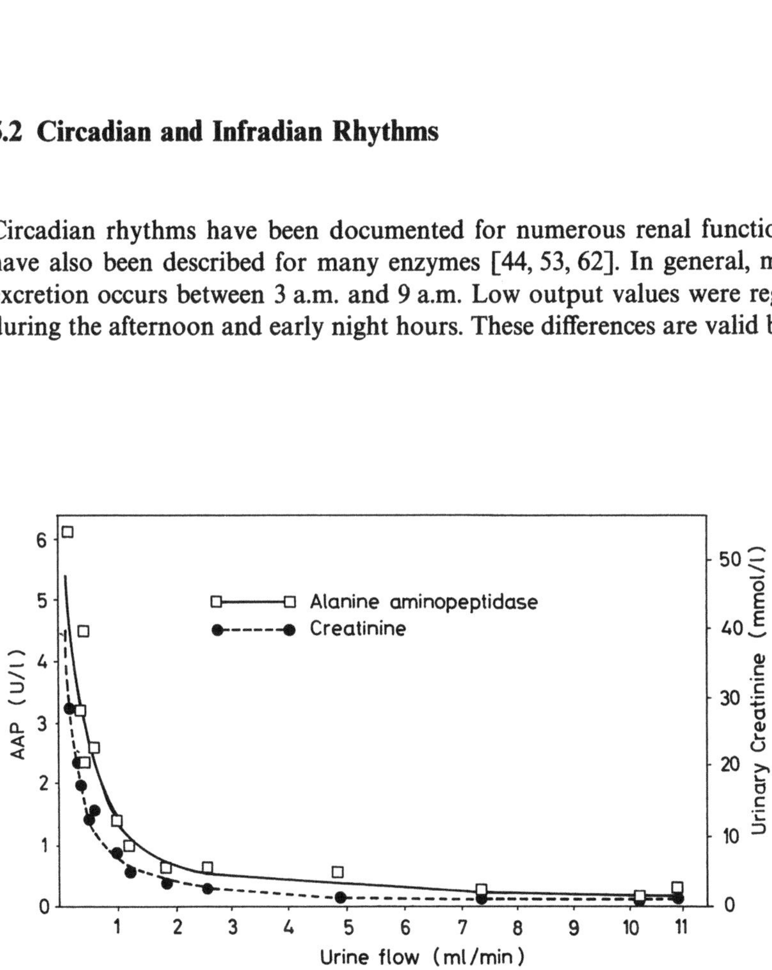

Fig. 3. Relation between volume-related alanine aminopeptidase (*AAP*) activity (□), urinary creatinine concentration (●) and urine flow rate (according to [11])

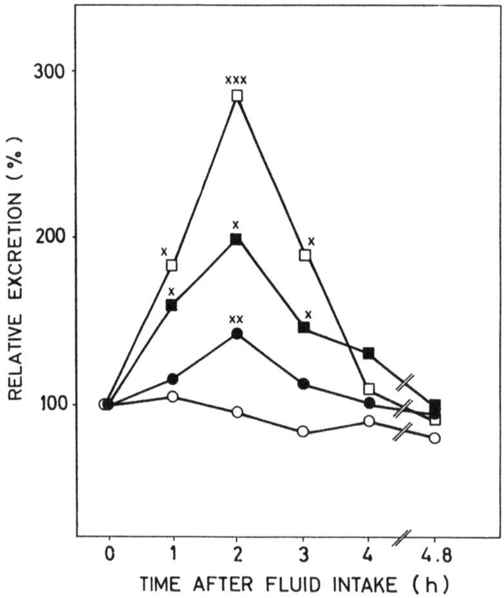

Fig. 4. Diuresis-dependent excretion related to urinary creatinine (U/mmol creatinine) of alanine aminopeptidase, alkaline phosphatase, gamma-glutamyltransferase, *N*-acetyl-*β*-D-glucosaminidase after fluid intake (according to [35]). Symbols and details as in Fig. 1

time-related and creatinine-related excretions; however, the differences increased if enzyme activity/creatinine ratios are used [44]. There is no statistically significant sex difference in the timing of the urinary circadian rhythms, but female subjects show a greater amplitude than male in urinary enzyme excretion [53]. These diurnal variations are also found in patients suffering from renal diseases, but the difference between maximum and minimum output is higher. The reasons for the diurnal variations of urinary enzyme excretions are unknown. Diurnal changes of cell permeability, lysosomal enzymatic activity, cell turnover and enzyme release have to be considered. Because the circadian time structure of the rhythm of renal function parameters like renal plasma flow, glomerular filtration rate or excretions of sodium and potassium does not correspond to that of enzymuria, it can be excluded that these renal function parameters are chronobiologically correlated in a direct way with changes of urinary enzyme output [44].

To avoid these changes of urinary enzyme excretion in the clinical interpretation of results, it is generally recommended to collect urine specimens within a fixed time period independently from the selected reference basis to express the enzyme excretion.

Beyond these well-established circadian rhythms of enzymuria, infradian rhythmic changes of urinary enzyme excretions are observed. This phenomenon was firstly reported in rats [24]. Grötsch et al. [24] described biological rhythms in the excretions of GGT, NAG and AAP with dominant periods in the infradian range of 7 and 9 days, respectively. In humans, the excretion of AAP, dipeptidylpeptidase IV and GGT followed an infradian rhythm with a mean period length of 10 days for the first two enzymes and 13 days for the last [11, 12]. These infradian biorhythms with changes in the range of 100% and more have to be considered in assessing the excretion of enzymes.

5.3 Age and Sex

True differences of urinary enzyme excretion given as time-related excretions in function due to age and sex could reflect (a) differences of kidney mass [25] from which the enzymes originate, (b) enzymatic changes in the kidney dependent on age and/or sex [76], (c) changes of the reabsorptive-catabolizing function of tubular cells which is responsible for processing low molecular weight enzymes like lysozyme and (d) the contribution of specific genital secretions to the urinary enzyme activities.

In practice, the influence of age and sex on the results of urinary enzyme activities essentially depends on the reference system (see Chap. 9) used to express the enzyme activity values (e.g. as volume activity, time-related excretion or creatinine-related excretion) and the urinary specimen used for measurement [52]. A statement on the dependence of particular enzyme's excretion on age and sex is only valid in connection with the information on the reference basis. Expression as volume activities differ in their age-dependent pattern from the expression of urinary enzyme activities given as the enzyme activity/creatinine ratio (Fig. 5). Whereas the activity/creatinine ratio decreases with increasing age in adolescence, this typical behaviour is not evident for volume activities [18]. It is obvious that the decrease is caused by increased output of urinary creatinine. An increased ratio is especially found in children below the age of 2 years in comparison with other age groups [1, 6, 19, 20, 41, 72].

Otherwise, in random day urinary specimens of healthy individuals between 20 and 80 years a slight but significant age-dependent increase of NAG/creatinine ratio was evident, whereas no such increase was realized with the morning urine specimens [9]. These differences could also be explained by the changed level of creatinine excretion [9].

The same issue has to be considered when the influence of sex is to be interpreted. Since males subjects excrete more creatinine than female, sex-dependent differences observed in time-related urinary enzyme excretion can disappear when urinary enzyme excretions are related to creatinine and vice versa [9, 36, 52]. It was e.g. found that men excreted significantly more GGT, trehalase and NAG than women if excretion was calculated on the basis of urinary time volume. However, if urinary creatinine was used as the reference system, the excretions of these three enzymes were higher in women than in men [52]. The AAP excretion in 24-h urine of men was found to be 1.2-fold higher

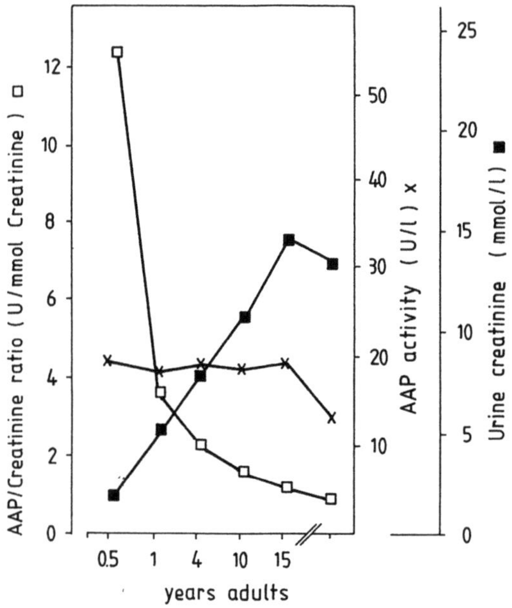

Fig. 5. Volume activity (U/l; *X*), activity/creatinine ratio (U/mmol creatinine; □) of alanine aminopeptidase (*AAP*) and urinary creatinine concentration (mmol/1; ■) in function of age (modified according to [36]). Values are given as medians

than in the respective samples of women [9], but this sex-dependent difference was not observed when the activity was related to urinary creatinine [33].

Since urinary enzyme activities are greatly scattered, some authors could not establish sex- or age-dependent variations in particular enzymes for which other authors demonstrated significant differences. Numerous studies have been done using either small numbers of subjects or dubious statistical procedures, leading to questionable conclusions.

5.4 Genetic Component

Paigen et al [63] tested the presence of a genetic component determining enzyme excretion rates by comparing the excretion of four lysosomal enzymes (β-glucuronidase, β-galactosidase, NAG, α-galactosidase) in monozygotic and dizygotic twins. Using calculations of heritability they found that half or more of the total variance of urinary enzyme excretion seemed to be genetic in origin. This appears to be true for both total enzyme excretion and the coordination of lysosomal enzyme output. Genetic polymorphism of some urine enzymes, e.g. deoxyribonuclease I [40], has been detected. These results are useful and should shed further light on the anthropological and forensic significance of urinary enzymes [60].

Various diseases of inborn error of metabolism increase or decrease the excretion rates of some enzymes and change their isoenzyme pattern (see Sect. 15.9).

5.5 Hormonal Influences and Nutrition

Some hormones like steroids and thyroid hormones are known to be effective by changing the transcription of specific genes. They increase the biosynthesis of messenger RNA and thus that of specific proteins [4]. Androgens and thyroid hormones have been mentioned as esential factors in influencing this transcription level in proximal tubular cells [51, 71]. Histochemical investigations of the renal proximal tubule of male and female rats after castration reveal the possible influence of androgens and oestrogens on the activity of lysosomal and brush border enzymes [21, 27, 42, 51, 57, 58]. GGT activity in the kidneys of castrated rats was lower than that in the kidneys of uncastrated or testosterone-substituted castrates. Enzymuria found in the castrated cows results from the decreased specific tissue activity [51]. Testosterone propionate, administered to female mice, markedly increased the urinary excretion of acid hydrolases. Analysis in mice receiving a large dose of testosterone propionate (10 mg daily) revealed a 250-fold increase in the urinary excretion of β-glucuronidase and a 12- to 25-fold increase in the excretion of four other acid hydrolases within 14 days. Testosterone administration in female mice elicits the formation of numerous myeloid bodies, a form of lysosomal residual body, in the kidney proximal tubule cells. These residual bodies are discharged, apparently by exocytosis, into the tubular lumen, releasing their complement of soluble hydrolases into the urine [21, 51] (see Chap. 3). Similar studies on the influence of the gonadal system on urinary enzyme excretion in humans are not available.

As these hormones show a circadian rhythm, circadian variations in protein synthesis in tubular cells and, consequently, the amount of released enzymes result. However, there is a time lag of about 4 h between the peaks of urinary enzyme activity (for example: GGT and AAP) and triiodothyronine and thyroxine [13].

This phenomenon could be interpreted as the latency period between effected synthesis and the release of enzymes. Thus, physiological changes in triiodothyronine levels could be partly responsible for altered enzyme excretion [13, 69, 71]. Orally ingested triiodothyronine leads to higher brush border and lysosomal enzyme output [10, 26].

In idiopathic hypoparathyroidism (IHP), 100 units of human parathyroid hormone (PTH) increased the urinary excretion of NAG from 0.029 ± 0.027 to 0.173 ± 0.035 U/l in two patients. This increase was not shown in patients suffering from pseudohypoparathyroidism (PHP) (a disorder characterized by target tissue resistance to PTH). PTH did not significantly increase the plasma levels of NAG in patients with IHP or with PHP. An increase of urinary excretion of NAG may be a new indicator of renal effects of PTH [55, 67].

Interestingly, urinary acid phosphatase secretion (origin: prostate) is an indicator of sexual arousal. The amount of acid phosphatase secreted by a man viewing a sexually arousing videotape is significantly greater than the amount secreted under conditions that evoke laughter, anxiety or anger [3].

Studies in dogs showed increased urinary enzyme excretion by adrenergic receptor stimulation. Increases in lysosomal and brush border enzymes during isoproterenol infusion were not due to blood pressure reduction [2]. Renal tubular β-receptors are localized mainly at the distal tubules along the nephron.

Only little is known about the influence of nutrition on urinary enzyme excretion. Kang et al. [37] investigated urinary enzyme excretion in rats fed with a diet containing sucrose, starch and/or fat. The excretion of NAG was significantly higher in rats fed on sucrose. In rats additionally fed a greater amount of fat, dietary sucrose produced a higher activity of the enzyme than did additional dietary starch. In healthy humans, NAG excretion increased significantly after an intake of sucrose up to 250 g per day. High sucrose intake leads to so-called sucrose nephrosis [31, 49] with an extreme enlargement of lysosomes. Mechanisms of enzyme release under this condition are discussed in Chap. 3.

Increased alkaline phosphatase activity in rats is found after feeding a low phosphorus diet [39, 48].

Urinary kallikrein level is related to dietary sodium intake. In rats, there was a positive correlation between urinary kallikrein and sodium excretion in acute experiments, whereas a long-term restriction of dietary sodium intake was characterized by an inverse relationship [3, 7]. It has been concluded that this phenomenon is probably caused by a simultaneous stimulation of aldosterone secretion because spironolactone as an antagonist prevents the increase of kallikrein excretion in cases of dietary sodium restriction [7].

In protein-caloric malnutrition (kwashiorkor) coupled with severe vitamin A deficiency, increased excretions of arylsulfatase A and B and of β-glucuronidase were found [4, 46]. The increased activities were attributed to the possible degeneration of the proximal tubules because a reduction of enzymuria occurred on recovery. A correlation seems to exist to vitamin A as in malnourished persons with or without vitamin A deficiency symptoms, arylsulfatase A level was significantly reduced, and that of arylsulphatase B was not altered [46].

5.6 Pregnancy

Changes in proximal tubular function and glomerular filtration rate are well established by the end of the first trimester of pregnancy. Renal size has increased. A number of researchers have investigated urinary enzyme excretion in terms of gestational age [8, 14, 29, 45, 61, 70]. Urinary NAG, AAP and GGT in pregnant women without complications [14, 23, 29, 61, 70] increased progressively through the first, second and third trimesters. The median values for the three markers in healthy third-trimester women were significantly higher than those for women in the first trimester [14]. Differently increasing rates within these periods are evident for various enzymes (Fig. 6). It has been shown that the activities of some enzymes tend to be higher in twin as compared with single pregnancies [56]. All these changes may be indicative of hypertrophy and/or hyperplasia of the nephron and part of the overall renal adaptation to normal pregnancy. Borenstein et al. [8] could not find urinary excretions of AP, aspartate aminotransferase (ASAT), leucine aminopeptidase, LDH and NAG to be dependent on gestational age.

About a year after an uncomplicated pregnancy, enzymuria normalizes [29]. Thus, changed enzyme excretions in uncomplicated pregnancy could be a reflection of an enhanced enzyme turnover.

5.7 Physical Activity

Exercise is usually associated with a reduction of effective renal plasma flow. This reduction is related to the intensity of exercise and may be caused by the combination of sympathetic nervous activity which constricts both afferent and efferent arterioles and the release of larger amounts of adrenaline and noradrenaline. The result is a decreased glomerular filtration rate and altered tubular functions. In consequence, changes of enzyme excretion result from an increased enzyme release from tubular cells and a reduced reabsorptive-catabolizing capacity of the tubulus apparatus (see Chap. 3). The following examples of changed enzyme excretion demonstrate these aspects. NAG and GGT activities

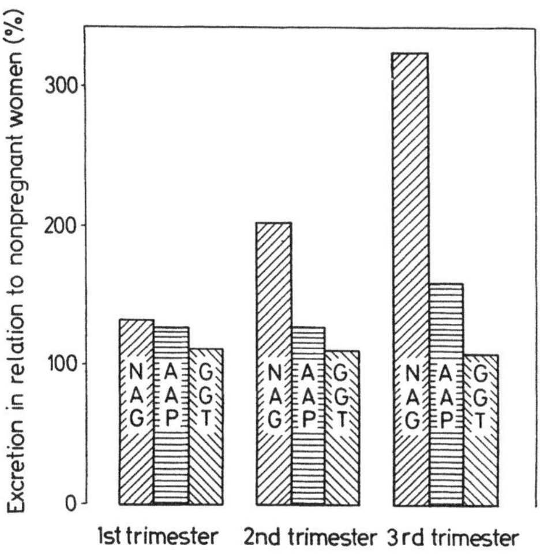

Fig. 6 Excretion of urinary enzymes alanine aminopeptidase, gamma-glutamyl transferase and *N*-acetyl-*β*-D-glucosaminidase (*AAP*, *GGT* and *NAG*) by pregnant women in terms of gestational age. Values are calculated from data of Cheung et al. [14] and are given in relation to data found in non-pregnant controls as percentage medians

in urine increased after a run for 1 hour in 90% of runners, showing a linear relationship to the 'run speed' (NAG: $r = 0.79$, GGT: $r = 0.52$). The urinary enzyme activities returned to the normal range in all cases within 12–36 h [16]. Whiting et al. [74] decribed significantly increased urine NAG levels before and immediately after a marathon race in 48 runners (pre: 121 ± 86, post: 250 ± 200 U/mg urinary creatinine). This post-race increase in enzyme activities possibly reflects the renal hemodynamic changes with tubular cell ischaemia during exercise.

In addition, low molecular mass proteins like lysozyme increase after exercise [65]. An increase of almost 50–60-fold was observed. Strenuous physical exercise leads to so-called post-exercise proteinuria, which differs from normal proteinuria at rest [66]. Post-exercise proteinuria is of mixed type as a result of both increased glomerular permeability and some failure of tubular reabsorption of low molecular mass proteins. Urinary amylase activity and amylase clearance were unchanged [64]. Because tubular damage is usually associated with increased excretion of lysozyme as well as amylase, these discrepancies between exercise and pathological conditions are at present difficult to explain.

5.8 Environmental and Occupational Influences

Environmental influences on urinary excretion can result from the effect of nephrotoxic substances which are found in the workplace or in polluted areas. The aspect of nephrotoxicity is in detail described in Chap. 14.

Table 1. Increased urinary enzyme excretion in workers exposed to nephrotoxic chemicals

Exposition	Enzymes	References
Cadmium	β-galactosidase	5
	NAG	22, 38
	trehalase	30
	leucine aminopeptidase	59
Lead	NAG, lysozyme	15
	NAG, β-glucuronidase	54
Uranium	catalase	17
Mercury	NAG	28, 54
Chlorinated hydrocarbon pesticides	NAG	43
Organic solvents	NAG, β-glucuronidase	54
Aromatic nitro-amino compounds	NAG	75
Styrene	NAG, GGT	50

NAG, *N*-acetyl-β-D-glucosaminidase; GGT, gamma-glutamyltransferase

Examples showing an influence of environmental pollution on urinary enzyme activity can be given. Japanese authors reported higher urinary trehalase activity in subjects with cadmium-induced itai-itai disease and their family members living in a cadmium-polluted area in the Jinzu river basi [59]. Bernard et al. [5] measured a moderately increased release of β-galactosidase in workers exposed to cadmium. Urinary activities of LDH, AP and catalase remained unchanged. Hyperenzymuria was observed mainly in workers with more than 25 years of exposure. When compared with appropriate control populations, workers exposed to lead or to organic solvents had significant increases in NAG activity [15, 75]. Table 1 surveys the findings. It can be concluded that exposure to environmental nephrotoxins at levels currently considered safe can produce alterations of urinary enzyme excretion.

References

1. Adelman RA, Zakauddin S (1980) Urinary enzyme activities in children and neonates receiving gentamicin therapy. Dev Pharmacol Ther 1:325–332
2. Akabane S, Kawamura M, Ito K, Ogino K (1984) Intrarenal beta-adrenergic stimulation enhances excretion of urinary lysosomal enzyme in anesthetized dogs. Nephron 36:265–269
3. Barcley AM, Little DM (1972) Urinary acid phosphatase secretion under different arousal conditions. Soc Psychophys Res 9:69–77
4. Begun A (1973) Urinary excretion of β-glucuronidase in protein-caloric malnutrition. Clin Chim Acta 46:229–234
5. Bernard A, Buchet JP, Roels, Masson P, Lauwerys R (1979) Renal excretion of proteins and enzymes in workers exposed to cadmium. Eur J Clin Invest 9:11–22
6. Bertotti A, Mangiarotti MA, Peratoner L, Stormi M, A'Avanzo R, De Marchi S, Barbone F (1987) Studio dellècrezione urinaria dell'alfa-glucosidasi, N-acetilglucosaminidase e alanina-aminopeptidasi nel neonato a termine sano. Minerva Pediatr 39:445–450
7. Boenner G (1988) Das renale Kallikrein-Kinin-System. Springer, Berlin Heidelberg New York
8. Borenstein R, Rozenman D, Lancet M, Bar-Khayim Y, Cohen N, Gertler A (1980) Significance of urinary enzymes in gestosis. Gynecol Obstet Invest 10:193–203
9. Burchardt U, Miesel B (1977) Alaninaminopeptidaseausscheidung mit dem Harn in Abhängigkeit von Alter und Geschlecht. Z Inn Med 32:319–322
10. Burchardt U, Peters JE, Thulin H, Hempel RD, Gründig CA (1974) Enzymausscheidung im Harn bei Hyperthyreose. Z Inn Med 29:719–722
11. Burchardt U, Hempel A, Höpfner JB, Hempel RD (1976) Alanineaminopeptidase- und Kreatininausscheidung unter verschiedenen Diuresezuständen beim Menschen. Z Inn Med 31:550–552
12. Burchardt U, Winkler K, Klagge M, Balschun D, Barth A (1988) Infradian biorhythms of enzymuria in man? J Clin Chem Clin Biochem 26:491–496
13. Cal J-P, Lemoigne F, Crockett R, Cambar J (1987) Circadian rhythm in gamma glutamyltranspeptidase and leucine aminopeptidase urinary activity in rats. Chronobiol Int 4:153–160
14. Cheung CK, Lao T, Swaminathan R (1989) Urinary excretion of some proteins and enzymes during normal pregnancy. Clin Chem 35:1978–1980
15. Coratelli P, Giannattasio M, Lomonte C, Marzolla R, Rana F, L'Abbate N (1988) Enzymuria to detect tubular injury in workers exposed to lead: a 12-month follow-up. Contrib Nephrol 68:207–211
16. De Pooli Vitali E, Gilli P, Squerzanti R, Nunzi L, Braga D, Ferinelli A (1984) Urinary enzyme activities (UEA) for gamma-glutamyl-transferase (γ-GT) and *N*-acetyl-glucosaminidase (NAG) in long distance runners. LAB-ABC Vaktiischrift vorde analytische, klinische chemie en biotechnologie 9:79

17. Dounce AL, Roberts E, Wills JH (1949) Catalasuria as a sensitive test for uranium poisoning. In: Voegtlin C, Hodge HC (eds) Pharmacology and toxicology of uranium compounds, vol 1. Mc Graw Hill, New York, pp 889–950
18. Dzialoszynski L, Dzialoszyńska A, Strzykala K, Sluzewski W (1975) Arylsulfatase A and B activity in urine of newborns and children up to the age of 12 years. Pediatr Pol 10:1263–1268
19. Ehrich JPH, Wurster U, Lütge M, Mertz I, Doehring E, Oemar B (1986) Enzymurie und Nierenerkrankungen im Kindesalter. Monatschr Kinderheilkd 134:832–839
20. Feldmann D, Flandrois C, Jardel T-M, Aymard F (1989) Circadian variations and reference intervals for some enzymes in urine of healthy children. Clin Chem 35:864–867
21. Girolami JP, Bascands JL, Prevot D, Pecher C, Manuel Y (1989) Urinary output modulation of alanine aminopeptidase, glutamyl transpeptidase and N-acetyl-β-D-glucosaminidase by castration and testosterone in male normal rats. Enzyme 41:181–186
22. Gompertz D (1993) Renal dysfunction in cadmium smelters: relation to in-vivo liver and kidney cadmium concentrations. Lancet ix:1185–1187
23. Goren MP, Sibai BM, El-Nazar A (1987) Increased tubular enzyme excretion in preeclampsia. Am J Obstet Gynecol 157:906–908
24. Grötsch H, Hropot M, Klaus E, Malerczyk V, Mattenheimer H (1985) Enzymuria of the rat: biorhythms and sex differences. J Clin Chem Clin Biochem 23:343–347
25. Haschen RJ (1977) Diagnostic significance of proteolytic enzymes. Z Med Lab Diagn 18:147–158
26. Hempel RD, Burchardt U, Haschen RJ (1977) Alaninaminopeptidaseausscheidung im Harn zur Verlaufskontrolle der Hyperthyreosetherapie. Dtsch Gesundheits Wes 32:22–24
27. Higashiyama N, Nishiyama S, Itoh T, Nakamura M (1983) Effect of castration of urinary N-acetyl-β-D-glucosaminidase levels in male beagles. Renal Physiol 6:226–231
28. Himeno S, Watanabe C, Suzuki T (1986) Urinary biochemical changes in workers exposed to mercury vapor. Ind Health 24:151–155
29. Hultberg B, Isaksson A, Krutzen E, Nilsson-Ehle (1989) Urinary excretion of N-acetyl-β-D-glucosaminidase in normal and complicated pregnancy. J Clin Chem Clin Biochem 27:487–489
30. Iwata K, Katoh T, Morikawa Y, Aoshima K, Nishijo M, Teranishi H, Kasuya M (1988) Urinary trehalase activity as an indicator of kidney injury due to environmental cadmium exposure. Arch Toxicol 62:435–439
31. Janigan DT, Trump BF (1960) Sucrose nephrosis: electron microscopic and histochemical observations. J Histochem Cytochem 8:385–396
32. Jösch W, Dubach UC (1967) Einfluß der Diurese auf die Enzymurie. Clin Chim Acta 15:325–330
33. Jung K, Scholz D (1980) An optimized assay of alanine aminopeptidase activity in urine. Clin Chem 26:1251–1254
34. Jung K, Schulze G (1986) Diuresis-dependent excretion of multiple forms of renal brush-border enzymes in urine. Clin Chim Acta 156:77–84
35. Jung K, Schulze G, Reinholdt C (1986) Different diuresis-dependent excretions of urinary enzymes: N-acetyl-β-D-glucosaminidase, alanine aminopeptidase, alkaline phosphatase, and γ-glutamyltransferase. Clin Chem 32:529–532
36. Jung K, Hempel A, Grützmann K-D, Hempel R-D, Schreiber G (1990) Age-dependent excretion of alanine aminopeptidase, alkaline phosphatase, γ-glutamyltransferase and N-acetyl-β-D-glucosaminidase in human urine. Enzyme 43:10–16
37. Kang SS, Price RG, Yudkin J, Worcester NA, Bruckdorfer KR (1979) The influence of dietary carbohydrate and fat on kidney calcification and the urinary excretion of N-acetyl-β-glucosaminidase (EC 3.2.1.30). Br J Nutr 41:65–71
38. Kawada T, Koyama H, Suzuki S (1989) Cadmium, NAG activity, and β_2-microglobulin in the urine of cadmium pigment workers. Br J Ind Med 46:52–55
39. Kempson SA, Dousa TP (1979) Phosphate transport across renal cortical brush border membrane vesicels from rats stabilized on a normal, high or low phosphate diet. Life Sci 24:881–888
40. Kishi K, Yasuda T, Awazu S, Mizuta S (1989) Genetic polymorphism of urine deoxyribonuclease I. Hum Genet 81:295–297
41. Klingmüller V, Bergmann E, Claaßen P (1984) Alaninaminopeptidase-Ausscheidung bei Kindern: Referenzbereiche und der Einfluß nierengängiger Kontrastmittel. J Clin Chem Clin Biochem 22:787–789
42. Koenig H, Goldstone A, Hughes C (1978) Lysosomal enzymuria in the testosterone-treated mouse. Lab Invest 39:329–341

43. Kossmann S, Wartalaska G (1984) Activity of some enzymes in the urine of workers producing chloronated hydrocarbon pesticides. Med Pr 25:403–407
44. Lakatua DJ, Blomquist CH, Haus E, Sackett-Lundeen L, Berg H, Swoyer J (1982) Circadian rhythm in urinary N-acetyl-β-glucosaminidase (NAG) of clinically healthy subjects. Am J Clin Pathol 76:69–77
45. Landon MJ, Bates D, Kirklsy M, Venables G (1978) Effect of anticonvulsant drugs on urinary excretion of γ-glutamyl transferase in women. Ann Clin Biochem 15:313–315
46. Latif KA, Amla J, Rao PPR (1979) Urinary excretion of arylsulfatases in malnourished vitamin A deficient children. Clin Chim Acta 96:131–138
47. Leonard DD (1974) Urinary beta-glucuronidase in man: relation to urinary specific gravity and creatinine concentration. Clin Nephrol 2:41–43
48. Levine BS, Ho K, Hodsman A, Kurokawa K, Coburn JW (1984) Early renal brush border membrane adaptation to dietary phosphorus. Miner Electrolyte Metab 10:222–227
49. Li MK, Kavanagh JP, Prendiville V, Buxton A, Moss DG, Black-lock NJ (1986) Does sucrose damage kidneys? Br J Urol 58:353–357
50. Malaguti Alberti L, Severini GC (1989) Urinary enzymes: early biochemical indicators in the biological monitoring of people exposed to styrene. Minerva Med 80:383–388
51. Malyusz M, Ehrens HJ (1983) Effect of castration and testosterone substitution on the urinary output of gamma-glutamyl-transpeptidase and of N-acetyl-beta-D-glucosaminidase of male rats with renovascular hypertension. Enzyme 29:93–99
52. Maruhn D, Fuchs I, Mues G, Bock KD (1976) Normal limits of urinary excretion of eleven enzymes. Clin Chem 22:1567–1574
53. Maruhn D, Strozyk K, Gielow L, Bock KD (1977) Diurnal variations of urinary enzyme excretion. Clin Chim Acta 75:427–433
54. Meyer BR, Rosenman K, Lerman Y, Drayer DE, Reidenberg MM (1984) Increased urinary enzyme excretion in workers exposed to nephrotoxic chemicals. Am J Med 76:989–998
55. Mizunashi K, Furukawa Y, Taguchi K, Kuwohara M, Miura R, Yumita S, Yoshinaga K (1989) Effects of parathyroid hormone on urinary excretion of N-acetyl-β-D-glucosaminidase in idiopathic hypoparathyroidism. Calcif Tissue Int 45:375–377
56. Mullan DP (1967) Urinary excretion of leucine aminopeptidase in pregnancy. J Clin Pathol 20:660–664
57. Nakamura M, Miyata K, Tamura T, Kawabata T (1981) Sex difference in urinary excretion of alanine aminopeptidase in the rat. Jpn J Nephrol 23:209–212
58. Nakamura M, Itoh T, Miyata K, Higashiayama N, Takesue H, Nishiyama S (1983) Difference in urinary N-acetyl-β-D-glucosaminidase activity between male and female beagle dogs. Renal Physiol 6:130–133
59. Nakano M, Aoshima K, Katoh T, Teranishi H, Kasuya M, Katoh T (1987) Severity of tubular brush border damage in cadmium-polluted area (Jinzu River Basin): clinical role of urinary trehalase. Environ Res 44:161–168
60. Navon R, Padeh B (1972) Urinary test for identification of Tay-Sachs genotypes. J Pediatr 80:1026–1030
61. Noble MCB, Landon MJ, Davison JM (1977) The excretion of gamma-glutamyltransferase in pregnancy. Br J Obstet Gynaccol 84:522–527
62. Paigen K, Peterson J (1978) Coordinacy of lysosomal enzyme excretion in human urine. J Clin Invest 61:751–762
63. Paigen K, Peterson J, Ward E (1984) A genetic component in human lysosomal enzyme excretion. Biochem Genet 22:517–527
64. Poortmans JR (1972) Effect of exercise on the renal clearance of amylase and lysozyme in humans. Clin Sci 43:115–120
65. Poortmans JR (1985) Postexercise proteinuria in humans. Facts and mechanisms. JAMA 253:236–240
66. Poortmans JR, Jeanloz RW (1968) Quantitative immunological determination of 12 plasma proteins excreted in human urine collected before and after exercise. J Clin Invest 47:386–393
67. Raab, W (1969) Renale Enzymausscheidung der Ratten nach Gabe von Hypophysenhinterlappenhormonen. Wien Klin Wochenschr 81:689–691
68. Schmidt E, Schmidt FW (1979) Enzyme diagnosis in disease of the liver and the biliary system. In: Schmidt E, Schmidt FW, Trautschold I, Friedel R (eds) Advances in clinical enzymology. Karger, Basel, p 239
69. Schulte-Wissermann H, Straub E, Funke PJ (1977) Influence of L-thyroxine upon enzymatic

activity in the renal tubular epithelium of the rat under normal conditions and in mercury induced lesions. Virchows Arch [B] 23:175–184
70. Skrha J, Perusicova J, Sperl M, Bendl J, Stolba P (1989) N-acetyl-β-glucosaminidase and albuminuria in normal and diabetic pregnancies. Clin Chim Acta 182:281–288
71. Somjen D, Ismail-Beigi F, Edelman JS (1981) Nuclear binding of T_3 and effect on QO_2 Na-K-ATPase, and alpha-GPDH in liver and kidney. Am J Physiol 240:E146-E154
72. Tessin I (1988) Enzymuria in neonates during treatment with tobramycin or ceftazidime. Pediatr Infect Dis J 7:142–143
73. Thiele KG (1973) γ-Glutamyltranspeptidase-Aktivität im Urin bei Gesunden und Nierenkranken. Klin Wochenschr 51:339–345
74. Whiting PH, Maughan RJ, Miller JDB (1984) Dehydratation and serum biochemical changes in marathon runners. Eur J Appl Physiol 52:183–187
75. Yoshida M, Sunaga M, Hara J, Katsumata M, Minami M (1989) Elevation of urinary N-acetyl-β-D-glucosaminidase and β-galactosidase activities in workers with long-term exposure to aromatic nitro-amino compounds. Bull Environ Contam Toxicol 43:1–8
76. Zabel K, Schiebler TH (1980) Histochemical, autoradiographic and electron microscopic investigations of the renal proximal tubule of male and female rats after castration. Histochem 69:255–276

Part 2

Methodological Fundamentals of the Activity Determination of Urinary Enzymes

6 Preanalytical Treatment of Urine

H. Mattenheimer and D. Maruhn

6.1 Interference Factors

Urine constitutes an unfavourable environment for enzymes that interferes with the quantitative measurement of enzyme excretion. In contrast to plasma enzymes, which are protected by a relatively stable "enzyme friendly" milieu in a closed system, enzymes from kidney cells lining the tubular lumen are released into an open system and encounter in the urine an "enzyme hostile" environment, which may alter the molecular structure of enzyme proteins and affect their catalytic activity: there is little or no protective protein, pH and urine volume vary, and so do metabolite and salt concentrations; urine contains variable concentrations of enzyme inhibitors and perhaps activators.

6.1.1 Hydrogen Ion Concentration

Urinary enzymes are relatively stable in the pH range 5.5–7.5. In model experiments, in which human kidney extract was added to heat-inactivated human urine, alanine aminopeptidase, alkaline phosphatase, γ-glutamyltransferase, N-acetyl-β-glucosaminidase (NAG), and lactate dehydrogenase (LDH) lost activity at lower pH values when kept at 37°C. The effect was most pronounced with LDH, while NAG was least affected [14]. Lysosomal hydrolases, α-galactosidase, β-galactosidase, and NAG were shown to lose activity in alkaline urine (pH $\geqslant$ 8) [18, 27]. The results from these *in vitro* experiments strongly suggest that the length of time the urine stays in the bladder may have a considerable effect on the activity of enzymes.

6.1.2 Enzyme Inhibitors

Urine contains endogenous enzyme inhibitors, mostly substances of low molecular mass that can be removed by gel filtration or dialysis. Identified inhibitors are listed in Table 1; those for other enzymes are still unknown, e.g., NAG [19] and γ-glutamyltransferase [34, 38], to name but two of the more commonly used enzymes.

Urinary Enzymes
K. Jung/H. Mattenheimer/U. Burchardt (Eds.)

Table 1. Endogenous inhibitors of urinary enzymes

Enzyme	Inhibitor(s)	Ref.
LDH	Peptides, 1760 and 2500 Da	[35]
	Urea	[43]
Alkaline phosphatase	Inorganic phosphates	[9]
	Pigment, 2100 Da	[41]
β-Glucuronidase	1.4-Saccharolactone	[16]
α-Glucosidase	Glucose	[3]
Arylsulfatase	Sulfate ions	[32]
Dipeptidase	ATP, ADP, AMP,	
(EC 3.4.13.1)	inorganic phosphates	[31]
AAP	Amino acids	[23]
N-acetyl-β-D-glucosaminidase	Urea	[3a]

AAP, alanine aminopeptidase; LDH, lactate dehydrogenase

Drugs have also been found to inhibit urinary enzymes. The list in Table 2 is certainly incomplete in regard to enzymes and interferences from other drugs and/or their metabolites.

The concentration of inhibitors varies considerably within and between individuals [6, 13, 16, 19, 34, 38]. Occasionally one finds that the activity of an enzyme decreases following gel filtration, and the possibility of enzyme activators in urine has been discussed, though none has so far been identified [6, 20].

Through advances in methodology some of the interferences can now be controlled; standardization of urine collection periods and preanalytical treatment of the sample are as important as optimization of assay methods.

Table 2. Drugs inhibiting urinary enzymes

Enzyme	Substance	Ref.
LDH	Ethionamide	[28]
	Nitrofurantoin	[28]
	Penicillin	[29]
	Salicylate	[10]
AP	Ethionamide	[28]
	Gentamicin	[33]
	Nitrofurantoin	[28]
	Sulfonamide	[29]
AAP	Isoxepac	[7]
NAG	Salicylate	[11]
	Mitomycin	[8a]
	Amikacin	[25a]

AAP, alanine aminopeptidase; AP, alkaline phosphatase; LDH, lactate dehydrogenase; NAG, *N*-acetyl-β-D-glucosaminidase

6.2 Collection and Storage of Samples

The collection of urine may be preceded by preparing the patient, in order to eliminate some factors that are known to cause transitory enzymuria. The patient should refrain from physical exercise (see Sect. 5.7), should not receive X-ray contrast media (see Sect. 18.1), and should not undergo major surgery (see Sect. 15.8). If medically justifiable, medication should be withheld for 24 h prior to the collection period.

Since the urine volume varies within relatively wide limits, it is not possible to establish a consistent reference volume for expressing enzyme activity. The activity should be measured in timed specimens and expressed in 'units per collection period'. Another option is to use urinary creatinine as reference quantity (see Chap. 9). Creatinine as reference is essential when measurements are made in random urine samples.

No agreement exists concerning the optimal urine collection period for enzyme measurements in man [20; see Chap. 9]. Circadian excretion rhythms have been observed for several enzymes and they appear to be different for individual enzymes [4, 17, 21, 38]. Some investigators, therefore, prefer the 24 h collection in order to obtain an average value, but periods as short as 2 h have also been recommended, e.g., between 6 a.m. and 8 a.m. Standardization of collection periods within a laboratory is, of course, advisable. Collections for 24 h should be made under toluene and/or with cooling of the urine. For the effect of urine flow on enzymuria see Sect. 5.1.

Collection periods for urine from laboratory animals housed individually in metabolic cages are usually 18 or 24 h; urine should be collected in cooled vessels or preserved by the addition of sodium azide. For long-term experiments, timed urinary enzyme excretion may be factored by body weight [8].

The stability of enzymes during storage of urine varies from enzyme to enzyme (Table 3). As a general rule, storage of crude (native) urine at room temperature should be limited to a few hours. At 4°C some enzymes appear to maintain their activity for several days, provided bacterial growth is prevented by addition of, e.g., sodium azide or thymol. Freezing of crude urine without appropriate additives (see Chap. 7) destroys the activity of some enzymes. However, freezing of urinary gel filtrates allows storage for several weeks without appreciable activity loss for most enzymes.

Long-term storage of native urine with preservation of enzyme activities is possible with the addition of 30% v/v ethylene glycol or glycerol [22]. In urines prepared accordingly, and stored at 4°C and -20°C, the activities of NAG, alanine aminopeptidase, γ-glutamyltransferase, and LDH were maintained for over 1 year. The urines were gel filtered prior to enzyme analysis.

Urines prepared and stored in the way described above are recommended for quality control [22].

Table 3. Stability of enzymes in urine

	Crude urine			Crude urine with additive*		Gel filtrate or dialyzate		
	RT	4°C	20°C	4°C	– 20°C	4°C	– 20°C	– 80°C
AAP	3 h [36]	3 d [24]	unst [24]	18 d [24]	18 d [24] 1 y [22]	3 d [36]	14 d [13, 24]	—
AP	3 h [36]	6 h [40]	unst [39]	—	—	1 d [5, 40]	3 m [36]	—
GGT	3 h [36]	5 h [30]	unst [37]	—	1 y [22]	8 d [38]	7 d [34]	—
LDH	—	6 h [40]	—	—	1 y [22]	1 d [40]	—	—
NAG	3 h [36]	⩽ 50 d [22]	1 y [22]	—	1 y [22]	3 m [36]	3 m [36]	—
Kallikrein	4 h [1]	—	6 w [1, 25]	—	—	—	—	1 y [1]
Ribonuclease	2 h [12]	5 d [12]	14 d [12]	—	—	—	—	—
Angiotensinase	—	10 d (Scherberich, personal communication)	—	—	—	—	—	—

AAP, alanine aminopeptidase; AP, alkaline phosphatase; GGT, γ-glutamyl-transferase; LDH, lactate dehydrogenase; NAG, *N*-acetyl-β-D-glucosaminidase; RT, room temperature; unst, unstable; h, hours; d, days; w, weeks; m, months; y, year

* 30% v/v glycerol or ethylene glycol

6.3 Preparation of Samples

Urine samples are prepared for enzyme analysis by centrifugation for removal of cell debris and sediment and by separation from inhibitors.

Inhibition may be overcome simply by diluting the urine. However, dilution is not to be recommended for several reasons:

1. The concentrations of inhibitors vary and the interference may not be eliminated by a standard dilution.
2. Dilution may further reduce the stability of enzymes and lead to denaturation.
3. Dilution reduces the analytical sensitivity.

The most commonly used procedures for inhibitor removal are gel filtration [42] and dialysis. Gel filtration is recommended because it is more practicable and in some instances more effective than dialysis [9]. Sephadex G50, fine (Pharmacia), was originally suggested for filtration [42]. Other investigators prefer Sephadex G 25, medium, which is available in small, ready-to-use columns (PD 10, Pharmacia) [6, 9, 15].

6.3.1 Recommended Procedure

- Prepare PD 10 columns according to manufacturer's instructions.
- Add 4 ml NaCl, 150 mmol/1, to column and drain completely.
- Centrifuge urine for 10 min at about 1000 g.
- Add 2.5 ml of the supernatant to the column and drain completely.
- Add 4 ml NaCl, 150 mmol/l.
- Collect the total eluate for enzyme assays (dilution factor: 1.6).
- Regenerate column with distilled water, followed by NaCl.

When conserved with sodium azide, the columns may be repeatedly used for 4–6 months. However, columns used for gel filtration of bilirubin- or gentamicin–containing urines should be discarded [20].

Gel filtration with PD 10 columns lasts 10–15 min and several columns can be operated simultaneously. The within-column and between-columns imprecision are well within acceptable limits [6].

References

1. Ash KO, Smith JB, Lynch M, Dadone M, Tolman KG, Williams RR (1987) Urinary kallikrein: assay validation and physiological variability. Clin Chim Acta 151:133–140
2. Berscheid G, Grötsch H, Hropot M, Klaus E, Mattenheimer H (1983) Enzymuria of the rat: the preparation of urine for enzyme analysis. J Clin Chem Clin Biochem 21:799–804
3. Bomback FM, Nakagawa S, Kumin S, Nitowski AM (1976) Altered lysosomal glycohydrolase activities in juvenile diabetes mellitus. Diabetes 25:420–427

3a. Bondiou MT, Bourbouze R, Bernard R, Percheron F, Paraz-Gonzalez N, Cabezas JA (1985) Inhibition of A and B N-acetyl-β-D-glucosaminidase urinary isoenzymes by urea. Clin Chim Acta 149:67–73

4. Burchardt U, Peters JE, Neef H, Thulin H, Gründig CA, Haschen RJ (1977) Der diagnostische Wert von Enzymbestimmungen im Harn. Z Med Labor Diagn 18: 190–212
5. Dubach UC, Padlina G (1966) Aktivität der alkalischen Phosphatase im Urin. Klin Wochenschr 44:180–186
6. Grötsch H, Mattenheimer H (1983) Urine. In: Bergmeyer HU, Bergmeyer J, Grassl M (eds) Methods of enzymatic analysis, vol. 3, 3rd edn. Chemie, Weinheim, pp 42–49
7. Grötsch H, Berscheid G, Grigoleit HG, Dagrosa EE, Hundt HKL (1981) Alanine aminopeptidase activity as an indicator of nephrotoxicity: drug interactions with enzyme activity in vitro. IRCS Med Sci 9:422
8. Grötsch H, Hropot M, Malerczyk V, Mattenheimer H (1985) Enzymuria of the rat: biorythms and sex differences. J Clin Chem Clin Biochem 23:343–347

8a. Guder WG, Heidland A (1986) Urine analysis. Report on the workshop conference of the German Society for Clinical Chemistry and the Society for Nephrology in Würzburg, October 25–26, 1985. J Clin Chem Clin Biochem 24:611–620 (results of Maruhn)

9. Hilliard SD, O'Donnell JF, Schenken S (1965) On the nature of the inhibition of urinary alkaline phosphatase. Clin Chem 11:570–574
10. Hines WJW, Smith MJH (1964) Inhibition of dehydrogenases by salicylate. Nature 201:192
11. Horak E, Hopfer SM, Sunderman FW Jr (1981) Spectrophotometric assay for urinary *N*-acetyl-β-D-glucosaminidase activity. Clin Chem 27:1180–1185
12. Jung K (1986) An optimized micromethod for determining the catalytic activity of serum ribonuclease. J Clin Chem Clin Biochem 24:243–250

13. Jung K, Scholz D (1980) An optimized assay of alanine aminopeptidase activity in urine. Clin Chem 26:1251–1254
14. Jung K, Pergande M, Schröder K, Schreiber G (1982) Influence of pH on the activity of enzymes in urine at 37°C. Clin Chem 28:1814
15. Knoll E, Wisser H, Rautenstrauch H (1980) Vergleichende Untersuchungen zur diagnostischen Wertigkeit von Diskelektrophorese der Urinproteine und N-Acetylglucosaminidaseausscheidung zur Erkennung von tubulären Nierenschädigungen bei chronischer Polyarthritis. J Clin Chem Clin Biochem 18:53–58
16. Kushinsky S, Chen VL (1967) Urinary inhibitors of β-glucuronidase. Enzymol Biol Clin 8:266–282
17. Lakatua DJ, Blomquist CH, Haus E, Sackett-Lundeen L, Berg H, Swoyer J (1982) Circadian rhythm in urinary *N*-acetyl-β-glucosaminidase (NAG) of clinically healthy subjects. Timing and phase relation to other urinary circadian rhythms. Am J Clin Pathol 78:69–77
18. Lockwood TD, Bosman HB (1979) The use of urinary *N*-acetyl-β-glucosaminidase in human renal toxicology. I. Partial biochemical characterization and excretion in humans and release from the isolated perfused rat kidney. Toxicol Appl Pharmacol 49:323–336
19. Maruhn D (1976) Rapid colorimetric assay of β-galactosidase and N-acetyl-β-glucosaminidase in human urine. Clin Chim Acta 73:453–461
20. Maruhn D (1983) Methodische Aspekte der Harnenzymologie. Z Gesamte Inn Med 38:557–564
21. Maruhn D, Strozyk K, Gielow L, Bock KD (1977) Diurnal variations of urinary enzyme excretion. Clin Chim Acta 75:427–433
22. Maruhn D, Wehling K, Metz U (1986) A stable liquid control material for urinary enzyme assays. Clin Chim Acta 160:119–122
23. Mattenheimer H, Frölke W, Grötsch H, Simane Z (1986) Identification of inhibitors of urinary alanine aminopeptidase. Clin Chim Acta 160:129–135
24. Mattenheimer H, Frölke W, Grötsch H, Maruhn D, Simane Z (1988) Recommendation for the measurement of "alanine aminopeptidase" in urine. J Clin Chem Clin Biochem 26:635–644
25. Meani A, Fabris C, Vianello D, Piccoli A, del Favero G, Farini R (1985) A kinetic assay for human urinary kallikrein. Enzyme 33:89–93
25a. Mozer J-L, Gibey R, Henry J-C (1989) Inhibition de l'activaté de la N-acetyl-β-glucosaminidase (NAG) urinaire et de ses isoenzymes par l'amicacine. C R Soc Biol 183:166–173
26. Nosal T, Kiser WS, Robitaille ML, King JW (1966) Urinary alkaline phosphatase and lactic dehydrogenase activity in the diagnosis of urologic neoplasms. Clin Chem 12:542
27. Paigen K, Peterson J (1978) Coordinacy of lysosomal enzyme excretion in human urine. J Clin Invest 61:751–762
28. Raab W (1972) Diagnostic value of urinary enzyme determinations. Clin Chem 18:5–25
29. Raab W, Hohenegger H (1968) Renale Enzymausscheidung bei Ratten nach hohen Dosen von Penicillin, Kanamycin und Sulfonamiden. Clin Chim Acta 20:95–103
30. Rambabu K, Jacob RT, Pattabiraman TN (1981) Variants of γ-glutamyl transferase in human urine. Clin Chim Acta 109:257–265
31. René AM, Kessler RS (1976) Colorimetric determination of renal dipeptidase in human serum and urine. IRCS Med Sci 4:111
32. Roy AB (1976) Sulfatases, lysosomes and disease. Aust J Exp Biol Med Sci 54:111–135
33. Sack K, Wilhelm J (1975) The diagnostic value of enzymuria, cell excretion, and proteinuria in experimental renal diseases. Urol Res 3:31–40
34. Salgo L, Szabo A (1982) γ-Glutamyl transpeptidase activity in human urine. Clin Chim Acta 126:9–16
35. Schoenenberger GA, Wacker WEC (1966) Peptide inhibitors of lactic dehydrogenase (LDH). II. Isolation and characterization of peptides I and II. Biochemistry 5:1375–1379
36. Schulze G, Goebel H (1988) Harnenzymdiagnostik – ein Vorschlag zur Definition der präanalytischen Bedingungen. Z Med Lab Diagn 29:381–386
37. Stokke O (1974) Preservation of γ-glutamyl transpeptidase in human urine. Clin Chim Acta 57:143–148
38. Szasz G (1970) γ-Glutamyltranspeptidase-Aktiviät im Urin. Z Klin Chem Klin Biochem 8:1–8
39. Van Breda Vriesman PJC (1968) Discussion remark. Curr Probl Clin Biochem 2:117
40. Wacker WEC, Amador E, Dorfman LE, Zimmerman TS (1965) Eine Methode zur Bestimmung und diagnostischen Anwendung der Lactatdehydrogenase und alkalischen Phosphatase im menschlichen Urin. Ärztl Lab 11:204–213

41. Walters CL (1958) A competitive alkaline phosphatase inhibitor from extracts of normal male urines. Enzymologia 20:33–56
42. Werner M, Maruhn D, Atoba M (1969) Use of gel filtration in the assay of urinary enzymes. J Chromatogr 40:254–263
43. Wilkinson JH (1968) Diagnostic significance of enzyme determinations in urine. Curr Probl Clin Biochem 2:207–227

7 Quality Assurance

Z.J. Simane

The quality assurance of urinary enzyme measurements is difficult because of lack of suitable control material. The few control urines commercially available are mostly prepared for special purposes; their enzymatic activities are either not declared or the preparations are enzymatically inactive. The usual commercial liquid or lyophilized material based on human or animal serum, although occasionally used [5], cannot replace a special control material for urinary enzymes because it does not contain some of the enzymes of interest and because the activities of a number of enzymes present in the control sera differ substantially from those of urine. Furthermore, the enzymes of control sera show kinetic characteristics which differ from those of urinary enzymes and would not meet the essential "similar behaviour" requirements of control materials [1].

Therefore, while tracing the development of practicable methods for urinary enzymes, some authors have also described the preparation and properties of control materials based on human urine. In this connection there have been two main problems: the poor stability of the enzymes in native urine, particularly when kept frozen, and the relatively low level of enzyme activity in normal human urine. The latter circumstance has been the reason for the use of urine from patients with renal disease for preparation of liquid control materials [4–6].

For stabilization of enzyme activities in liquid control materials against freeze denaturation an addition of albumin or albumin buffer [3, 4, 7, 9] or of cryoprotectants such as ethylene glycol [4, 6, 8], glycerol [6, 7], sodium glutamate [9] or dimethyl sulfoxide (DMSO) [7], alone or in combination, has been recommended.

It was found that the enzymes in gel filtrates from urines were more stable against low temperatures and could be kept frozen for months without considerable loss of their activities. Such preparations were used successfully as control materials [2, 5]. Moreover, a preparation based on lyophilized gel filtrate from human urine enriched by an extract from rat or pig kidney was used as a reliable quality assurance material (Simane, unpublished data).

Currently, quality assurance data are available for the following urinary enzymes:

Alanine aminopeptidase [2, 4, 6, 7]
Alkaline phosphatase [4]
Arylsulfatase [5]

Urinary Enzymes
K. Jung/H. Mattenheimer/U. Burchardt (Eds.)

γ-Glutamyltransferase [4, 6, 7]
β-Galactosidase [5]
α-Glucosidase [5]
β-Glucuronidase [5]
LDH [6]
Lysozyme [4]
NAG [2, 4, 5–7]

At present, the incontestable need for a suitable control material for quality assurance in urinary enzyme analysis cannot be covered from commercial sources. Thus, the analyst has to prepare such materials in his own laboratory. The control urines mentioned above [2, 4, 5–7] (Simane, unpublished data) seem to yield satisfactory results, no matter whether liquid or lyophilized and whether stabilized by additives or by gel filtration. However, a direct comparison of the reported data would be difficult because the enzyme stability has been evaluated by different methods. The choice of a suitable preparation will depend, therefore, primarily on available materials and techniques.

References

1. Bretaudiere JP, Dumont G, Rej R, Bailly M (1981) Suitability of control materials. General principles and methods of investigation. Clin Chem 27:798–805
2. Diener U, Knoll E, Langer B, Rautenstrauch H, Ratge D, Wisser H (1981) Urinary excretion of *N*-acetyl-β-D-glucosaminidase and alanine aminopeptidase in patients receiving amikacin or *cis*-platinum. Clin Chim Acta 112:149–157
3. Hafkenscheid JCM, van der Ven-Jongekrijg J (1983) Stability test of six enzymes for internal quality control. Enzyme 29:239–249
4. Jung K, Grützmann KD (1988) Quality control material for activity determinations of urinary enzymes. Clin Biochem 21:53–57
5. Maruhn D (1979) Preparation of urine for enzyme determinations by gel filtration. Curr Probl Clin Biochem 9:22–29
6. Maruhn D, Wehling K, Metz U (1986) A stable control material for urinary enzyme assays. Clin Chim Acta 160:119–122
7. Mueller PW, MacNeil ML, Steinberg KK (1986) Stabilization of alanine aminopeptidase, gamma glutamyltranspeptidase, and *N*-acetyl-β-D-glucosaminidase activity in normal urines. Arch Environ Contam Toxicol 15:343–347
8. Pope WT, Caragher TE, Grannis GF (1979) An evaluation of ethylene glycol-based liquid specimens for use in quality control. Clin Chem 25:413–418
9. Tamiya T, Okahashi N, Sakuma R, Aoyama T, Akahane T, Matsumoto JJ (1985) Freeze denaturation of enzymes and its prevention with additives. Cryobiology 22:446–456

8 Methods for Measuring Urinary Enzyme Activities

8.1 Membrane-Bound Enzymes

8.1.1 Alanine Aminopeptidase

H. Mattenheimer, K. Jung, and H. Grötsch

8.1.1.1 *Introduction*

Aminopeptidases (α-aminoacylpeptide hydrolases, EC 3.4.11) comprise a group of exopeptidases, which split single amino acids from the N-terminus of peptide chains. Two, namely the cytosolic (EC 3.4.11.1) and the microsomal forms (EC 3.4.11.2), are of relatively broad substrate specificity. The cytosolic enzyme, the "classic" leucine aminopeptidase [6], hydrolyzes L-peptides with an N-terminal leucine residue. The recommended substrates are L-leucinamide [4, 5] and L-leucine hydrazide [7]. Chromogenic substrates like L-leucine-4-nitroanilide or L-leucine-β-naphthylamide are not hydrolyzed. The pH optimum of the cytosolic enzyme is between 9–10; it is activated by Mg^{2+} and Mn^{2+} ions and inhibited by Co^{2+} ions [19]. The microsomal enzyme, which concerns us here, acts preferentially on peptides with N-terminal alanine, and the name alanine aminopeptidase (AAP) was introduced [18]; L-alanine-β-naphthylamide [17] and L-alanine-4-nitroanilide [21] are the recommended substrates. L-Leucine derivatives like L-leucine-β-naphthylamide [3] and L-leucine-4-nitroanilide [20] are also split, although at a slower rate. The enzyme, then, is termed either AAP or leucine aminopeptidase (LAP), depending on which substrate is being used. The terminology becomes confusing, particularly when one considers that, strictly speaking, the name "leucine aminopeptidase" should be reserved for the cytosolic enzyme. L-Alanine-4-nitroanilide is the substrate of choice for the determination of AAP because of its relative specificity (L-leucine-4-nitroanilide and L-alanine-β-naphthylamide are also split by cystyl aminopeptidases), higher hydrolysis rate and better solubility as compared with L-leucine-4-nitroanilide. The main advantages of the nitroanilides over the naphthylamides are the possibility of continuous recording of the release of 4-nitroaniline and the easy adaptation of the assay to automated analysers.

Aminopeptidases present in urine hydrolyse a variety of substrates, among them the 4-nitroanilides of L-alanine and L-leucine, as well as L-alaninamide and L-leucinamide. A comparison of purified microsomal and cytosolic enzyme preparations with urine suggests, but does not prove, the following for human

Urinary Enzymes
K. Jung/H. Mattenheimer/U. Burchardt (Eds.)

urine: (a) a microsomal (membrane-bound) enzyme hydrolyzes all four substrates; (b) a leucinamide-splitting enzyme, most likely of cytosolic origin, is present in varying proportions; (c) an enzyme specific for the hydrolysis of L-alaninamide may exist [14].

The main aminopeptidase in urine is AAP, which is released from the brush border membrane of the proximal tubule (see Chap. 3). AAP isolated from human kidney is a glycoprotein of 240 kDa, with two subunits of 120 kDa, each containing one zinc atom; carbohydrate amounts to about 20% with a high proportion of sialic acid [13]. In human urine two multiple forms are found, a soluble and a particulate AAP, which can be separated by ultracentrifugation and electrophoresis [9]. Both variants have nearly identical kinetic characteristics in regard to K_m and dependence on pH and substrate concentration. The particulate variant is the more heat-sensitive of the two [9].

With phenylalanine, one of the natural inhibitors of alanine aminopeptidase in urine, it was possible to postulate the occurrence of two alanine aminopeptidases in human urine [14, 15]. Whether these are identical with the soluble and particle-bound forms separated by ultracentrifugation [9] has yet to be investigated.

8.1.1.2 *Assay*

L-Alanine-4-nitroanilide + H_2O → L-alanine + 4-nitroaniline.

The increase in absorbance of the liberated 4-nitroaniline is directly measured at 405 nm.

Three optimized methods have been described with L-alanine-4-nitroanilide as substrate [2, 10, 16]; they differ in reaction temperature, buffer and buffer concentration, substrate concentration, and particularly in the use of either gel filtered [10, 16] or native [2] urine samples. Flandrois et al. [2] suggested reaction conditions (300 mmol/l TRIS buffer, pH 7.8; 5.8 mmol/l L-alanine-4-nitroanilide) under which gel filtration would no longer be required. However, the influence of urinary inhibitors on AAP is not fully eliminated [8], and we strongly suggest the use of gel-filtered samples.

Optimized assay conditions for AAP vary with the reaction temperature; two methods are described here: method I at 25°C for urine of humans and some laboratory animals [16], and method II at 37°C, so far developed and tested for human urine only [10].

The activity of AAP is inhibited by higher substrate concentrations [16]; substrate inhibition is slightly stronger at 25°C than at 37°C. At 25°C, inhibition in human urine begins at substrate concentrations exceeding 0.8 mmol/l, and in rat and dog urine exceeding 0.3 and 0.5 mmol/l respectively; these concentrations lie in the range of the K_m values. Assay conditions were therefore chosen to give the lowest possible K_m. The K_m value depends on the type of buffer, and it increases with pH and temperature. The recommended assay conditions [16] are: triethanolamine buffer 70 mmol/l, pH 7.6; L-alanine-4-nitroanilide concentrations depending on the species (see above); incubation temperature 25°C.

At 37°C, 50 mmol/l TRIS buffer, pH 7.8 and 2 mmol/l L-alanine-4-nitroanilide were found to be optimal reaction conditions for human urine [10].

Collection, storage and preparation of samples are as described in Chap. 6.

8.1.1.3 *Method I*

Reagents and Solutions. All reagents must be of analytical quality, and deionized water must be used.

1. Triethanolamine buffer, 100 mmol/l, pH 7.60: dissolve 1.856 g triethanolamine hydrochloride in about 90 ml water, adjust pH to 7.60 (glass electrode) with 100 mmol/l NaOH (approximately 7.6 ml) and add water to 100 ml.
2. Substrate solution (human urine), L-alanine-4-nitroanilide, 8.0 mmol/l: dissolve 19.64 mg L-alanine-4-nitroanilide hydrochloride in about 7 ml water, adjust pH to 7.6 with diluted NaOH and add water to 10 ml. For rat and dog urine prepare solutions containing 3.0 and 5.0 mmol/l, respectively.

Measurement.
Prepare incubation mixture in cuvettes (d = 1.0 cm) and read or record absorbance changes (ΔA) in a spectrophotometer or spectrum-line photometer at 405 nm. Incubation temperature 25°C, incubation volume 500 μl, sample volume fraction 0.2.

Pipette successively			Final concentration
Triethanolamine buffer	(1)	350 μl	70 mmol/l
Sample		100 μl	
Substrate	(2)	50 μl	0.8 mmol/l
Mix; read or record absorbance for 3 min (longer with low activities)			

For rat and dog urine the appropriate substrate solutions must be used, with final concentrations of 0.3 and 0.5 mmol/l, respectively.

ΔA/min should not exceed 0.133 with human urine, 0.05 with rat urine and 0.083 with dog urine. Then, 150 mmol/l NaCl is recommended for dilution.

Calculation

$$\text{AAP U/l} = \Delta A/\Delta t \times 10^3/9.9 \times 500/100 = \Delta A/\Delta t \times 505 (\times \text{dilution factor})$$

where 9.9 is the millimolar absorption coefficient of 4-nitroaniline at 405 nm. The dilution factor encompasses the dilution by gel filtration, by ethylene glycol (in case of urine storage), and dilution because of high activity.

8.1.1.4 *Method II*

Reagents and Solutions. All reagents must be of analytical grade, and deionized water must be used.

1. TRIS/HCl buffer, 59 mmol/l, pH 7.85: dissolve 715 mg of tris(hydroxymethyl) aminomethane in about 90 ml water, adjust pH to 7.85 at 37°C with 1 mol/l HCl, and add water to 100 ml. This solution is stable for at least 1 year at 20°C or 4 weeks at 4°C.
2. Substrate solution (L-alanine-4-nitroanilide, 26 mmol/l; TRIS/HCl buffer, 59 mmol/l, pH 7.80), dissolve 63.9 mg of L-alanine-4-nitroanilide hydrochloride in 10 ml TRIS buffer solution (1). Hydrochloride reduces the pH of TRIS buffer solution (1) from 7.85 to 7.80 in the final reaction mixture. The substrate solution is stable for at least 2 weeks at 4°C.

Measurement. Prepare incubation mixture in cuvettes (d = 1.0 cm) and read or record absorbance changes in a spectrophotometer at 405 nm. Incubation temperature 37°C, incubation volume 650 μl, sample volume fraction 0.154.

Pipette successively	Sample (μl)	Blank (μl)	Final concentration
TRIS buffer (1)	500	500	50 mmol/l
Sample	100	—	
Water	—	100	
Mix; preincubate for 10 min at 37°C			
Substrate solution (2)	50	50	2 mmol/l
Mix; read or record absorbance for 3 min (longer with low activities)			

ΔA/min should not exceed 0.250; samples with higher activity should be diluted with 0.15 *M* NaCl.

Calculation

$$\Delta A/\min_{sample} - \Delta A/\min_{blank} = \Delta A/\min_{corr}$$

$$\text{AAP U/1} = \Delta A/\min_{corr} \times 10^3/9.9 \times 650/100 \times \text{DF}$$

$$= \Delta A/\min_{corr} \times 675 \times \text{DF}$$

where 9.9 is the millimolar absorption coefficient of 4-nitroaniline at 405 nm and DF the dilution factor (dilution by gel filtration, by addition of ethylene glycol in case of urine storage and with highly active samples).

The blank value, registering spontaneous hydrolysis of the substrate, is generally low with suitable batches of L-alanine-4-nitroanilide ($\leq$ 0.4 U/l) and may be neglected with routine AAP determinations.

8.1.1.5 *Separation of Multiple Forms*

Ultracentrifugation and electrophoresis are suitable approaches for the separation of particulate and soluble variants of brush border enzymes [9].

Ultracentrifugation. A sample of 4.0 ml of gel-filtered urine is centrifuged at 105 000 × *g* at 4°C for 1 h. Then, 3.0 ml of the supernatant, containing the soluble form of AAP, are carefully removed. AAP activity is measured in the gel filtrate (GF) and the supernatant (S), and the activity of the particulate AAP is calculated as follows:

$$\text{particulate AAP(U/l)} = \text{activity (GF)} - \text{activity (S)}$$

or

$$\text{particulate AAP (\%)} = [\text{activity(GF)} - \text{activity(S)}] \times 100/\text{activity (GF)}$$

In the urine of 20 healthy adults 46.7% ± 9.8% (mean ± s) of the AAP was found to be particle bound [9].

Electrophoresis. Separation by electrophoresis [11] is performed with agarose gel (10 g/l in barbital buffer, pH 8.6, 85 mmol/l) on 7.5 × 11 cm glass plates. Urine as gel filtrate samples are concentrated 50–100-fold (e.g. microcollodion bags, Sartorius, FRG); then, 4 μl are applied in punched holes (0.5 × 5 mm). Electrophoresis is carried out at 50 mA/gel for about 120 min and a running distance of 4.5 cm. Bands with AAP activity are stained by incubating the gels in a solution containing 50 mmol/l phosphate buffer, pH 7.2, 1 mmol/l L-alanine-β-naphthylamide and 1 mmol/l tetrazotized di-2-anisidine (fast blue B). The particulate activity of AAP is found near the application point, while the soluble activity is localised more towards the front. Figure 1 shows the particulate and soluble variants of AAP in comparison with alkaline phosphatase (AP) and gamma-glutamyltransferase (GGT).

8.1.1.6 *Reference Values*

Method I. Reference values established with method I for various animal species are summarized in Chap. 17. Data on humans are currently available for men only and were supplied by Dr. Simane. In 247 urine samples from 103 volunteers AAP expressed in U/mmol creatinine was: $\bar{x} = 0.3$, $s = 0.1$, range 0.07–1.01, median 0.28.

Method II. Urinary excretion of AAP was determined in morning random urine of 429 subjects ranging from 5 days to 58 years of age [12]. The distribution of the AAP/creatinine ratios was log normal in the different age groups; the ratios decreased significantly with increasing age. Sex-related differences were found in age groups over 6 years. Table 1 shows the upper reference limits calculated as 97.5th percentiles.

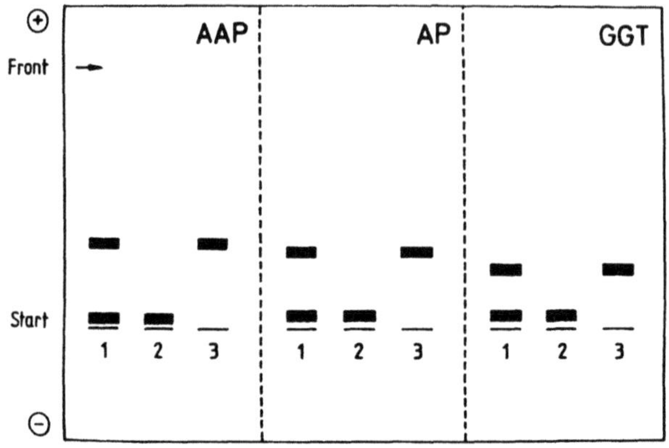

Fig. 1. Electrophoretic analysis of urine, particulate and soluble variants of alanine aminopeptidase, alkaline phosphatase and gamma-glutamyltransferase (*AAP*, *AP* and *GGT*). *1*, Urine; *2*, pellet after ultracentrifugation; *3*, supernatant after ultracentrifugation. Electrophoresis was performed on agarose, according to [11]

Table 1. Upper reference limits of age-dependent alanine aminopeptidase (AAP)/creatinine ratios in human urine established with method II at 37°C

Age group	Sex and number	U/mmol creatinine (97.5th percentiles)[a]
1 (5 days – 6 months)	M (n = 46) + F (n = 21)	39.5
2 (7 months – 2 years)	M (n = 11) + F (n = 6)	26.6
3 (3 – 6 years)	M (n = 48) + F (n = 37)	6.1
4 (4 – 12 years)	M (n = 75)	4.5
	F (n = 46)	3.3
5 (13 – 18 years)	M (n = 44)	5.0
	F (n = 32)	2.5
6 (23 – 58 years)	M (n = 30) + F (n = 33)	2.2

[a] The percentiles were calculated on the basis of a log normal distribution. Data from men and women were combined when sex-related differences were not observed [12]

References

1. Farr W, Rehfeld N, Reichelt D, Haschen HJ, (1968) Vergleichende Untersuchungen zur Bestimmung der Aminosäurerylamidase im menschlichen Serum. Z Med Labortechn 9:78–86
2. Flandrois C, Lahet C, Feldman D, Gabastou A, Maire I (1988) Urinary alanine aminopeptidase assay improved as results of multivariate response-surface analysis. Clin Chem 34:954–957
3. Goldbarg JA, Rutenburg AM (1958) The colorimetric determination of leucine aminopeptidase in urine and serum of normal subjects and patients with cancer and other diseases. Cancer 11:283–291

4. Hafkenscheid JCM (1984) Aminopeptidases and amino acid arylamidases: introduction. In: Bergmeyer HU, Bergmeyer J, Grassl M (eds.) Methods of enzymatic analysis, 3rd edn. Vol. V. Verlag Chemie, Weinheim, Deerfield Beach FL, Basel, pp 11–15
5. Hafkenscheid JCM, Kohler BEM (1985) A continuous method for the determination of leucine aminopeptidase in human serum with L-leucinamide as substrate. J Clin Chem Clin Biochem 23:395–398
6. Hanson H, Hütter HJ (1966) Zur Darstellung von Leucinaminopeptidase aus Schweinenieren, Hoppe Seyler's Z Physiol Chem 347:118–126
7. Haschen RJ, Farr W, Reichelt D (1968) Photometrische Bestimmung der klassischem Leucin-aminopeptidase in Blutplasma und–serum. J Clin Chem Clin Biochem 6:11–18
8. Jung K, Jähnchen G (1990) How should we measure activity of alanine aminopeptiadse in urine? Clin Chem 36:177–178
9. Jung K, Pergande M, Wischke UW (1984) Characterization of particulate and soluble variants of the brush-border enzymes alanine aminopeptidase, alkaline phosphatase and gamma-glutamyltransferase in human urine. Biomed Biochim Acta 43:1357–1364
10. Jung K, Scholz D (1980) An optimized assay of alanine aminopeptidase activity in urine. Clin Chem 26:1251–1254
11. Jung K, Wischke UW (1984) Electrophoretic variants of alanine aminopeptidase, alkaline phosphatase, and gamma-glutamyltransferase in urine. Clin Chem 30:856–859
12. Jung K, Hempel A, Grützmann KD, Hempel RD, Schreiber G (1990) Age dependent excretion of alanine aminopeptidase, alkaline phosphatase, γ-glutamyltransferase and N-acetyl-β-D-glucosaminidase in human urine. Enzyme 43:10–16.
13. Kao YJ, Starnes WI, Behal FJ (1978) Human kidney alanine aminopeptidase: physical and kinetic properties of a sialic acid containing glycoprotein. Biochemistry 17:2990–2994
14. Mattenheimer H (1988) Aminopeptidases in urine. In: Blackmore DJ, Eckersall PD, Evans GO, Sommer H, Stonard MD, Woodman DD (eds.) Animal clinical biochemistry, Cambridge University Press, Cambridge, New York, New Rochelle, Melbourne, Sidney, pp 209–218
15. Mattenheimer H, Frölke W, Grötsch H, Simane Z (1986) Identification of inhibitors of urinary alanine aminopeptidase. Clin Chim Acta 160:129–135
16. Mattenheimer H, Frölke W, Grötsch H, Maruhn D, Simane Z (1988) Recommendation for the measurement of "alanine aminopeptidase" in urine. J Clin Chem Clin Biochem 26:635–644
17. Peters JE, Schneider I, Haschen RJ (1972) Bestimmung der L-Alanyl-Peptidhydrolase (Alanin-Aminopeptidase, Aminosäure-Arylamidase) im menschlichen Harn. Clin Chim Acta 36:289–301
18. Rehfeld N, Peters JE, Giesecke H, Beier L, Haschen RJ (1967) Untersuchungen über Aminosäure-arylamidasen. II. Reinigung und Characterisierung der Aminosäure-arylamidase Isoenzyme aus Leber und Duodenalschleimhaut des Menschen. Acta Biol Med Germ 19:819–830
19. Sanderink GJ, Artur Y, Siest G (1988) Human aminopeptidases: a review of the literature. J Clin Chem Clin Biochem 26:795–807
20. Tuppy H, Wiesbauer U, Wintersberger E (1966) Aminosäure-p-nitroanilide als Substrate für Aminopeptidasen und andere proteolytische Fermente. Hoppe Seyler's Z Physiol Chem 329:278–288
21. Wachsmuth ED, Fritze I, Pfleiderer G (1966) An aminopeptidase occurring in pig kidney. I. An improved method of preparation. Physical and enzymic properties. Biochemistry 5:169–174

8.1.2 Alkaline Phosphatase

K. Jung

8.1.2.1 *Introduction*

Alkaline phosphatase [AP, orthophosphoric-monoester phosphohydrolase (alkaline optimum), EC 3.1.3.1] is a zinc-containing glycoprotein which splits many phosphate esters including naturally occurring and synthetic substrates.

Three different forms of AP coded from distinct genes are found in human organs [8, 11]: the placental, intestinal and liver-bone-kidney type. The last one has been called the "tissue-unspecific" type. They can be distinguished from one another on the basis of kinetic parameters, physico-chemical and immunological properties.

The human kidney contains as the major form the tissue-unspecific type which is the intrinsic constituent of the luminal plasma membrane in the proximal tubule [13]. In contrast to this membrane-bound type, the intestinal one is only found in the cytosol and amounts to about 4% [12].

The total AP activity in the urine of healthy people consists of about 70% of the liver-like fraction and about 20% of intestinal-like fraction [12]. Urinary AP molecules have a relative molecular mass of 75–124 kDa [1, 9, 14]. This high molecular mass makes it unlikely that AP represents serum enzyme cleared by the kidney, and it can be accepted that AP is derived from kidney tissue [2, 3]. It has been suggested that human kidney AP could be a tetramer, and urinary AP consists of its active dimeric or trimeric subunits [3].

In addition to the immunological differentiation, urinary AP activity can be separated into soluble and particulate fractions by ultracentrifugation and electrophoresis as described for AAP ([6]; see Sect. 8.1.1.5).

8.1.2.2. *Assay*

$$\text{4-nitrophenylphosphate} + H_2O \xrightarrow{AP} \text{4-nitrophenol} + \text{phosphate}$$

(hydrolysis)

$$\text{4-nitrophenylphosphate} + \text{buffer-ROH} \xrightarrow{AP} \text{4-nitrophenol} + \text{R-phosphate}$$

(transphosphorylation)

AP splits off (as do transphosphorylases) the phosphate group from 4-nitrophenylphosphate. The increase in absorbance of the liberated 4-nitrophenol under alkaline conditions is measured at 405 nm.

Numerous methods of determining AP activity have been described [10], but no special investigations have been made to optimize reaction conditions for urinary AP activity. The latter is generally measured by adapted methods for determining serum AP activities. Because low levels are found in human urine, methods with high analytical sensitivity are needed to guarantee a reliable determination. The substrate 4-nitrophenylphosphate and the transphosphorylating buffers like diethanolamine and 2-amino-2-methyl-1-propanol form the reaction conditions for these methods. The rate of AP action in the presence of the transphosphorylating buffer is two- to sixfold greater than in the presence of non-activating buffer such as glycine. Thus, I recommend a method which uses diethanolamine as phosphate-accepting buffer [15]. This method is more sensitive than procedures with other phosphate-accepting buffers.

Final reaction conditions are: 10 mmol/l 4-nitrophenylphosphate; 1 mol/l diethanolamine, pH 9.8; 0.5 mmol/l Mg^{2+}; incubation temperature 37°C.

Collection, storage and preparation of samples are described in Chap. 6.

Reagents and Solutions. All reagents must be of analytical quality; diethanolamine must be free of monoethanolamine (< 5 mmol/l in final reaction mixture) and 4-nitrophenylphosphate must be free of inorganic phosphate (< 0.05 mmol/l in final reaction mixture) [4, 5]. Deionized water must be used.

1. Mg^{2+} solution (32.5 mmol/l): dissolve 801 mg $MgSO_4 \cdot H_2O$ in 100 ml water.
2. Diethanolamine/Mg^{2+} solution (1.3 mol/l diethanolamine, pH 9.8; 0.65 mol/l Mg^{2+}): mix 125 ml diethanolamine with about 800 ml water, adjust pH to about pH 10.0 with 1 mol/l HCl, add 20 ml Mg^{2+} solution (1) and then adjust pH to 9.8 at 37°C. Dilute solution to exactly 1 l in a volumetric flask with water. This solution is stable for at least 1 year at – 20°C or 4 weeks at 4°C.
3. 4-Nitrophenylphosphate solution (130 mmol/l): dissolve 965 mg of 4-nitrophenylphosphate (disodium salt 6 H_2O) in 20 ml water. This solution is stable for at least 4 weeks at – 20°C. It is advisable to freeze the solution in small portions for daily use.

Measurement. Prepare incubation mixture in cuvettes (d = 1.0 cm) and read or record absorbance changes in a spectrophotometer at 405 nm. Incubation temperature 37°C. Incubation volume 650 μl, sample volume fraction 0.154.

Pipette successively	Sample assay (μl)	Reagent blank (μl)	Final concentrations
Buffer/Mg^{2+} solution (2)	500	500	Diethanolamine 1 mol/l
Sample	100	—	
Water	—	100	Mg^{2+} 0.5 mmol/l
Mix; preincubate for 10 min at 37°C			
Nitrophenylphosphate solution (3)	50	50	10 mmol/l
Mix, read or record absorbance for 5 min			

ΔA/min should not exceed 0.300. For higher activities, 154 mmol/l NaCl is used for sample dilution.

Calculation. Correct readings for sample by blank readings and calculate AP activity according to the following equation:

$$\mathrm{U/l} = \frac{\Delta \mathrm{A/min} \times 10^3 \times 0.650 \times \mathrm{DF}}{19.25 \times 0.1} = \Delta \mathrm{A/min} \times 337.7 \times \mathrm{DF}$$

Table 1. The 97.5th percentiles of alkaline phosphatase (AP)/creatinine ratios as upper reference limits in function of age (according to [7]). The percentiles were calculated on the basis of log normal distribution

Age group	Sex and number	AP (U/mmol creatinine) (97.5th percentile)
1 (5 days–6 months)	M ($n = 46$) + F ($n = 21$)	17.5
2 (7 months–2 years)	M ($n = 11$) + F ($n = 6$)	13.7
3 (3–6 years)	M ($n = 48$) + F ($n = 37$)	5.3
4 (7–12 years)	M ($n = 75$) + F ($n = 46$)	3.5
5 (13–18 years)	M ($n = 44$) + F ($n = 32$)	3.2
6 (23–58 years)	M ($n = 30$) + F ($n = 33$)	1.6

where 19.25 is the millimolar absorption coefficient of 4-nitrophenol at 405 nm and 37°C and DF the dilution factor (dilution by gel filtration and/or by addition of ethylene glycol to urine sample in case of storage).

8.1.2.3 *Reference Ranges*

Using the described method urinary excretion of AP was determined in morning random urine samples from 429 subjects of various ages (5 days to 58 years) [7]. AP/creatinine ratios were log normally distributed in the different age groups and significantly decreased with increasing age (Table 1). Sex-related differences were not observed. Table 1 contains the upper reference limits calculated as 97.5th percentiles.

8.1.2.4 *Separation of Multiple Forms*

Urinary AP activity can be separated into particulate and soluble fractions by ultracentrifugation and electrophoresis [6]. In healthy adults, about 45% of AP activity is particle bound [6]. The separation methods correspond to the procedures described for the separation of AAP forms (see Sect. 8.1.1.5). After electrophoresis, activity bands are stained by incubating gels in 500 mmol/l 2-amino-2-methyl-1,3-propandiol buffer (pH 9.70) and 2 mmol/l α-naphthylphosphate. The subsequent coupling is carried out with 1 mmol/l tetrazotized di-2-anisidine (fast blue B) in 100 mmol/l acetic acid.

References

1. Butterworth PJ (1968) Human kidney and urinary alkaline phosphatases. Biochem J 107:467–472
2. Cornell A, Hodson AW, Latner AL (1979) Immunological identity of urinary alkaline phosphatase in health and disease. Clin Chim Acta 97:217–223

3. Hodson AW (1983) Separation of alkaline phosphatase isoenzymes from human kidney and comparison with alkaline phosphatases from other human tissues, urine, and Escherichia coli. Clin Chem 29:100–106
4. Jung K, Pergande M (1980) Influence of inorganic phosphate on the activity determination of isoenzymes of alkaline phosphatase in various buffer systems. Clin Chim Acta 102:215–219
5. Jung K, Pergande M, Reichmann G, Sitte A, Egger E (1978) Influence of monoethanolamine on activity measurements of the isoenzymes of alkaline phosphatase. J Clin Chem Clin Biochem 16:223–224
6. Jung K, Pergande M, Wischke UW (1984) Characterization of particulate and soluble variants of the brush-border enzymes alanine aminopeptidase, alkaline phosphatase and gamma-glutamyltransferase in human urine. Biomed Biochim Acta 43:1357–1364
7. Jung K, Hempel A, Grützmann KD, Hempel RD, Schreiber G (1990) Age-dependent excretion of alanine aminopeptidase, alkaline phosphatase, γ-glutamyltransferase and N-acetyl-β-D-glucosaminidase in human urine. Enzyme 143:10–16
8. Mc Kenna MJ, Hamilton TA, Sussman HH (1979) Comparison of human alkaline phosphatase isoenzymes. Structural evidence for three protein classes. Biochem J 181:67–73
9. Miura M, Koyama I, Matsuzaki H, Sakagishi Y, Ikezawa H, Komoda T (1988) Organ specific properties for human urinary alkaline phosphatases. Clin Chim Acta 171:63–74
10. Moss DW (1977) The scientific foundation of clinical enzymology. Naturwissenschaften 64:253–259
11. Moss DW (1986) Multiple forms of acid and alkaline phosphatase: genetics, expression and tissue specific modification. Clin Chim Acta 161:123–135
12. Pfleiderer G, Baier M, Mondorf AW, Stefanescu T, Scherberich JE, Müller H (1980) Change in alkaline phosphatase isoenzyme pattern in urine as possible markers for renal disease. Kidney Int 17:242–249
13. Scherberich JE, Falkenberg FW, Mondorf AW, Müller H, Pfleiderer G (1976) Biochemical and immunological studies on isolated brush border membranes of human kidney cortex and their membrane surface proteins. Clin Chim Acta 55:179–197
14. Stinson RA, Seargant LE (1981) Comparative studies of pure alkaline phosphatase from five human tissues. Clin Chim Acta 110:261–272
15. The Committee on Enzymes of the Scandinavian Society for Clinical Chemistry and Clinical Physiology (1974) Recommended methods for the determination of four enzymes in blood. Scand J Clin Lab Invest 33:291–306

8.1.3 Gamma-glutamyltransferase

Z.J. Simane

8.1.3.1 *Introduction*

Gamma-glutamyltransferase [GGT; (γ-glutamyl)-peptide-amino acid γ-glutamyltransferase; EC 2.3.2.2] was first isolated and characterized by Hanes et al. [3, 4] in the early 1950s. GGT from human or animal kidney is a glycoprotein (90–120 kDa, depending on the isolation technique used) which is bound to the plasma membrane and contains hexoses, hexosamine and sialic acid [14, 23]. The amphiphilic molecule consists of two unequal subunits which can be released from the membrane by proteinases or detergents. The light subunit, a carbohydrate-rich and hydrophilic glycopeptide, is located on the outer membrane surface and carries the catalytic centre. The heavy hydrophobic fragment seems to contain a peptide of 3–19 kDa which mediates the anchorage of the molecule in the lipid layer of the brush border membrane [6, 22, 23]. GGT species isolated from various tissues have the same molecular mass and possess the same catalytic activity; they differ, however, in their isoelectric point, which probably depends on the sialic acid content [11].

GGT is one of the key enzymes of the γ-glutamyl cycle. Its functions are to catalyze the cleavage of glutathione into glutamic acid and cysteinylglycine and, acting as transpeptidase, the transfer of the γ-glutamyl moiety to acceptor amino acids. Reduced glutathione is resynthesized in the γ-glutamyl cycle and participates then as coenzyme in certain intracellular enzyme reactions [12].

GGT occurs abundantly in human and animal organs of high secretory or absorptive capacity such as the kidney, liver, pancreas and intestine. Its highest activity has been found in the kidney, located in the brush border of the proximal tubuli and Henle's loop [5]. The entire urinary GGT activity in healthy individuals originates from the kidney epithelium and is released from it by physiological cell disintegration [18]. The GGT activity is two- to sixfold higher than that of serum [18, 24]. Two multiple forms of the enzyme, i.e. a soluble and a particulate fraction, were detected in human urine. They can be separated from each other by ultracentrifugation, electrophoresis, gel filtration or affinity chromatography [9]. In hyperenzymuria caused by glomerular injury a serum fraction of the enzyme can occur additionally. Both enzyme components of kidney origin have nearly the same K_m values and show an identical dependence on pH and substrate concentration. They differ only in their stability at 56°C: the soluble fraction does not show any loss of activity at this temperature whereas the particulate variant is inactivated under the same conditions to 86% $\pm$ 10% of the initial value [9]. The difference of both forms in this respect may correspond with the cytoplasmic or microsomal origin of the enzyme [2].

8.1.3.2 *Assay*

GGT regulates the transfer of γ-glycyl between a number of donor and acceptor substrates; for analytical use a synthetic donor, L-glutamyl-4-nitroanilide [13] or the more soluble L-γ-glutamyl-3-carboxy-4-nitroanilide [21], and glycylglycine as acceptor are mostly preferred.

The estimation of urinary GGT was first described by Orlowski and Szewczuk [15]; the first systematic study of analytical conditions in urine was presented by Szasz [20]. He found that there are only negligible differences between optimum reaction conditions for urinary and serum enzymes, if at all. Therefore, methods may be applied for urine which are commonly used for serum estimation. To this end, a number of procedures were published which differ from each another in donor substrate and its concentration, reaction temperature and composition of buffer [10]. Opinion is divided as to the necessity of discharging the interfering urine components by dialysis or gel filtration prior to the assay. Some authors consider the pretreatment not to be necessary [10, 17], whereas Szasz [20] reports on the presence of heat-resistant and dialysable inhibitors in urine. In my own experiments no differences were found between GGT activities of crude and gel-filtered human, canine and rat urine (unpublished data).

In the following text two methods for urinary GGT estimation are presented: the procedure derived from the paper of Szasz et al. [21] preferred in the FRG and the standard method of the European Committee for Clinical Laboratory Standards [1], which is a modification for routine use of the recommended provisional IFCC Reference Method [7].

Both procedures described yield significantly different values with variations in reaction conditions (temperature, substrate concentration, buffer). Furthermore, it must be considered that the temperature converting factors for some species (dog, rat) are not identical with those recommended for human urine (Table 1).

Collection, storage and preparation of samples are as described in Chap. 6.

Table 1. Temperature converting factors (mean ± SD) for urinary gamma-glutamyltransferase as determined in human, Wistar rat and canine samples using the presented methods I and II

	Temperature ratio					
	30° vs 25°C		37° vs 30°C		37° vs 25°C	
	Crude urine	Gel-filtered urine	Crude urine	Gel-filterd urine	Crude urine	Gel-filtered urine
Humans (male, $n = 20$)						
Method I						
Mean	1.34	1.36	1.42	1.35	1.91	1.84
SD	0.06	0.08	0.05	0.05	0.05	0.11
Method II						
Mean	1.40	1.36	1.38	1.40	1.92	1.90
SD	0.04	0.04	0.04	0.04	0.04	0.06
Wistar rats (male, $n = 20$)						
Method I						
Mean	1.46	1.53	1.48	1.47	2.16	2.24
SD	0.03	0.05	0.02	0.06	0.06	0.04
Method II						
Mean	1.58	1.53	1.53	1.49	2.42	2.27
SD	0.03	0.04	0.03	0.03	0.07	0.07
Beagle dogs (male, $n = 20$)						
Method I						
Mean	1.30	1.25	1.61	1.71	2.09	2.14
SD	0.10	0.10	0.07	0.10	0.16	0.20
Method II						
Mean	1.42	1.37	1.70	1.54	2.42	2.11
SD	0.08	0.11	0.06	0.08	0.17	0.20

8.1.3.3 *Method I* [24]

Assay at 25°C:

$$\text{L-}\gamma\text{-glutamyl-3-carboxy-4-nitroanilide} + \text{glycylglycine} \xrightarrow{\text{GGT}}$$

$$\text{L-}\gamma\text{-glutamyl- glycylglycine} + \text{3-carboxy-4-nitroaniline}$$

The absorption maximum of 3-carboxyl-4-nitroaniline is at 380 nm; however, the donor substrate interferes significantly up to 410 nm. Therefore, the absorption of 3-carboxy-4-nitroaniline liberated by GGT is measured at Hg 405 nm or higher wavelengths (405–410) and not at its maximum.

The variables of the procedure are described by Persijn et al. [16]. The optimal pH value was found to be 8.25 (at 25°C) using TRIS buffer 100 mmol/l. TRIS ions are inhibitory above 200 mmol/l. The concentration of the acceptor substrate, glycylglycine, was set at 100 mmol/l. Under these conditions, the concentration of the donor substrate L-γ-glutamyl-3-carboxy-4-nitroanilide had to be set at 2.9 mmol/l.

In the present method the optimum conditions of measurement were modified concerning substrate concentration, buffer and pH to obtain an identical catalytic activity with the previously frequently used method of Szasz [19] in which the substrate was γ-glutamyl-4-nitroanilide. This does not play any role if the method is employed for urinary GGT evaluation. Nevertheless, the fact that a number of kits commercially available are based on this procedure seems to justify its description.

Reagents and Solutions. All reagents must be of analytical grade, and deionized water must be used.

1. TRIS/glycylglycine solution of TRIS 120 mmol/l, pH 8.25, and glycylglycine 120 mmol/l: dissolve 1.452 g TRIS in about 90 ml water, during stirring add 1.584 g glycylglycine and adjust pH to 8.25 (glass electrode) with 2 mol/l HCl. As preservative 100 mg sodium azide may be used. Add water to 100 ml.
2. L-γ-Glutamyl-3-carboxy-4-nitroanilide (35.1 mmol/l): dissolve 121.6 mg L-γ-glutamyl-3-carboxy-4-nitroanilide, monoammonium salt, in 10 ml water (with 1 mg sodium azide per ml). Measure pH value (glass electrode) and adjust to 7–7.5.
3. Reagent mixture (TRIS, 110 mmol/l, pH 8.25, 25°C; glycylglycine, 10 mmol/l; L-γ-glutamyl-3-carboxy-4-nitroanilide, 3.2 mmol/l): mix 20 ml solution (1) and 2 ml solution (2).

When kept at 2°–8°C, solution (1) is stable for 3 months, solution (2) for 3 weeks and the reagent mixture for 10 days.

Measurement. Given wavelength Hg 405 nm, light path 10 mm, final volume 2.2 ml, volume fraction $\phi = 0.091$ and 25°C (thermostated in cuvette holder), measure against air. A reagent blank is not necessary. Before starting the assay, adjust the temperature of solutions to the measuring temperature.

Pipette successively		Final concentration	
Reagent mixture (3)	2.0 ml	L-γ-glutamyl-3-carboxy-4-nitroanilide glycylglycine TRIS	2.9 mmol/l 100 mmol/l 100 mmol/l
Wait for 5 min, until temperature of 25°C is reached			
Sample	0.2 ml		
Mix, read or record absorbance for 3 min (longer with low activities)			

Calculation

$$\text{GGT U/l} = \Delta A/\Delta t \times 1158\,(\times \text{dilution factor})$$

ΔA/min should not exceed 0.200, and 0.15 mol/l NaCl is recommended for dilution.

The dilution factor encompasses the dilution by gel filtration, by ethylene glycol (in case of urine storage) and dilution because of high activity.

Some filter photometers do not allow measurement exactly at Hg 405 nm. In such a case, it is recommended to estimate the absorption coefficient of the reaction product 3-carboxy-4-nitroaniline under these special conditions and to consider it for the calculation.

8.1.3.4 *Method II* [1]

Assay at 37° or 30°C: the principle is the same as in method I.

Optimized assay conditions for human serum [7] include pH 7.9 (at 30°C), concentration of acceptor substrate glycylglycine (simultaneously acting as buffer substance) 150 mmol/l, concentration of donor substrate 6 mmol/l. The procedure can be run both at 30° and 37°C.

Reagents and Solutions. All reagents must be of analytical grade, and deionized water must be used. To prevent microbial contamination, the use of sterilized containers is recommended.

1. *N*-Glycylglycine-NaOH buffer 183.3 mmol/l, pH 7.9 (30°C): dissolve 24.2 g *N*-glycylglycine in about 800 ml water, adjust pH to 7.90 at 30°C with 2 mol/l NaOH (glass electrode), and add water to 1000 ml. (At 37°C the buffer has pH 7.70; this has only a negligible influence on the reaction rate.)
2. L-γ-Glutamyl-3-carboxy-4-nitroanilide 66.0 mmol/l: dissolve 2.286 g of L-γ-glutamyl-3-carboxy-4-nitroanilide, monoammonium salt, monohydrate, in about 90 ml water and add water to 100 ml.
3. Reagent mixture (L-γ-glutamyl-3-carboxy-4-nitroanilide 6.6 mmol/l, *N*-glycylglycine 165 mmol/l), pH 7.9 at 30°C: mix 10 ml solution (2) and 90 ml

solution (1). The absorption of the mixture at 410 nm against a water blank should not exceed 0.500.

4. Sodium chloride solution 154 mmol/l: dissolve 9.00 g sodium chloride in a total final volume of 1000 ml water.

When kept at 4°C or − 20°C, solutions (1) and (2) are stable for 1 month and 9 months, respectively. It is recommended to prepare solution (3) fresh on each day it is used.

Measurement. Given wavelength 410 nm, light path 10 mm, final volume 2.2 ml, volume fraction $\phi = 0.091$ and temperature 30° or 37 °C (thermostated in cuvette holder), measure against reagent blank. Adjust the solutions to reaction temperature before starting the assay.

Pipette successively	Sample	Blank	Final concentration
Reagent mixture (3)	2.0 ml	2.0 ml	L-γ-glutamyl-3-carboxy-4-nitroanilide 6 mmol/l *N*-glycylglycine 150 mmol/l
Equilibrate temperature to 30 °C			
Sample (urine) Solution (4)	0.2 ml	 0.2 ml	Volume fraction $\phi = 0.091$
Mix, read or record absorbance for up to 5 min			

Calculation.

$$\text{GGT U/l} = \Delta A/\Delta t \times 1391 \ (\times \text{dilution factor})$$

ΔA/min should not exceed 0.215, and 154 mmol/l NaCl is recommended for dilution.

8.1.3.5 *Separation of Multiple Forms*

The particulate and soluble fractions of GGT can be separated analogously to other brush border enzymes. The procedure is described in Sect. 8.1.1.5.

In urine samples of 20 ostensibly healthy adults, the percentage of total activity in the particulate fraction was found to be 75.7% ± 3.25% [8].

According to Fedyanin et al. [2], the cytoplasmic and microsomal fractions of urinary GGT can be separated by heating the sample to 58 °–60 °C for 30 min. Using this procedure, the mean percentage of the stable (cytoplasmic) fraction was found to be 54%.

References

1. European Committee for Clinical Laboratory Standards (1988) ECCLS Standards for enzyme determinations 1–4. ECCLS Document ISSN 1011–6265, no 2, pp 67–74
2. Fedyanin JP, Golub TV, Karaeva LN, Figurnov VA (1988) Measurements of the blood and urine gamma-glutamyl transpeptidase isoenzymes. Lab Delo 1:14–16
3. Hanes CS, Hird FJR, Isherwood FA (1950) Synthesis of peptide in enzymic reactions involving glutathione. Nature 166:288–292
4. Hanes CS, Hird FJR, Isherwood FA (1952) Enzymic transpeptidation reactions involving γ-glutamyl peptides and α-amino-acyl peptides. Biochem J 51:25–35
5. Heinle H, Wendel A, Schmidt U (1977) The activities of the key enzymes of the γ-glutamyl cycle in microdissected segments of the rat nephron. FEBS Lett 73:220–224
6. Hughey RP, Curthoys NP (1976) Comparison of the size and physical properties of γ-glutamyltranspeptidase purified from rat kidney following solubilization with papain or with Triton X-100. J Biol Chem 251:7863–7870
7. IFCC Expert Panel on Enzymes (1983) IFCC methods for measurement of enzymes, pt 4. International Federation of Clinical Chemistry (IFCC) method for γ-glutamyltransferase, EC 2.3.2.2, stage 2, draft 2. J Clin Chem Clin Biochem 21:633–646
8. Jung K, Wischke UW (1984) Electrophoretic variants of alanine aminopeptidase, alkaline phosphatase and γ-glutamyltransferase in urine. Clin Chem 30:856–859
9. Jung K, Pergande M, Wischke UW (1984) Characterization of particulate and soluble variants of brush-border enzymes alanine aminopeptidase, alkaline phosphatase and gamma-glutamyltransferase in human urine. Biomed Biochim Acta 43:1357–1364
10. Liese W, Jung K, Köhler A, Grützmann KD, Peters JE (1988) Grundlagen der Enzymaktivitätsbestimmungen in der Klinischen Chemie. VEB Verlag Volk und Gesundheit, Berlin, pp 151–155, 187–188
11. Matsuda Y, Tsui A, Katunuma N (1980) Studies on the structure of γ-glutamyltranspeptidase. J Biochem 87:1243–1248
12. Nemesanszky E, Lott JA (1985) Gamma-glutamyltransferase and its isoenzymes: progress and problems. Clin Chem 31:797–803
13. Orlowski M, Meister A (1963) γ-Glutamyl-ρ-nitroanilide: a new convenient substrate for determination and study of L- and D-γ-glutamyltranspeptidase activities. Biochim Biophys Acta 73:679-681
14. Orlowski M, Meister A (1965) Isolation of γ-glutamyl transpeptidase from hog kidney. J Biol Chem 240:338–347
15. Orlowski M, Szewczuk A (1962) Determination of γ-glutamyl transpeptidase activity in human serum and urine. Clin Chim Acta 7:755–760
16. Persijn JP, van der Slik W (1976) A new method for the determination of γ-glutamyltransferase in serum. J. Clin Chem Clin Biochem 14:421–427
17. Rambabu K, Pattabiraman TN (1976) Studies on γ-glutamyl transpeptidase of human urine. Clin Chim Acta 73:251–255
18. Rosalki SB (1975) Gamma-glutamyl transpeptidase. Adv Clin Chem 17:53–107
19. Szasz G (1969) A kinetic photometric method for serum γ-glutamyl transpeptidase. Clin Chem 15:124–136
20. Szasz G (1970) γ-Glutamyltranspeptidase-Aktivität im Urin. Z Klin Chem Klin Biochem 8:1–8
21. Szasz G, Weimann G, Stähler F, Wahlefeld AW, Persijn JP (1974) New substrates for measuring γ-glutamyl transpeptidase activity. Z Klin Chem Klin Biochem 12:228
22. Tate SS, Meister A (1977) Affinity labeling of γ-glutamyl-transpeptidase and location of the γ-glutamyl binding site on the light subunit. Proc Natl Acad Sci USA 74:931–935
23. Tate SS, Ross ME (1977) Human Kidney γ-glutamyl transpeptidase: Catalytic properties, subunit structure, and localization of the γ-glutamyl binding site on the light subunit. J Biol Chem 252:6042–6045
24. Wahlefeld AW, Bergmeyer HU (1984) γ-Glutamyltransferase: routine method. In: Bergmeyer HU, Bergmeyer J, Grassl M (eds) Methods of enzymatic analysis, vol 3, 3rd edn. Verlag Chemie, Weinheim, pp 352–356

8.1.4 Aminopeptidase A (Angiotensinase A)

J.E. Scherberich

8.1.4.1 *Introduction*

Aminopeptidase A (Angiotensinase A, EC 3.4.11.7) is a membrane-bound enzyme which splits off N-terminal amino acid/aspartic acid from peptides, especially angiotensin II [1, 2]. Histochemical studies using α-L-glutamic acid-2-naphthylamide showed that aminopeptidase A is localized in glomerular endothelia and podocytes and, to lesser extent, in the brush border membrane of the proximal tubule [2, 5, 10]. Splitting of the substrate of aminopeptidase A is inhibited by angiotensin II, but not by angiotensin III, in a competitive manner; thus, aminopeptidase A is defined as an angiotensinase (= angiotensinase A), a fact which has been supported by peptide degradation studies using glomeruli isolated from human kidney [8, 11].

The molecular mass of angiotensinase A after solubilization of epithelial membranes from human kidney is approximately 210 kDa. It is activated by 5-10 mmol/l Ca^{2+} and inhibited by ethylenediaminetetraacetic acid (EDTA), while Mg^{2+}, Mn^{2+}, and Zn^{2+} ions have no effect. The pH optimum of the enzyme from human kidney particles is 7.4; the Michaelis–Menten constant for α-L-glutamyl-4-nitroanilide is 0.16 mmol/l and 0.25 mmo/l in the presence of 0.05 mmol/l angiotensin II [3–5, 9].

Angiotensinase A was enriched from urine more than 20-fold using lectin-affinity chromatography and fast-protein liquid chromatography [7]. Its molecular mass was 146 kDa, and the pI was 4.8. Negative staining of the enzyme isolated from the urine of patients with tubular damage revealed the presence of small knobs of 4–7 nm in diameter which may form pseudofilamentous aggregates of 30–50 nm [12].

Aminopeptidase is excreted in urine of normal persons and, at an increased rate, in patients with arterial hypertension and renal diseases [3, 5, 6].

8.1.4.2. *Assay*

$$\text{L-}\alpha\text{-Glutamyl-4-nitroanilide} + H_2O \xrightarrow{\text{Angiotensinase}}$$

$$\text{4-nitroaniline} + \text{glutamic acid}$$

The increase in absorbance of the liberated 4-nitroaniline is measured at 405 nm.

At the reaction temperature 37 °C, 1.6 mmol/l L-α-glutamyl-4-nitroanilide, 50 mmol/l TRIS/HCl buffer, and pH 7.7 are optimal reaction conditions. Aminopeptidase A determinations in tissue samples should be made in the presence of 62.5 mmol/l Ca^{2+}. The addition of Ca^{2+} is not recommended for assays in urine samples.

Collection, Storage, and Preparation of Samples. Native urine specimens over a 24-h collection period or, alternatively, aliquots of the second morning urine (mid-stream urine aliquot of the spontaneously voided urine in a time between 8.00 and 10.00 h) are used for measurements. Storage of urine specimens for up to 10 days at 4 °C does not change the angiotensinase activity.

Reagents and Solutions.
1. TRIS/HC1 buffer (60 mmol/l, pH 7.70): dissolve 0.727 g of tris(hydroxymethyl)aminomethane in about 90 ml water, adjust to pH 7.70 at 37 °C with 1 mol/l HCl, and add water to 100 ml.
2. L-α-Glutamyl-4-nitroanilide (16.6 mmol/l): dissolve 94.8 mg of L-α-glutamyl-4-nitroanilide by adding 0.5 ml of 2 mol/l HCl and add water to 20 ml. This solution is stable at 4 °C for 4 weeks.

Measurement. Prepare incubation mixture in cuvettes (d = 1.0 cm) and read or record absorbance changes in a spectrophotometer at 405 nm. The incubation temperature is 37 °C, incubation volume 1.0 ml, and sample volume fraction 0.1.

Pipette successively	Sample assay (μl)	Reagent blank (μl)	Final concentrations
TRIS buffer (1)	800	800	TRIS buffer, 50 mmol/1
Sample	100	—	
Water	—	100	
Mix; preincubate for 10 min at 37 °C			
L-α-Glutamyl-4-nitroanilide solution (2)	100	100	1.6 mmol/l
Mix, read or record absorbance for 10 min			

A/min should not exceed 0.250. For higher activities, 0.154 mol/l NaCl solution is used for sample dilution.

Calculation. Correct readings for sample by blank readings and calculate angiotensinase activity according to the instruction given for the estimation of alanine aminopeptidase:
Angiotensinase U/1 = ΔA/min × 1010

8.1.4.3 *Reference Ranges*

Using the described method, urinary excretion of angiotensinase A was determined in second morning urine samples of 68 healthy adults (34 women, 34 men; aged 18–65 years). Volume activity amounted to 1.59 ± 0.68 U/l (arithmetic mean ± SD and a median of 0.962 U/l. The angiotensinase/creatinine ratio (U/mmol creatinine) was not significantly different between urine specimens from men and women and showed the following figures: arithmetic mean ± SD, 0.109 ± 0.051; median, (lower and upper quartile): 0.092 (0.062, 0.155).

In contrast to other urinary enzymes, e.g., gamma-glutamyltransferase, angiotensinase is negatively correlated with the creatinine clearance [6].

References

1. Glenner GG, Mc Millan PHJ, Folk LF (1962) A mammalian specific for the hydrolysis of N-terminal L-glutamyl and aspartyl residues. Nature 194:867
2. Kugler P, Wolf G, Scherberich JE (1985) Histochemical demonstration of peptidases in the human kidney. Histochemistry 83:337–342
3. Scherberich JE, Stuckhardt C,Wolf G, Kugler P, Schoeppe W (1986) Angiotensinase A of human kidney:histo/-biochemical studies of an enzyme possibly involved in the regulation of intrarenal renin-angiotensin system. Nieren Hochdruckkr 15:386
4. Scherberich JE, Wolf G, Stuckhardt C, Schoeppe W (1987) Characterization and clinical significance of angiotensinase-A of human kidney. Nephrol Dial Transplant 2:387
5. Scherberich JE, Wolf G, Stuckhardt C, Kugler P, Schoeppe W (1989) Characterization and clinical role of glomerular and tubular proteases from human kidney. Adv Exp Med Biol 240:275-282
6. Scherberich JE, Wolf G, Albers C, Nowack A, Stuckhardt C, Schoeppe W (1989) Glomerular and tubular membrane antigens reflecting cellular adaptation in human renal failure. Kidney Int 36 (Suppl 27):38–51
7. Scherberich JE, Wiemer J, Herzig C, Fischer P, Schoeppe W (1990) Isolation and partial characterization of angiotensinase A and aminopeptidase M from urine and human kidney by lectin affinity chromatography and high-performance liquid chromatography. J Chromatogr 521:279–289
8. Wolf G, Stuckhardt C, Scherberich JE (1986) Biochemical and immunohistochemical studies on angiotensinase A in human kidney and renal adenocarcinoma. Immunbiol 173:445–446
9. Wolf G, Scherberich JE, Stuckhardt C, Schoeppe W (1987a) Characterization of angiotensin receptors from human kidney. Prot Biol Fluids 35:289-292
10. Wolf G, Scherberich JE, Stuckhardt C, Schoeppe W (1987b) Angiotensin-degrading aminopeptidase A (APA) of human kidney:biochemical and histochemical characterizaton. Acta Endocrinol 114:283
11. Wolf G, Scherberich JE, Schoeppe W (1988) Glomeruli isolated from human kidney split angiotensin II: further evidence that renal "aminopeptidase A" (APA) is an angiotensinase. Z Anal Chem 330:429–430
12. Wiemer J, Scherberich, JE, Schoeppe W (1991) Purification and characterization of angiotensinase A (ATA) from human urine. Nieren- und Hochdruckkr. 20:460

8.2 Lysosomal Enzymes

8.2.1 N-Acetyl-β-D-Glucosaminidase; [2-Acetamido-2-deoxy-β-D-glucoside-acetamidodeoxyglucohydrolase, EC 3.2.1.30]

Z.J. Simane

8.2.1.1 *Introduction*

A β-glucosaminidase, which catalyzes the hydrolysis of *N*-acetyl-D-glucosaminopyranoside from β-glycosidic linkages, was discovered in almond emulsin by Helferich and Iloff [9]. Its activity could not be distinguished from that of *N*-

acetyl-β-galactosaminidase (EC 3.2.1.53) and the term 'hexosaminidase' was therefore used to describe either activity [27, 34].

N-Acetyl-β-D-glucosaminidase (NAG) has a molecular mass of 150–160 kDa and occurs in lysosomes of many tissues and in body fluids [3,8,28]. It is particularly abundant in organs where high rates of mucoid turnover might be expected, e.g., testis and spleen. Robinson and Stirling [27] found that the NAG activity of human tissue extracts could be attributed to two components, termed A and B, which could be resolved by starch-gel electrophoresis or anion-exchange chromatography, or distinguished by their different heat stabilities. The isoelectric points of forms A and B were found to be pH 5.4 and 7.9 respectively [29]. Both isoenzyme types are characterized by the structure of their subunits. The A form consists of three nonidentical polypeptide chains, α, β_a and β_b, whereas the B form consists of two $\beta_a\beta_b$ units. Two types of NAG A have been described [5, 12, 30]. The tissue A form, present in liver, kidney and spleen, can be distinguished from the A^s form, present in serum, on the basis of its behaviour on DEAE-cellulose chromatography and gel electrophoresis. The two major NAG forms, A and B, are both present in the renal cortex [33] and the urine. A renal origin of the activity excreted by urine was demonstrated in rat by comparison of the release from the isolated perfused rat kidney with excretion from the in situ organ. The rate of NAG excretion in humans is directly proportional to the corresponding rate of release by the tubular epithelium. The urinary activity is not associated with cells; it is almost entirely soluble [17].

The urine of healthy humans contains both the A and B forms, and a minor component, M, which was found only in male urine and is in its properties similar to the A form. The B form generally comprised 15%–25% of the total NAG activity [7]. Other authors [5, 31] normally detected only the A form in human urine, while NAG B activity appeared only as a consequence of renal tissue lesions. There are only traces of the intermediate forms I_1 and I_2 which are considered to be the incompletely solubilized A form.

Thus, the elevation of urinary NAG activity in disease is accompanied by a change in the relative activities of the multiple forms present. The activity of both NAG A and B increased in urine from patients with renal disease; the higher the NAG activity excreted, the greater the proportion of the B form. In contrast, glomerular damage was accompanied by the presence of the serum A^s form in the urine [5].

8.2.1.2 *Methods*

The early procedures developed for the estimation of the NAG activity have been limited by a lack of suitable substrates. Those with an alkyl N-acetyl-β-glucosaminide as substrate [20] depend on the relative nonspecific estimation of reducing sugar liberated. Phenyl derivatives, and in particular 4-nitrophenyl-*N*-acetyl-β-glucosaminide, were used by numerous authors [1, 3, 7, 8, 10, 11, 13, 14, 17, 19, 20, 26, 28] in methods depending on the measurement of the liberated phenol. Recently, a kinetic rate assay of urinary NAG with the substrate 2-chloro-4-nitrophenyl-*N*-acetyl-β-D-glucosaminide has been described [18].

4-Nitrophenol has a relatively low molar absorptivity of 18.6×10^3 l $\times$ $mol^{-1} \times cm^{-1}$ at 405 nm and, unfortunately, the color of the released phenolic chromophore (λ_{max} 405 nm) is similar to that of urine which may result in high background readings. Furthermore, the yellow color may be mimicked by some urine constituents such as bilirubin or hemoglobin as well as by some drugs. In addition, NAG inhibitors can occur in urine samples [10]. There are two conceivable ways of avoiding such interference: (a) discharging disturbing compounds prior to the assay by gel filtration, and (b) diluting the sample to such an extent that interference is negligible. The first Method [10, 14, 19] is practicable even if it makes the analysis more time consuming. It is the method of choice, if more enzymes have to be measured in one sample or if a high concentration of disturbing substances (e.g., drugs in animal studies) might be expected.

Leaback and Walker [15] described the synthesis and properties of methylumbelliferyl N-acetyl-β-D-glucosaminide which they used as a fluorogenic substrate for an extremely sensitive NAG assay. This principle was also used for the estimation of urinary NAG activity [13, 25] and the procedure was adapted for an automated flow system [31] and for a centrifugal analyzer [24]. A modification using a microtiterplate technique was also described [16]. Utilizing the high sensitivity of the fluorescence method, the urine samples can first be so diluted as to eliminate the influence of inhibitors without the need for previous gel filtration. However, fluorometry is not a widely used technique in clinical chemistry, and so it is expensive if employed solely for urinary NAG.

More recently, two new substrates for colorimetric measurement of the NAG activity were described. Yuen et al [35, 36] reported the synthesis of 2-methoxy-4-(2′-nitrovinyl)-phenyl-2-acetamido-2-deoxy-β-D-glucopyranoside (MNP-GlcNAc) and a corresponding colorimetric assay for urinary NAG. Because of the high molar absorptivity of MNP (26.8×10^3 l $\times mol^{-1} \times cm^{-1}$ at 505 nm), the effect of urinary inhibitors could be minimized by diluting the urine in the assay procedure. A good correlation was found with assays using methylumbelliferyl and 4-nitrophenyl glycosides. The MNP method was also adapted for automated flow system and for a reagent kit.

A further chromogenic substrate for estimation of urinary NAG, 3-cresolsulfonphthalein *N*-acetyl-β-D-glucosaminide (MPC-NAG), was described by Noto et al. [21]. This substrate yields an aglycone with a molar absorptivity of 40.67 $\times 10^3$ l $\times mol^{-1} \times cm^{-1}$ at 580 nm. The high reagent substrate concentration relative to the K_m for NAG (approximately seven K_m) decreases the potential for analytical interference by urinary inhibitors. The assay was automated for various analyzer models [6,22].

In the following sections two methods for urinary NAG are presented. The first procedure is modified from that described by Maruhn [19] in which 4-nitrophenyl-*N*-acetyl-β-D-glucosaminide is used as substrate, while the second assay described by Noto et al [21] is based on the cleavage of 3-cresolsulfonphthaleinyl-*N*-acetyl-β-D-glucosaminide. There is also a kit commerically available which is based on the latter method [4].

8.2.1.3 Method I [19]

$$\beta\text{-4-nitrophenyl-2-acetamido-2-deoxy-D-glucoside} + H_2O \xrightarrow{\text{NAG}}$$

$$\text{4-nitrophenol} + N\text{-acetamido-2-deoxy-D-glucose}$$

Optimized Assay Conditions include pH 4.15, using citrate buffer, 100 mmol/l; concentration of the substrate 4-nitrophenyl-2-acetamido-2-deoxy-D-glucoside 10 mmol/l, temperature 37 °C.

Collection, Storage and Preparation of Samples. As described in Chap. 6.

Reagents and Solutions. All reagents must be of analytical grade, and deionized water must be used.

1. Citrate buffer, pH 4.15, 100 mmol/l. Solution A: dissolve 2.1014 g citric acid, monohydrate, in water. Add water to 100 ml. Solution B: dissolve 2.941 g trisodium citrate-2-hydrate in water. Add water to 100 ml. Mix solution A and solution B until pH 4.15 is reached (glass electrode).
2. AMP buffer, pH 10.25, 0.75 mol/l: dissolve 6.65 g 2-amino-2-methyl-*1*-propanol in 25 ml water and adjust pH to 10.25 (glass electrode) with 2 mol/l HCl. Add water to 100 ml.
3. Reagent mixture. Under gentle warming, dissolve 250 mg β-4-nitrophenyl-2-acetamido-2-deoxy-D-glucoside in 73 ml citrate buffer. The solution may be dispensed in portions of 0.2 ml each in disposable tubes and kept frozen at − 20 °C until use.

Stability of solutions.
Kept at 2 °C to 8 °C, the buffer solutions (1) and (2), above are stable for 3 months. The frozen reagent mixture (3) is stable for at least 6 months.

Procedure. Assay conditions should be as follows: Wavelength Hg 405 nm; light path 10 mm, final volume 0.6 ml; volume fraction $\phi = 0.33$; 37 °C (thermostatted cuvette holder). Measure against water, use sample blank and reagent blank. Before starting the assay, adjust the temperature of solutions to the measuring value.

Pipette successively	Analysis sample (S)	sample blank (SB)	Reagent blank(RB)
Reagent mixture (solution 3)	200	—	200
Sample	200	200	—
NaCl 0.15 mol/l	—	—	200
Incubate in a water bath (37 °C) for 15 min			
AMP buffer (solution 2)	200	200	200
Mix; read absorbance within 30 min			

Calculation. NAG U/l = [A(S) − A(SB) + A(RB)] × 10.81 (× dilution factor). The absorbance of the sample [A(S)] should not exceed 1.000. NaCl 0.15 mol/l is recommended for dilution.

The dilution factor encompasses the dilution by gel filtration, the dilution by ethylene glycol (in case of urine storage) and dilution necessary because of high activity.

8.2.1.4 Method II [21]

$$\text{3-cresolsulfonphthaleinyl-2-acetamido-2-deoxy-D-glucoside} \xrightarrow[Na_2CO_3]{NAG}$$
$$\text{sodium 3-cresolsulfonylphthalein} + \text{2 acetamido-2-deoxy-D-glucose}$$

Optimized Assay Conditions are pH 4.9 (at 37 °C), using citrate buffer, 50 mmol/l, pH 4.75. Concentration of the substrate 3-cresol-sulfonphthaleinyl-2-acetamido-2-deoxy-D-glucoside 2.75 mmol/l, temperature 37 °C.

Collection, Storage and Preparation of Samples. As described in Chap. 6, however, native urine samples without gel filtration can be used.

Reagents and Solutions. All reagents must be of analytical grade, and deionized water must be used.

1. Citrate buffer, pH 4.75, 50 mmol/l. Solution A: dissolve 1.057 g citric acid, monohydrate, in water. Add water to 100 ml. Solution B: dissolve 1.47 g sodium citrate-2-hydrate in water. Add water to 100 ml. Mix solutions A and B until pH 4.75 is reached (glass electrode).
2. Stop reagent: dissolve 3.18 g sodium carbonate anhydrous in 100 ml water.
3. Substrate solution: dissolve 185 mg of 3-cresolsulfonphthaleinyl 2-acetamido-2-deoxy-D-glucoside and 61.8 mg sodium tetraborate in 100 ml of citrate buffer (1).

Stability of Solutions. If kept at between 4 °C and 8 °C, the solutions remain stable for at least 1 month.

Procedure. Assay conditions should be as follows: Wavelength 580 nm; light path 10 mm, final volume 3.05 ml; volume fraction $\phi = 0.016$; 37 °C (thermostatted cuvette holder). Measure against reagent blank.

Pipette successively: (in ml)	Sample (S)	Reagent blank (RB)
Substrate solution (solution 3)	1.00	1.00
Incubate for 5 min at 37 °C; add		
Sample	0.05	—
Water	—	0.05
Mix and incubate for exactly 15 min at 37 °C; add		
Stop reagent (Solution 2)	2.00	2.00
Mix, allow to stand for 10 min, and record absorbance within 50 min.		

Calculation. NAG U/l = 100 × A(S) (× dilution factor). The dilution factor encompasses the dilution by gel filtration, if so prepared, and the dilution by ethylene glycol (in case of urine storage).

8.2.1.5 *Multiple Forms*

As compared with the measurement of total urinary NAG, the estimation of its multiple forms can yield more detailed information about the nature of a renal lesion [2, 7, 23]. Ion-exchange chromatography is the most frequently used method for separating the enzyme forms A and B [2, 5, 7, 12, 27, 32]; for rapid routine determination an adaptation of this procedure, carried out in test tubes, appears to be advantageous [23]:

A 0.1-ml urine sample and 0.3 ml of 10 mmol/l sodium phosphate buffer, pH 7, are added to 2 ml of 20% diethylaminoethanol (DEAE) Trisacryl M gel suspension equilibrated with the phosphate buffer. The suspension is gently stirred for 15 min and then centrifuged. The NAG-B activity is determined on a 0.1-ml sample of supernatant (corresponding to a final 20-fold dilution of the urine sample), The NAG A activity is calculated by substracting NAG B value from the total activity of the urine sample. The activities may be measured using Method II (above).

References

1. Borooah J, Leaback DH, Walker PG (1961) Studies on glucosaminidase. II. Substrates for N-acetyl-β-glucosaminidase. Biochem J 78:106–110
2. Bourbouze R, Dubois M, Gluckman JC, Legrain M (1987) Excretion of urinary N-acetyl-β-D-glucosaminidase isoenzymes after renal transplantation in the rat. J Clin Chem Clin Biochem 25:76–76
3. Conchie J, Findlay J, Levvy GA (1959) Mammalian glycosidases. Distribution in the body. Biochem J 71:318–325
4. Cooper EH, Forbes MA (1984) Evaluation of a new test for N-acetyl-β-D-glucosaminidase in urine. Clin Chem 30:593–594
5. Ellis BG, Tucker SM, Thompson AE, Price RG (1975) Presence of serum and tissue forms of N-acetyl-β-glucosaminidase in urine from patients with renal disease. Clin Chim Acta 116:195–202
6. Goren MP, Wright RK, Osborne S (1986) Two automated procedures for N-acetyl-β-D-glucosaminidase determination evaluated for detection of drug-induced tubular nephrotoxicity. Clin Chem 32:2052–2055
7. Grebner EE, Tucker J (1973) Human urinary N-acetyl-β-hexosaminidases. Biochim Biophys Acta 32.1:228–233
8. Hahn L (1945) On the mucopolysaccharide-splitting enzyme system of the mammalian testis. Ark Kemi Mineral Geol 21A:1–13
9. Helferich B, Iloff A (1933) Über Emulsin. XIII. Darstellung und fermentative Spaltung von Glykosiden des N-Acetyl-glucosamins und der 2-Desoxyglucose. Z Physiol Chem 221:252–258
10. Horak E, Hopfer SM, Sunderman FW Jr (1981) Spectrophotometric assay for urinary N-acetyl-β-D-glucosaminidase activity. Clin Chem 27:1180–1185

11. Hultberg B, Ockerman PA (1972) Artificial substrates in the assay of acid glycosidases. Clin Chim Acta 39:49–58
12. Ikonne JU, Ellis RB (1973) N-Acetyl-β-D-hexosaminidase component A. Different forms in human tissues and fluids. Biochem J 135:457–462
13. Jung K, Nesener E (1988) Bestimmung der N-Acetyl-β-D-glucosaminidase-Activität im Urin mit einem kolorimetrischen und einem fluorimetrischen Test. Z Med Lab Diagn 29:254–262
14. Knoll E, Wisser H, Rautenstrauch H (1980) Vergleichende Untersuchung zur diagnostischen Wertigkeit von Diskelektrophorese der Urinproteine und N-Acetylglucosaminidaseausscheidung zur Erkennung von tubulären Nierenschädigungen bei chronischer Polyarthritis. J Clin Chem Clin Biochem 18:53–58
15. Leaback DH, Walker PG (1961) Studies on glucosaminidase 4. The fluorimetric assay of N-acetyl-β-glucosaminidase. Biochem J 78:151–156
16. Linko-Löppönen S, Mäkinen M (1985) A microtiter plate assay for N-acetyl-β-D-glucosaminidase using a fluorogenic substrate. Anal Biochem 148:50–53
17. Lockwood TD, Bosmann HB (1979) The use of urinary N-acetyl-β-glucosaminidase in human renal toxicology. Partial biochemical characterization and excretion in human and release from isolated perfused rat kidney. Toxicol Appl Pharmacol 49:323–336
18. Makise J, Saito E, Obuchi M, Kanayama M, Ichikawa K, Harakawa K, Yoshida K (1988) Kinetic rate assay of urinary N-acetyl-β-D-glucosaminidase with 2-chloro-4-nitrophenyl-N-acetyl-β-D-glucosaminide as substrate. Clin Chem 34:2140–2143
19. Maruhn D (1976) Rapid colorimetric assay of β-galactosidase and N-acetyl-β-glucosaminidase in human urine. Clin Chim Acta 73:453-461
20. Neuberger A, Pitt Rivers RV (1939) The hydrolysis of glucosaminidases by an enzyme in Helix pomatia. Biochem J 33:1580–1590
21. Noto A, Ogawa Y, Mori S, Yoshioka M, Kitakaze T, Hori T, Nakamura M, Miyake T (1983) Simple rapid spectrophotometry of urinary N-acetyl-D-glucosaminidase with use of a new chromogenic substrate. Clin Chem 29:1713–1716
22. Orsonneau JL, Blanchardie P, Lustenberger P, Bernard S (1987) Measurement of urine N-acetyl-β-D-glucosaminidase with a centrifugal analyzer. Clin Chem 33:616
23. Paraire M, Bourbouze R, Baumann FC, Percheron F (1983) Differential assay of A and B isoenzymes in urinary N-acetyl-β-D-glucosaminidase. Clin Chim Acta 129:233–238
24. Powell SC, Scaro J, Wilson E, Shihabi ZK (1983) Assay of urinary N-acetyl-β-glucosaminidase in a centrifugal analyzer. Clin Chem 29:1717–1719
25. Price RG, Dance N, Richards B, Catell WR (1970) The excretion of N-acetyl-β-D-glucosaminidase and β-galactosidase following surgery to the kidney. Clin Chim Acta 27:65–72
26. Pugh D, Leaback DH, Walker PG (1957) Studies on glucosaminidase. N-acetyl-β-glucosaminidase in rat kidney. Biochem J 65:464–469
27. Robinson D, Stirling JL (1968) N-acetyl-β-glucosaminidases in human spleen. Biochem J 107:321–327
28. Roseman S, Dorfman A (1951) α-Glucosaminidase. J. Biol Chem 191:607–620
29. Srivastava SK, Awasthi YC, Yoshida A, Beutler E (1973) Studies on human β-D-N-acetylhexosaminidases. I. Purification and properties. J Biol Chem 249:2043–2048
30. Swallow DM, Stokes DC, Corney G, Harris H (1974) Difference between the N-acetyl hexosaminidase isoenzymes in serum and tissue. Ann Hum Genet 37:287–302
31. Tucker SM, Boyd PJR, Thompson AE, Price RG (1975) Automated assay of N-acetyl-β-glucosaminidase in normal and pathological human urine. Clin Chim Acta 62:333–339
32. Tucker SM, Pierce RJ, Price RG (1980) Characterisation of human N-acetyl-β-D-glucosaminidase isoenzymes as an indicator of tissue damage in disease. Clin Chim Acta 102:29–40
33. Wiktorowicz JE, Awasthi YC, Kurosky A, Srivastava SK (1977) Purification and properties of human kidney-cortex hexosaminidases A and B. Biochem J 165:49–53
34. Woollen JW, Heyworth R, Walker PG (1961) Studies on glucosaminidase. III. Testicular N-acetyl-β-glucosaminidase and N-acetyl-β-galactosaminidase. Biochem J 78:111–116
35. Yuen CT, Price RG, Chattagoon L, Richardson AC, Praill PFG (1982) Colorimetric assays for N-acetyl-β-D-glucosaminidase and β-D-galactosidase in human urine using newly-developed ω-nitrostyryl substrates. Clin Chim Acta 124:195–204
36. Yuen CT, Kind PRN, Price RG, Praill FG, Richardson AC (1984) Colorimetric assay for *N*-acetyl-β-D-glucosaminidase (NAG) in pathological urine using the ω-nitrostyryl substrate: the development of a kit and the comparison of manual procedure with the automatic fluorimetric method. Ann Clin Biochem 21:295–300

8.3 Cytosolic Enzymes: Lactate Dehydrogenase

A.S. Bhargava and P. Slabke

8.3.1 Introduction

The cytosolic enzyme lactate dehydrogenase (LDH, EC 1.1.1.27) catalyses the reversible reduction of pyruvate to lactate. This enzyme is known to occur in five different molecular forms (LDH_1–LDH_5 isoenzymes). The maximum activity of the enzyme LDH occurs in kidney when compared with other organs such as heart, skeletal muscle, pancreas, spleen, liver and lung on a wet organ weight basis in humans [22] and dogs [23]. The measurement of LDH activity in urine and its increase in several renal diseases as an indicator of renal cellular injury was first described by Rosalki and Wilkinson in 1959 [13] using a method developed by Wroblewski and La Due for serum analysis [23]. Several inhibitors of LDH activity are known to exist in human urine [1, 14, 15]. Furthermore, urea in concentrations which might be reached in urine has been shown to inhibit LDH activity in kidney homogenates [21].

In 1965 Güttler and Clausen [9] described a method for the determination of urinary LDH activity without dialysis of human urine samples, based on the reduction of pyruvate to lactate. Later, Dubach 1966 [6] and Raab 1967 [12] dialysed urine samples before measuring LDH activities in humans and rats.

Since the use of gel filtration or dialysis to remove inhibitors in urine is generally practised before measuring LDH activities (see Chap. 6, collection, storage and preparation of samples), established methods for the determination of serum LDH activity are also recommended for urine samples as the method for LDH activity in urine has not yet been optimized. However, an optimal technique is not to be expected as different ratios of individual isoenzymes of LDH present in the sample reach their maximum activity under different reaction conditions [5]. Furthermore, the activity is also inhibited by high concentrations of the substrate pyruvate [20].

Various methods for the determination of total LDH activities which differ in temperature, buffer solution and concentrations of pyruvate as substrate and of NADH have been recommended by several national scientific societies [3–5, 17].

8.3.2 Assay

LDH catalyzes the following reaction:

$$\text{pyruvate} + \text{NADH} + H^+ \rightarrow \text{lactate} + \text{NAD}^+$$

The LDH activity is determined by the rate of decrease of NADH in the above reaction. NADH is measured by the absorption at 334 nm.

The assay conditions correspond to the recommendation of the German Society for Clinical Chemistry 1972 [5] for normal human serum as described by Bergmeyer and Berndt [2]: phosphate buffer 50 mmol/l pH 7.5; pyruvate 0.6 mmol; NADH 0.18 mmol/l; incubation temperature 25°C.

Collection, storage and preparation of samples are described in Chap. 6.

Reagents and Solutions. All reagents must be of analytical quality, and re-purified water must be used.

1. Phosphate/pyruvate solution (52 mmol/l phosphate, pH 7.5; 0.63 mmol/l pyruvate): dissolve 778 mg dipotassium phosphate and 100 mg monopotassium phosphate in about 90 ml water, adjust to pH 7.5 (glass electrode) with 100 mmol/l NaOH or HCl, add water to make 100 ml and then add 6.9 mg sodium pyruvate.
2. Nicotinamide adenine dinucleotide, reduced form (11.3 mmol/l β-NADH): dissolve 14 mg $NADH–Na_2$ and 15 mg sodium bicarbonate in 1.5 ml water.

Measurement. Prepare the incubation mixture in cuvettes (d = 1.0 cm) according to the following scheme, and read or record absorbance changes in a spectrophotometer or spectrum line photometer at 334 nm, incubation temperature 25 °C, incubation volume 3.15 ml, sample volume 0.1 ml (sample volume fraction 0.032).

Pipette successively			Final concentration
Phosphate/pyruvate solution	(1)	3.00 ml	50 mmol/l; 0.6 mmol/l
NADH solution	(2)	0.05 ml	0.18 mmol/l
Sample		0.10 ml	
Mix; read absorbance and start stopwatch. Repeat the reading after exactly 1, 2 and 3 min or monitor the reaction on a recorder			

Calculation.

$$\text{LDH U/l} = \Delta A/\Delta t \times 10^3/6.18 \times 3.15/0.1 = \Delta A/\Delta t \times 5097 \ (\times \text{dilution factor})$$

where 6.18 is the millimolar absorption coefficient of NADH at 334 mm. The dilution factor encompasses the dilution by gel filtration, and by dialysis and dilution necessary because of high activity. ΔA/min should not exceed 0.100. For dilution, 0.15 mol/l NaCl is recommended.

8.3.3 LDH Isoenzymes

The diagnostic utility of LDH isoenzyme patterns in urine is still under discussion. The urine of healthy persons, rats and dogs was found to contain predominatly LDH_1–LDH_3 [7, 7, 21], reflecting mainly the isoenzyme patterns of cortical tubules, despite considerable species differences in the distribution of the isoenzymes between cortical, medullary and papillary structures [11, 18, 19]. Several methods such as gel electrophoresis [6], diethylaminoethyl cellulose column chromatography [8], spectrophotometry based on different stabilities of individual isoenzymes at alkaline pH [16] and high-performance liquid chromatography [10] have been used for these determinations.

References

1. Bartels H (1966) Zur Bestimmung der Lactatdehydrogenasektivität im Urin. Schweiz Med Wochenschr 96:1304–1305
2. Bergmeyer HU, Berndt E (1974) Lactat-Dehydrogenase. UV-Test mit Pyruvat und NADH. In: Bergmeyer HU (ed) Methoden der enzymatischen Analyse, vol. 1, 2nd edn. Verlag Chemie, Weinheim, pp 607–612
3. Commission Enzymologie de la Societe Francaise de Biologie Clinique: Recommendations for determining the catalytic concentration of lactate dehydrogenase in human serum at 30°C (1982) Ann Biol Clin 40:160–162
4. Committee on Enzymes, the Scandinavian Society for Clinical Chemistry and Clinical Physiology: Recommended methods for the determination of four enzymes in blood (1974) Scand J Clin Lab Invest 33:291-306
5. Deutsche Gesellschaft für Klinische Chemie: Recommendations of the German Society for Clinical Chemistry (1972) Z Klin Chem Klin Biochem 10,6:281–291
6. Dubach UC (1966) On the origin of lactic dehydrogenase isoenzymes in urine. Helv Med Acta 2:139–150
7. Emanuelli G, Cestonaro G, Anfossi G, Calcamuggi G, Gatti C, Marcarino C (1982) Urinary enzyme excretion and renal lactate dehydrogenase isoenzyme pattern in acute $HgCl_2$ nephropathy of rat. Enzyme 27:89–98
8. Gelderman AH, Gelboin HV, Peacock AC (1965) Lactic dehydrogenase isozymes in urine from patients with malignancies of the urinary bladder. J Lab Clin Med 65:132–142
9. Güttler F, Clausen J (1965) Urinary lactate dehydrogenase activity. Determination of the urinary LDH isozyme pattern as a supplement to the measurement of total urinary LDH activity. Enzymol Biol Clin 5:55–64
10. Matsukura H, Suzuki Y, Takai R, Okada T, Naiki S, Sakuragawa N (1987) Chromatographic profiles of urinary isoenzymes in healthy children. Clin Chim Acta 169:209–216
11. Mattenheimer H (1971) Enzymes in the urine. Med Clin North Am 55:1493–1508
12. Raab W (1967) Experimentelle Untersuchungen zur Steigerung der Laktatdehydrogenase-Aktivität der Harnes. Experientia 23:883
13. Rosalki SB, Wilkinson JH (1959) Urinary lactic dehydrogenase in renal disease. Lancet 2:327–328
14. Schoenenberger GA, Wacker WEC (1965) Isolation and Characterization of a lactic dehydrogenase (LDH) inhibitor of human urine. J Clin Invest 44:1096
15. Schoenenberger GA, Wacker WEC (1966) Peptide inhibitors of lactic dehydrogenase (LDH). II. Isolation and characterization of peptides I and II. Biochemistry 5:1375–1379
16. Takizawa N, Yamauti J, Marui Y, Hayashi C, Yamashita J, Horio T (1983) Spectrophotometric method for selective assay of the five isoenzymes of human lactate dehydrogenase, based on their different stabilities at alkaline pH. Clin Chem 29, 11:1941-1945

17. Talner LB, Rushmer HN, Coel MN (1972) The effect of renal artery injection of contrast material on urinary enzyme excretion. Invest Radiol 7:311–322
18. Thiele KG, Mattenheimer H (1966) Die Isoenzyme der Lactat-Dehydrogenase in der Rattenniere. Z Klin Chem 4:232–234
19. Thiele KG, Mattenheimer H (1968) The isoenzymes of lactate dehydrogenase in the nephron of the healthy human kidney. Z Klin Chem Klin Biochem 6:132–138
20. Vassault A (1983) Lactate dehydrogenase. UV-method with pyruvate and NADH. In: Bergmeyer HU, Bergmeyer J and Graßl M (eds.) Methods of enzymatic analysis, 3rd edn. Vol. III. Verlag Chemie, Weinheim, Deerfield Beach Fl, Basel, pp 118–126
21. Wilkinson JH (1968) Diagnostic significance of enzyme determinations in urine. In: Dubach UC (ed.) Enzymes in urine and kidney. Verlag Hans Huber Publishers Berne and Stuttgart, pp 207–227
22. Wroblewski F (1957) The clinical significance of lactic dehydrogenase activity in the milieu interieur. Scand J Clin Lab Invest 10,31: 230–253
23. Wroblewski F, La Due JS (1955) Lactic dehydrogenase activity in blood. Proc Soc Exp Biol Med 90:210–213

8.4. Kallikrein

K. Jung

8.4.1 Introduction

Two groups of kallikrein can be distinguished: the glandular or tissue group (tissue kallikrein, EC 3.4.21.8) and the plasma one [3]. The glandular form occurs in the pancreas, salivary glands, kidney and the respective body fluids, pancreatic juice, saliva and urine. Immunologic and biochemical studies on the glandular form have demonstrated a high degree of similarity if not identity of the kallikrein from these different organs and body fluids. The protein parts are identical. Differences have been found, however, in the carbohydrate parts [3, 9].

Renal kallikrein can be found in the highest concentrations in the distal tubules [12] (see Chap. 2). It is localized near the lumen as an ectoenzyme, in the endoplasmic reticulum and at the basolateral membrane [5, 14]. According to this distribution, renal kallikrein is released into the urine, lymph and venous blood. Thus, urinary kallikrein stems from the kidney. The glomerular filtration of glandular kallikrein does not contribute to the urinary kallikrein because it is completely catabolized in the proximal cells [10].

Urinary kallikrein has similar properties to renal kallikrein [11, 14]. Several preparations of human urinary kallikrein have been isolated [3], and the presence of at least two multiple forms of kallikrein in urine have been considered [14]. Three multiple types purified by isoelectric focussing were characterized [13]. The isoelectric point lies near 4.0. However, it is uncertain whether this heterogeneity is caused by a differing amino acid or glycoprotein composition. Depending on the carbohydrate content, the relative molecular mass of the single-chain α-kallikrein lies between 32 and 40 kDa [7] while the double-chain β-kallikrein is little less as it arises by limited proteolysis from α-kallikrein [4]. Urine and renal tissue also contain an inactive enzyme, prokallikrein, which can be activated by trypsin.

Urinary kallikrein belongs to the class of serin proteinases. It releases the kinin kallidin (lys-bradykinin) from the renal kininogen that is similar, if not identical, to the plasma low molecular mass kininogen. In contrast to the plasma kallikrein, urinary and renal kallikreins are inhibited by soybean trypsin inhibitor whereas all kallikrein types are inhibited by aprotinin.

Three kinds of procedures to measure kallikrein exist [6, 14]:

1. Indirect determination by measurement of the kinin generated by bioassay (the liberated kinin is measured by its ability to induce contraction of an isolated smooth muscle preparation), enzyme immunoassay, radioimmunoassay or high-performance liquid chromatography.
2. Direct determination of the protein part of kallikrein by radioimmunoassay or enzyme immunoassay [2, 5, 6, 14].
3. Determination of the esterase activity of kallikrein [1, 6, 14]. Synthetic substrates such as α-*N*-tosyl-L-arginine methyl ester, benzoyl-L-arginine ethylester or D-valyl-L-leucyl-L-arginine-4-nitroanilide have been used to measure its catalytic activity. The correlation between synthetic substrate assays and enzyme immunoassays are close to 1 [5]. The tripeptide substrate D-valyl-L-leucyl-L-arginine-4-nitroanilide offers a simple and convenient way of determining kallikrein also in a non-specialized laboratory [1, 2, 6]. The method described thereafter is based on this substrate.

The determination of urinary kallikrein is of interest as the kallikrein-kinin system is probably involved in the regulation of important renal functions, and impaired renal kallikrein activity may be a factor in the maintenance of essential hypertension [2, 6].

8.4.2 Assay

D-valyl-L-leucyl-L-arginine-4-nitroanilide + H_2O $\longrightarrow$ D-valyl-L-leucyl-L-arginine + 4-nitroaniline

The increase in absorbance of the liberated 4-nitroaniline is measured at 405 nm. The interference action of urokinase on this tripeptide is compensated for by a blank containing aprotinin which inhibits kallikrein and does not affect urokinase.

Amundsen et al. [1] introduced the method based on this chromogenic tripeptide substrate and investigated the influence of various reaction conditions. The activity of kallikrein depends especially on the substrate concentration; optimal concentrations are reached exceeding 0.5 mmol/l (Fig. 1). The K_m value is 30 μmol/l [11]. There is a broad pH range to give maximal activities (Fig. 2). Buffer concentrations within the range of 25–200 mmol/l do not

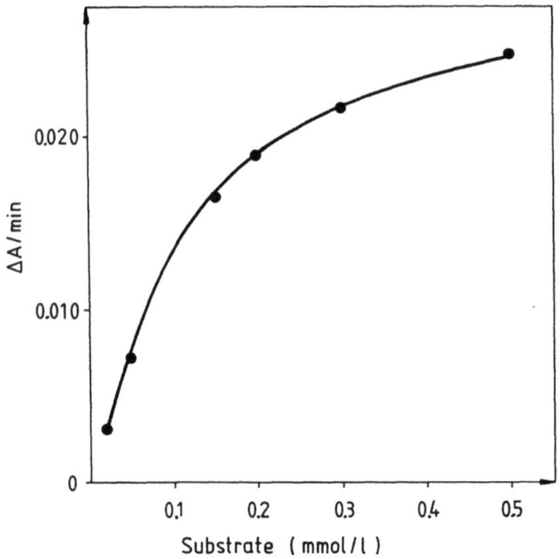

Fig. 1. Kallikrein activity depending on the concentration of D-valyl-L-leucyl-L-arginine-4-nitroanilide

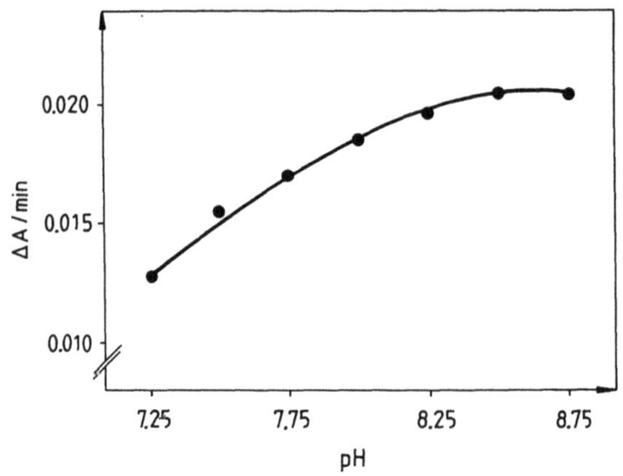

Fig. 2. Kallikrein activity depending on the pH of the reaction mixture

influence the activity. Considering these reaction conditions, the influence of substrate concentration on the linearity of the reaction (Fig. 3) and the cost of chromogenic substrate, final concentrations of 0.3 mmol/l chromogenic substrate, 100 mmol/l TRIS/HCl buffer and pH 8.5 are recommended for the test.

Because of the low activity of kallikrein in urine a high proportion of sample in the reaction mixture is needed for a reliable determination. Thus, to avoid turbidity and pH instability in the reaction mixture, gel-filtered samples should be used. In addition, gel filtration removes urinary inhibitors of kallikrein [8]. *Collection, storage and preparation of samples* are as described in Chap. 6. Urine samples stored at 4 °C and − 20 °C did not lose their kallikrein activity over 7 days and 14 months, respectively [2].

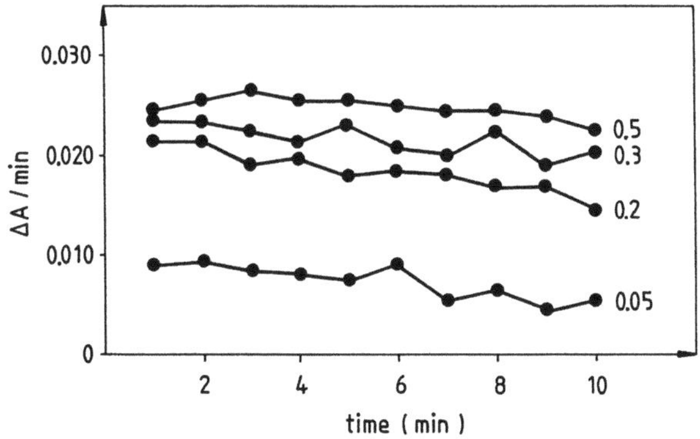

Fig. 3. Linearity of the assay depending on the substrate concentrations. Numbers indicate the concentration of D-valyl-L-leucyl-L-arginine-4-nitroanilide used (0.05 – 0.5 mmol/l)

Reagents and Solutions. All reagents must be of analytical grade, and deionised water must be used.

1. TRIS/HCl buffer (1.2 mol/l, pH 8.5): dissolve 14.54 g of TRIS is about 80 ml water, adjust to pH 8.5 with 1 mol/l HCl and dilute with water to 100 ml.
2. Aprotinin solution (kallikrein inhibitor, 600 000 units/1; TRIS/HCl buffer, 1.2 mol/l): dissolve 24 000 kallikrein inhibitor units in 100 ml of solution (1). Trasylol (Bayer AG, Leverkusen, FRG) can be used as aprotinin.
3. D-Valyl-L-leucyl-L-arginine-4-nitroanilide solution (3.6 mmol/l): dissolve 20.9 mg of substrate S-2266 (Kabi Diagnostica, Stockholm, Sweden or Feinbiochemica, Heidelberg, FRG) in 10 ml water.

Measurement. Prepare incubation mixture (reaction solutions plus gel-filtered urine samples) in cuvettes (d = 1.0 cm) and read absorbance changes at 405 nm, incubation temperature 37 °C, final volume 600 μl, sample volume fraction 0.83.

Pipette successively	Sample assay (μl)	Blank assay (μl)	Final concentration
TRIS/HCl buffer (1)	50	—	TRIS/HCl 100 mmol/l
Aprotinin solution (2)	—	50	aprotinin 20 000 KIU/1
Sample	500	500	
Mix; preincubate for 10 min at 37 °C			
Substrate solution (3)	50	50	0.3 mmol/l
Mix and read absorbance after 10 min			

ΔA/min should not exceed 0.02. For higher activities, 154 mmol/l NaCl solution is used for sample dilution.

Calculation. Correct readings for sample by blank readings and calculate kallikrein activity according to the following equation:

$$U/l = \frac{\Delta A/\min \times 10^3 \times 0.600 \times DF}{9.9 \times 0.500} = \Delta A/\min \times 121.2 \times DF$$

where 9.9 is the millimolar absorption coefficient of 4-nitroaniline at 405 nm and DF the dilution factor by gel filtration.

8.4.3 Reference Ranges

The kallikrein activity measured with the described method in random morning urine samples of 24 healthy adults (12 females, 12 males) amounted to between 0.22 and 2.79 U/l (mean value $\pm$ SD = 1.07 $\pm$ 0.59 U/l). The time-related excretion rates were between 13 and 92 mU/h (38.5 + 18.4 mU/h) and the excretion rates related to creatinine between 21.7 and 182 U/mol creatinine (80.9 $\pm$ 39.8 U/mol creatinine). Using a similar method (final concentration of D-valyl-L-leucyl-L-arginine-4-nitroanilide of 0.15 mmol/l) Bönner [2] found kallikrein excretion rates per 24 h between 0.42 $\pm$ 0.05 and 1.01 $\pm$ 0.13 U (arithmetic means $\pm$ SEM) depending on sex and the respective age group investigated.

References

1. Amundsen E, Putter J, Friberger P, Knos M, Larsbraten M, Claeson G (1979) Methods for the determination of glandular kallikreins by means of a chromogenic tripeptide substrate. Adv Exp Med Biol 120 A:83–95
2. Bönner G (1988) Das renale Kallikrein-Kinin-System. Springer, Berlin Heidelberg New York
3. Fiedler F (1979) Enzymology of glandular kallikreins. In: Erdös EG (ed) Bradykinin, kallidin, and kallikrein. Springer, Berlin Heidelberg New York, pp 103–161 (Handbook of experimental pharmacology, vol 25, suppl)
4. Fiedler F, Gebhard W (1980) Isolation and characterization of native single-chain porcine pancreatic kallikrein, another possible precursor of urinary kallikrein. Hoppe-Seyler's Z Physiol Chem 361:1661–1671
5. Franke M, Rohrschneider S, Geiger R (1982) Enzyme immunoassay of human urinary kallikrein. J. Clin Chem Clin Biochem 20:621–626
6. Geiger R (1985) Kallikrein. In: Bergmeyer HU (ed) Methods of enzymatic analysis, vol 5. Verlag Chemie, Weinheim, pp 129–143
7. Geiger R, Stuckstedte U, Fritz H (1980) Isolation and characterization of human urinary kallikrein. Hoppe-Seyler's Z Physiol Chem 361:1003–1016
8. Lögering HJ, Paar D, Maruhn D (1983) Kallikreinausscheidung im Nativharn und Harn nach Gelfiltration bei Normalpersonen. Med Welt 34:690–691
9. Margolius HS (1984) The kallikrein-kinin-system and the kidney. Annu Rev Physiol 46:309–326
10. Mills IH, Paterson CL, Ward PE (1975) The role of the kidney in the inactivation of injected ^{125}J-kallikrein. J Physiol (Lond) 251:281–286
11. Nustad K (1970) The relationship between kidney and urinary kininogenase. Br J Pharmacol 39:73–86
12. Omata K, Carretero OA, Scicli AG, Jackson BA (1982) Localization of active and inactive kallikrein (kininogenase activity) in the microdissected rabbit nephron. Kidney Int 22:602–607

13. Pierce JV, Nustad K (1972) Purification of human and rat urinary kallikrein. Fed Proc 31:623A
14. Ward PE, Margolius HS (1979) Renal and urinary kallikreins. In: Erdös EG (ed) Bradykinin, kallidin, 2nd kallikrein. Springer, Berlin Heidelberg New York, pp 525–542 (Handbook of experimental pharmacology, vol 25, suppl)

8.5 Low Molecular Mass Proteins with Enzymatic Properties

8.5.1 Lysozyme

K. Jung

8.5.1.1 *Introduction*

Human lysozyme (mucopeptide N-acetylmuramoyl hydrolase, EC 3.2.1.17) consists of a basic polypeptide chain of 130 amino acids with an isoelectric point of about pH 10.5. Since its relative molecular mass is only around 15 kDa, it belongs to the class of low molecular mass proteins [12, 16].

Lysozyme catalyzes the hydrolysis of the peptidoglycan layer of the bacterial cell wall. It splits the $\beta(1-4)$ glycosidic linkage between *N*-acetylmuramic acid and *N*-acetyl-D-glucosamine and forms dimers consisting of these two saccharides. Chitin, an oligosaccharide made up of components of *N*-acetyl-D-glucosamine, is also cleaved, but the hydrolysis rate is only about a tenth of that of polysaccharides from the bacterial cell wall [9]. Six reactive sites are described in the lysozyme molecule necessary for the formation of an enzyme-substrate complex and the subsequent catalysis.

Lysozyme occurs in neutrophilic granulocytes, monocytes and macrophages, and it is found in large quantity in many organs, including kidney, lung, spleen and bowel [10]. It is released from leucocytes into the serum and then freely filtered through the glomerular wall of the kidney into the ultrafiltrate. During the passage through the nephron, lysozyme is absorbed by the proximal tubular cells and is catabolized by lysosomes [2, 3]. This is the so-called reabsorptive-catabolic function of the kidney. Since this process is very efficient, only small concentrations of lysozyme (about a tenth of the serum concentration) are found in the urine of healthy people. Urinary lysozyme increases (1) when the mentioned reabsorptive-catabolic function of proximal tubule cells is impaired (e.g. kidney diseases) and (2) when the ultrafiltrate, due to an excessive endogenous lysozyme synthesis (monocytic leukaemia, inflammatory bowel disorder) contains more lysozyme than the tubule cells are capable of processing (overflow lysozymuria).

Lysozyme activity can be determined by immunological methods like electroimmunodiffusion [8], radioimmunoassays [15, 17] and enzyme immunoassays [1, 13], or by fluorimetric [9], colorimetric [9, 11], turbidimetric [e.g. 4, 5, 7], lysoplate [6] and HPLC [14] methods. Radiommunoassays, enzyme immunoassays and high-performance liquid chromatography methods are very sensitive; concentrations up to 1 μg/l can be measured [1, 13–15]. However, these methods are either cumbersome or require sophisticated equipment. Thus, turbidimetric methods are preferred in routine practice because of their simplicity and rapidity. These methods for determining lysozyme activity are based on the ability of lysozyme to reduce the turbidity of a bacterial suspension by hydrolyzing the glycosidic bonds in the polysaccharides of the cell wall. The change of turbidity can be measured either nephelometrically or photometrically [9]. The method described thereafter is based on this principle [4].

8.5.1.2 *Assay*

Lysozyme splits the glycosidic bonds on cell wall components of a suspension of lyophilised *Micrococcus lysodeikticus*. Through this, the turbidity of the suspension is reduced, which can be recorded at 365 nm. The decrease of turbidity parallels the enzyme activity in the sample.

Optimal analytical factors such as pH, kind of buffer, buffer concentration, ionic strength and incubation temperature of the assay mixture are interdependent [7]. At the reaction temperature of 37 °C, 70 mmol/l of phosphate buffer, pH 6.2 and 250 mg/l of *Micrococcus lysodeikticus* cells may be considered the optimum reaction conditions (Fig. 1) [4]. The analytical sensitivity and linearity of the reaction are better if the decrease of turbidity is monitored at a shorter wavelength. Measurement at 365 nm is recommended.

Various available commercial preparations of *Micrococcus lysodeikticus* and even different batches obtained from the same producer differ in their susceptibilty to hydrolysis by lysozyme [4, 5, 7]. Hereby, the sensitivity and linearity of the assay are influenced. It must be concluded that the declaration of lysozyme activity based on absorbance units is not suitable [5]. The absorbance changes must be related to a lysozyme standard. Different standards (egg-white, bovine and human lysozyme) are available which differ in their catalytic activity, e.g. human lysozyme is 2–12 times more active than lysozyme from egg white, but most reported values, as in the following method, are given as concentrations equivalent to egg-white lysozyme.

Collection, storage and preparation of sample are as described in Chap. 6.

Reagents and Solutions. All reagents must be of analytical quality, and deionized water must be used.

1. Phosphate buffer solution (73 mmol/l, pH 6.2): dissolve 8.085 g KH_2PO_4 and 2.384 g $Na_2HPO_4 \cdot 2H_2O$ in about 950 ml water, adjust to pH 6.2 with 100 mmol/l HCl and fill up to 1000 ml with water. The solution is stable for 4 weeks at 4 °C.
2. *Micrococcus lysodeikticus* substrate (260 mg/l): suspend 13 mg of freeze-dried *Micrococcus lysodeikticus* (e.g. from Boehringer Mannheim, FRG) in 50 ml phosphate buffer solution (1). The solution is stable for 5 days at 4 °C and 2 weeks at − 20 °C.

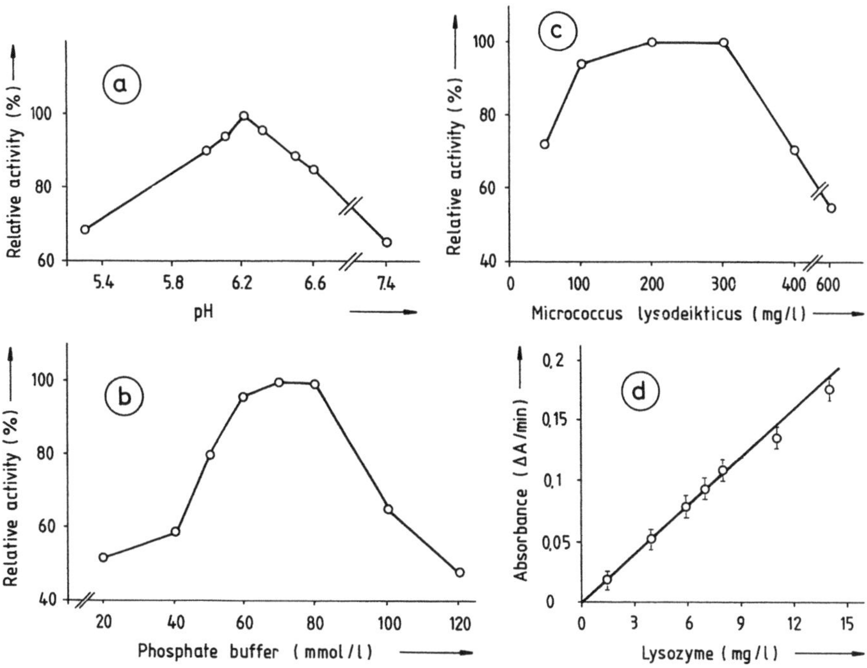

Fig. 1. **a** Effect of pH **b**, buffer concentration and **c** *Micrococcus* concentration on lysozyme activity, and **d** calibration curve

3. Lysozyme standard solutions: dissolve 5 mg egg-white lysozyme (e.g. from Worthington, Freehold, USA) in 50 ml phosphate buffer solution (1) and mix 0.1, 0.2, 0.4, 0.6 and 0.8 ml of this stock solution with 9.9, 9.8, 9.6 and 9.2 ml water, respectively. These solutions correspond to lysozyme concentrations of 1, 2, 4, 6, 8 and 10 mg/l, respectively, and should be used within 4 h

Measurement. Prepare incubation mixture in cuvettes of a holder with thermostat (d = 1.0 cm) and monitor absorbance changes on a recorder in a spectrophotometer at 365 nm against air, incubation temperature 37 °C, incubation volume 520 μl, sample volume fraction 0.0385. Centrifuged urine (not gel filtrate) is used as sample.

Pipette successively	Sample assay (μl)	Standard assay (μl)	Final concentrations
Micrococcus suspension (2)	500	500	phosphate 70 mmol/l
Preincubate for 10 min at 37 °C			*Micrococcus* 250 mg/l
Sample	20	—	
Standard	—	20	
Mix; monitor absorbance for 90 s on a recorder			

$\Delta A/\text{min}$ should not exceed the value obtained with the standard solution of 10 mg/1. For higher activities, water is used for sample dilution.

It is necessary to measure standard solutions for the construction of a calibration curve only if newly prepared *Micrococcus* suspensions are used.

Calculation. Calculate $\Delta A/\text{min}$ for standard solutions and sample from the linear part of the recorded reaction curve, construct a calibration curve and read the concentration of lysozyme in the sample from it (Fig. 1d).

8.5.1.3 *Reference Ranges*

The lysozyme excretion was measured with the described method in urine samples collected between 8 and 12 h a.m. from 58 healthy adults (26 female, aged between 17 and 54 years; 32 male, aged between 17 and 53 years). The median values and the 99th percentile as cut-off values are given in Table 1.

Table 1. Urinary lysozyme excretion in healthy adults (according to [4])

Reference basis	Sex	Percentiles	
		50 (median)	99
μg/l	M	92	335
	F	205	480
μg/4 h	M	13	70
	F	42	104
μg/mmol creatinine	M	5.8	38.4
	F	18.9	64.2

References

1. Brouwer J, van Leeuwen-Herberts T, Otting-van de Ruit M (1984) Determination of lysozyme in serum, urine, cerebrospinal fluid and feces by enzyme immunoassay. Clin Chim Acta 142:21–30
2. Christensen EL, Maunsbach AB (1974) Intralysosomal digestion of lysozyme in renal proximale tubule cells. Kidney Int 6:396–407
3. Cojocel C, Baumann K (1983) Renal handling of endogenous lysozyme in the rat. Renal Physiol 6:258–265
4. Diego J, Jung K (1982) Methodische Untersuchungen zur Lysozymbestimmung im Urin mit einem kinetischen Verfahren. Z Med Lab Diagn 23:13–19
5. Gorin G, Wang SF, Papapavlou L (1971) Assay of lysozyme by its lytic action on *M. lysodeikticus* cells. Anal Biochem 39:113–127
6. Gupta DK, von Figura K, Hasilik A (1987) Conditions for a reliable application of the lysoplate method in the determination of lysozyme. Clin Chim Acta 165:73–82
7. Houser MT (1983) Improved turbidimetric assay for lysozyme in urine. Clin Chem 29:1488–1493
8. Johannson BG, Malmquist J (1971) Quantitative immunochemical determination of lysozyme (muramidase) in serum and urine. Scand J Clin Lab Invest 27:255–261

9. Klass HJ, Hopkins J, Neale G, Peters TJ (1977) The estimation of serum lysozyme: a comparison of four assay methods. Biochem Med 18:52–57
10. Mason DY, Taylor CR (1975) The distribution of muramidase (lysozyme) in human tissue. J Clin Pathol 28:124–132
11. Nanjo F, Sakai K, Usui T (1988) p-Nitrophenyl penta-N-acetyl-β-chitopentasoide as a novel synthetic substrate for the colorimetric assay of lysozyme. J Biochem 104:255–258
12. Osserman EF, Canfield RE (1974) Lysozyme. Academic, New York
13. Porstmann B, Jung K, Schmechta H, Evers U, Pergande M. Porstmann T, Kramm HJ, Krause, H (1989) Measurement of lysozyme in human body fluids: comparison of various enzyme immunoassay techniques and their diagnostic application. Clin Biochem 22:349–355
14. Severini G, Aliberti LM (1987) Diagnostic significance of urinary enzymes: development of a high performance liquid chromatographic method for the measurement of urinary lysozyme. Clin Chim Acta 163:97–103
15. Shima K, Meisei H, Fukuda M, Tanaka A (1986) Determination of urinary lysozyme for potential detection of tubular dysfunction in diabetic nephropathy. Clin Chem 32:1818–1822
16. Stelzner A, Werle M (1989) Die immunologische und klinische Bedeutung des Lysozyms. Z Klin Med 44:261–264
17. Thomas MJ, Russo A, Craswell P, Ward M, Steinhardt I (1981) Radioimmunoassay for serum and urinary lysozyme. Clin Chem 27:1223–1226

8.5.2 Ribonuclease

K. Jung

8.5.2.1 *Introduction*

Ribonucleases (RNAase, ribonucleate 3′-pyrimidino-oligonucleotidohydrolase, EC 3.1.4.22) are enzymes which are ubiquitously distributed in human organs. They catalyze the depolymerization of RNA. On the basis of their different pH optima, they are often classified into basic and acidic forms. This classification should be avoided because the type of buffer used for the determination decisively influences the pH optimum of RNAse. Characterization of RNAase by substrate and immunological specificity has been suggested instead [1, 8].

According to this suggestion, two distinct classes of RNAses have been established, the pancreas-type RNAase isolated from the pancreas, kidney, lung, brain, heart, serum and urine, and the liver-spleen type, isolated from liver, spleen, granulocytes, serum and urine [8]. In human urine, these two major RNAases were purified [2] and termed RNAase C and RNAase U according to their different splitting activity against poly C and poly U. RNAase C exhibited a strong preference for poly C. This enzyme is 100- and 15-fold less active against RNA and poly U, respectively. In contrast, RNAase U hydrolyzes poly C very poorly and is 30- and three-fold more active against RNA and poly U, respectively. There is no antigenic cross-reaction between the two RNAase types [7].

Poly C-specific RNAase activity constitutes the main part of RNAase activity in human urine. This enzyme consists of a polypeptide chain of 126 amino acids and about 30% carbohydrates [4]. It can be resolved into multiple components by column electrofocussing and polyacrylamide gel electrophoresis whereas RNAase U as the minor component focusses as a single component [2].

Prior treatment of RNAase C with neuraminidase results in a single peak of enzyme, indicating that the charge heterogeneity is the consequence of a varying content of sialic acid residues. The variability of the carbohydrate portion caused by different isolation procedures may explain the range in molecular mass (14–44 kDa) observed [4, 9, 12]. The RNAase C apoprotein (without carbohydrates) has a relative molecular mass of 14 kDa [2].

Comparative studies of urinary and serum RNAase C [3, 10, 12] suggest that this enzyme reaches the urine as a component of filtered plasma and behaves as a low molecular mass protein (see Chap 3). Because the urinary RNAase does not entirely correspond to the serum forms, it is supposed that the serum RNAase is changed during the passage through the nephron [12]. Kidney tissue and the urinary tract have to be considered as additional sources of RNAase [12].

Various methods have been employed for RNAase determination: spectrophotometric methods with different substrates (RNA from yeast, transfer RNA from yeast and *Escherichia coli*, different polynucleotides), methods using radiolabelled substrates and radioimmunoassays [1,, 3, 5, 6, 10, 12, 13]. In routine diagnostics, spectrophotometric methods are of especial interest. In comparison with naturally occurring substrates, synthetic substrates offer advantages for the standardisation of methods. Polycytidylic acid (poly C) [11, 13] has been recommended for this purpose. It enables the determination of the main RNAase form in urine. The method described hereafter is based on this substrate [5].

8.5.2.2 *Assay*

The determination of RNAase is based on the principle that poly C is hydrolysed by this this enzyme during a fixed time incubation. After precipitation of undegraded substrate by perchloric acid-lanthanum nitrate solution, the absorbance increase of supernatant due to the release of acid-soluble oligonucleotides is measured at 278 nm. The activity is calculated using the absorption coefficient for cytidine 3′-monophosphate.

Reaction conditions given here are based on optimum conditions for serum pancreas-type RNAase which are also acceptable for the measurement of the poly C-specific form of RNAase in urine [5]. The method is optimized with regard to all analytical variables which influence both the reaction rate of RNAase (type of buffer, buffer concentration, concentration of poly C, incubation time) and the measurement of the reaction products [5].

Recommended final reaction concentrations are 0.3 mM poly C, 200 m*M* imidazole/HCl buffer, pH 7.0, and 50 m*M* NaCl. Optimal concentrations for the precipitation procedure, guaranteeing sufficient precipitation and minimal decomposition of unreacted substrate, are 160 m*M* perchloric acid and 4 m*M* lanthanum nitrate [5]. The method shows a linear response for RNAase activity up to a net absorbance of about 0.900.

Collection, storage and preparation of samples are as described in Chap. 6. Samples stored at room temperature, 4°C or —20°C did not lose their RNAase activity over 5 days [5]. However, it is recommended that samples should be stored at − 20°C when not tested within 3 days.

Because high activities of RNAase are found, the urine samples must be diluted (1:2500 with normal urine) preferably with 150 mmol/l NaCl solution.

Reagents and Solutions. All reagents must be of analytical quality, and deionized water must be used. All glassware and plastic ware (tubes, tips) must be free from ribonuclease (sterilization or boiling, 15 min).

1. Imidazole/HCl buffer (300 mmol/l, pH 7.0): dissolve 2.044 g imidazole in about 80 ml water, adjust to pH 7.0 with 1 mol/l HCl and dilute with water to 100 ml. Autoclave this solution. This buffer is stable at least 1 year at − 20°C and 4 weeks at 4°C. Only imidazole not containing UV-absorbing components should be used; the absorbance of this solution should be < 0.100.
2. Substrate solution (poly C, 0.45 mmol/l; imidazole/HCl buffer, 300 mmol/l, pH 7.0): dissolve 25 mg poly C in imidazole/HCl buffer (1), pipette 50 μl portions in microcentrifuge tubes and store at − 20°C. The solution is stable for at least 2 weeks.
3. Stop solution (perchloric acid, 200 mmol/l; lanthanum nitrate, 5 mmol/l): dissolve 541 mg $La(NO_3)_3 \cdot 6\,H_2O$ in 200 ml water, add 4.3 ml perchloric acid (1.67 kg/1) and dilute to 250 ml with water. This solution is stable at least 4 weeks at 4°C.
4. Diluent (sodium chloride solution, 150 mmol/l): dissolve 87.7 g sodium chloride in water and dilute to 1 l. Autoclave this solution.

Measurement. Prepare incubation mixture at 37°C by adding diluted samples to substrate solution (1) and read absorbance in the supernatant after precipitation at 278 nm in cuvettes of 1 cm light path. Stop solution must be used ice-cold. Incubation volume is 75 μl, sample volume fraction 0.333, final volume after stopping 375 μl.

Pipette successively into tubes	Sample (μl)	Reagent blank (μl)	Final concentrations
Substrate solution (2)	50	50	poly C 0.3 mmol/l imidazole 300 mmol/l
Preincubate 10 min at 37°C			
Diluted sample	25	—	
Diluent (4)	—	25	
Mix; incubate 15 min at 37°C			
Stop solution (3)	300	300	lanthanum nitrate 4 mmol/l perchloric acid 160 mmol/l
Mix; place the mixture in an icebath for 30 min, then centrifuge for 12 min at 4000 × *g* at 4°C. Remove 250μl of supernatant and read the absorbance within 15 min			

If the corrected absorbance exceeds 0.900, a greater dilution of samples must be used.

Calculation. Calculate RNAase activities according to the general equation:

$$\mathrm{kU/l} = \frac{\Delta A \times \text{final volume (ml)} \times 10^3 \times \mathrm{DF}}{\varepsilon \times \text{reaction time(min)} \times \text{sample volume(ml)} \times \text{light path(cm)}}$$

where ΔA = absorbance of the sample assay minus absorbance of the reagent blank

DF = dilution factor for urine sample

ε = 13.0, millimolar absorption coefficient of cytidine 3′-monophosphate

If DF and reaction time are 2501 and 15, respectively, the following formula will result:

$$\mathrm{kU/l} = \frac{\Delta A \times 0.375 \times 10^3 \times 2501}{13 \times 15 \times 0.025 \times 1} = \Delta A \times 192.4$$

8.5.2.3 *Reference Ranges*

The RNAase activity was measured with the described method in urine samples collected between 8 and 12 h a.m. in 81 healthy adults (43 men, 38 women). The values for all the conventional reference bases (kU/l, kU/time, kU/mmol creatinine) were normally distributed. The median values and 97.5th percentile as cut-off values are given in Table 1.

Table 1. Urinary RNAase activity in healthy adults. The percentiles were calculated on the basis of a normal distribution of the values. Only the time-related excretion differed between women and men and is separately indicated

Reference basis		Percentiles	
		50 (arithmetic mean)	97.5
kU/l		46.3	102
kU/4 h	women	5.7	10.7
	men	9.6	17.3
kU/mmol creatinine		5.43	6.56

References

1. Corbishley TP, Johnson PJ, Williams R (1984) Serum ribonuclease. In: Bergmeyer HU (ed) Methods of enzymatic analysis, vol. 4 Verlag Chemie, Weinheim, pp 133–143

2. Cranston JW, Perini F, Crisp ER, Hixson CV (1980) Purification and properties of ribonucleases from human urine. Biochim Biophys Acta 616:239–258
3. Cranston JW, Hoover HC, Crisp ER (1981) Immunological studies on ribonuclease C from human urine. Clin Chem 27:1362–1367
4. Iwama M, Kunihiro M, Ohgi K, Irie M (1981) Purification and properties of human urine ribonucleases. J Biochem 89:1005–1016
5. Jung K (1986) An optimized micromethod for determining the catalytic activity of serum ribonuclease. J Clin Chem Clin Biochem 24:243–250
6. Lee CC, Trotman CNA, Tate WP (1983) A ribonuclease assay: separation of product from undegraded aminoacyltransfer RNA substrate. Anal Biochem 135:64–68
7. Neuwelt EA, Schmukler M, Niziak MS, Jewett PB, Levy CC (1977) The immunological characterization of several human ribonucleases by using primary binding tests. Biochem J 163:419–426
8. Neuwelt EA, Boguski MS, Frank JJ, Procter-Appich K, Levy C (1978) Possible sites of origin of human plasma ribonucleases as evidenced by isolation and partial characterization of ribonclease from several human tissues. Cancer Res 38:88–93
9. Rabin EZ, Weinberger V (1975) The isolation, purification, and properties of a ribonuclease from normal human urine. Biochem Med 14:1–11
10. Rabin EZ, Weinberger V, Tattrie B (1977) Ribonuclease activity in human serum, cerebrospinal fluid, and urine. Clin Chim Acta 78:235–242
11. Reddi KK (1978) Serum ribonuclease of normal persons and patients with renal impairment. Clin Biochem 11:133–134
12. Thomas JM, Hodes ME (1981) Isoenzymes of ribonuclease in human serum and urine. I. Methodology and a survey of a control population. Clin Chim Acta 111:185–197
13. Zimmerman SB, Sandeen G (1965) A sensitive assay for pancreatic ribonuclease. Anal Biochem 10:444–449

8.6 Other Enzymes

H. Mattenheimer and D. Maruhn

The detailed descriptions of methods in the preceding sections cover the more commonly used enzymes. As evidenced throughout this book, however, several investigators have applied other enzymes; the assay conditions for some of these are briefly delineated in this section. More detailed information may be obtained from the literature cited.

8.6.1 Adenylate Kinase (EC 2.7.4.3)

$$\text{AMP} + \text{ATP} \xrightleftharpoons{\text{AK}} 2\ \text{ADP}$$

$$2\ \text{ADP} + 2\ \text{PEP} \xrightleftharpoons{\text{PK}} 2\ \text{ATP} + 2\ \text{pyruvate}$$

$$2\ \text{pyruvate} + 2\ \text{NADH} + 2\ \text{H}^+ \xrightleftharpoons{\text{LDH}} 2\ \text{L-lactate} + 2\ \text{NAD}^+$$

Assay conditions [5]: triethanolamine buffer pH 7.6, 50 mmol/l; AMP 1.3 mmol/l; ATP 1.3 mmol/l; phosphoenolpyruvate (PEP) 0.3 mmol/l; $MgSO_4$ 13 mmol/l; KCl 43 mmol/l; NADH 0.23 mmol/l; dithiothreitol 1 mmol/l; pyruvate kinase 2.7 kU/l; lactate dehydrogenase 15 kU/l.

8.6.2 Alanine Aminotransferase (EC 2.6.1.2)

L-alanine + 2-oxoglutarate $\rightleftharpoons_{ALAT}$ L-glutamate + pyruvate

pyruvate + NADH + H^+ $\rightleftharpoons_{LDH}$ lactate + NAD^+

Assay conditions [10]: TRIS buffer pH 7.3, 100 mmol/l; L-alanine 500 mmol/l; 2-oxoglutarate 15 mmol/l; NADH 0.18 mmol/l; pyridoxal phosphate 0.11 mmol/l; lactate dehydrogenase $\geq$ 1200 U/1.

8.6.3 Angiotensin I Converting Enzyme (EC 2.4.15.1)

Glycyl-L-histidyl-glycine → glycine + L-histidyl-glycine

The amount of histidyl-glycine formed is measured in a color reaction in an alkaline solution with *O*-phthalaldehyde as chromogen.

Assay conditions [2]: TRIS-HCl buffer pH 8.0, 818 mmol/l; glycyl-L-histidyl-glycine 0.91 mmol/l.

8.6.4 Aspartate Aminotransferase (EC 2.6.1.1)

L-aspartate + 2-oxoglutarate $\rightleftharpoons_{ASAT}$ oxaloacetate + L-glutamate
2-oxaloacetate + NADH + H^+ $\rightleftharpoons_{MDH}$ L-malate + NAD^+

Assay conditions [17]: TRIS buffer pH 7.8, 91 mmol/l; L-aspartate 240 mmol/l; NADH 0.16 mmol/l; 2-oxoglutarate 12 mmol/l; pyridoxal phosphate 0.11 mmol/l; lactate dehydrogenase 420 U/l (25°C); malate dehydrogenase 930 U/l (25°C).

The mitochondrial enzyme (m-ASAT) may be determined by an immunochemical method [7, 16].

8.6.5 Arylsulfatase A (EC 3.1.6.1)

Potassium-4-nitrocatechol sulphate → 4-nitrocatechol + sulphate

The amount of 4-nitrocatechol formed is measured in alkaline solution at 510 nm.

Assay conditions [15]: acetate buffer pH 5.0, 250 mmol/l; sodium pyrophosphate 0.25 mmol/l; NaCl 855 mmol/l; potassium-4-nitrocatechol sulfate 5 mmol/l. The enzyme may also be determined by a radioimmunoassay [10a].

8.6.6 Dipeptidyl Aminopeptidase IV (EC 3.4.14.5)

Glycyl-proline-4-nitroanilide → glycyl-proline + 4-nitroaniline

The amount of liberated 4-nitroaniline is measured at 405 nm.

Assay conditions [19]: TRIS-HCl buffer pH 8.7, 50 mmol/l; glycyl-proline-4-nitroanilide 5 mmol/l.

8.6.7 *β*-Galactosidase (EC 3.2.1.23)

4-nitrophenyl-*β*-D-galactopyranoside → 4-nitrophenol + *β*-D-galactose

The amount of 4-nitrophenol liberated is measured in alkaline solution at 405 nm.

Assay conditions [12]: citrate buffer pH 4.0, 50 mmol/l; 4-nitrophenyl-*β*-D-galactopyranoside 2.5 mmol/l.

8.6.8 *α*-Glucosidase (EC 3.2.1.20)

4-nitrophenyl-*α*-D-glucopyranoside → 4-nitrophenol + *α*-D-glucose

The amount of 4-nitrophenol liberated is measured in alkaline solution at 405 nm.

Assay conditions [6]: acetate buffer pH 4.2, 130 mmol/l; 4-nitrophenyl-*α*-D-glucopyranoside 10 mmol/l; KCl 10 mmol/l.

8.6.9 *β*-Glucuronidase (3.2.1.31)

4-nitrophenyl-*β*-D-glucuronide → 4-nitrophenol + *β*-D-glucuronate

The amount of 4-nitrophenol liberated is measured in alkaline solution at 405 nm.

Assay conditions [11]: acetate buffer pH 4.5, 100 mmol/l; 4-nitrophenyl-*β*-D-glucuronide 10 mmol/l.

8.6.10 Glutamate Dehydrogenase (EC 1.4.1.2)

$$\text{2-oxoglutarate} + \text{NADH} + NH_4^+ \rightleftharpoons \text{L-glutamate} + NAD^+ + H_2O$$

Assay conditions [18]: triethanolamine buffer pH 7.6, 50 mmol/l; NADH 0.2 mmol/l; ADP 1 mmol/l; EDTA 2.5 mmol/l; ammonium acetate 100 mmol/l; lactate dehydrogenase 2 kU/1; 2-oxoglutarate 7 mmol/l. The pH of the buffer depends on assay temperature: 7.6 at 37°C. 7.8 at 30°C, 8.0 at 25°C.

8.6.11 Glutathione Transferase (EC 2.5.1.18)

1-chloro-2,4-dinitrobenzene + glutathione → *S*-(2-chloro-4-nitrophenyl)-glutathione + HCl

Assay conditions [9]: phosphate buffer pH 7.5, 100 mmol/l; 1-chloro-2,4-dinitrobenzene 1.0 mmol/l; glutathione 5.0 mmol/l. Condition for assay at 25°C; measurement of change in absorbance at 345 nm. The enzyme may also be determined by a radioimmunoassay [1, 3].

8.6.12 Malate Dehydrogenase (EC 1.1.1.37)

$$\text{oxaloacetate} + \text{NADH} + \text{H}^+ \rightleftharpoons \text{L-malate} + \text{NAD}^+$$

(oxaloacetate is instable and, therefore, generated in the aspartate aminotransferase reaction)

$$\text{L-aspartate} + \text{2-oxoglutarate} \xrightarrow{\text{ASAT}} \text{L-glutamate} + \text{oxaloacetate}$$

Assay conditions [4, 8]: phosphate buffer pH 7.4, 91 mmol/l; L-aspartate 38 mmol/l; 2-oxoglutarate 1.2 mmol/l; NADH 0.22 mmol/l; aspartate aminotransferase 3.3 kU/l.

8.6.13 Trehalase (EC 3.2.1.28)

$$\alpha,\alpha\text{-trehalose} \rightleftharpoons 2\ \text{D-glucose}$$

Liberated glucose is measured with a glucose oxidase-peroxidase reaction.

Assay conditions [13, 14]: phosphate buffer pH 6.2, 80 mmol/l; trehalose 160 mmol/l.

References

1. Bäckman L, Appelkvist EL, Ringdén O, Dallner, G (1988) Glutathione transferase in the urine: a marker for post-translplant tubular lesion. Kidney Int 33:571–577
2. Baggio B, Favaro S, Cantaro S, Bertazzo L, Frunzio A, Borsatti A (1981) Increased urine angiotensin I converting enzyme activity in patients with upper urinary tract infection. Clin Chim Acta 109:211–218
3. Bass MM, Kirsch RE, Tuff SA, Campbell JA, Sauncers JS (1979) Radioimmunoassay measurement of urinary ligandin excretion in nephrotoxin-treated rats. Clin Sci 56:419–426
4. Bergmeyer HU, Bernt E (1983) Malate dehydrogenase, oxaloacetate to malate reaction. In: Bergmeyer HU, Bergmeyer J, Grassl M (eds.) Methods of enzymatic analysis, 3rd edn. Vol. 3. Verlag Chemie, Weinheim, Deerfield Beach FL, Basel, pp 171–175
5. Brolin SE (1983) Adenylate kinase (myokinase), UV-method. In: Bergmeyer HU, Bergmeyer J, Grassl M (eds.) Methods of enzymatic analysis, 3rd edn. Vol. 3. Verlag Chemie, Weinheim, Deerfield Beach FL, Basel, pp 541–545
6. Ceriotti F, Barenghi M, Pontillo M, Murone M, Bonini PA (1985) Optimization of the urinary α-glucosidase determination. J Clin Chem Clin Biochem 23:845–850
7. Dietrich B, Haschen RJ, Hütter HJ, Pantscheva-Haschen R (1988) Die mitochondriale Aspartataminotransferase in der Differenzierung von Hyperenzymurien. Z Med Lab Diagn 29:245–253
8. Fent K, Mayer E, Zbinden G (1988) Nephrotoxicity screening in rats: a validation study. Arch Toxicol 61:449–358

9. Habig WH, Pabst MJ, Jakoby WB (1974) Glutathione-S-transferases: the first step in mercapturic acid formation. J Biol Chem 249:7130–7139
10. Hørder M, Rej R (1983) Alanine aminotransferase, routine UV-method. In: Bergmeyer HU, Bergmeyer J, Grassl M (eds.) Methods of enzymatic analysis, 3rd edn. Vol. 3. Verlag Chemie, Weinheim, Deerfield Beach FL, Basel, pp 445–450
10a. Laidler PM, Ryder KW, Glick MR, Oel TO, Van Etten RL (1985) Radioimmunoassay for arylsulfatase A in urine. Clin Chem 31:391–396
11. Maruhn D, Fuchs I, Mues G, Bock KD (1976) Normal limits of urinary excretion of eleven enzymes. Clin Chem 22:1567–1574
12. Maruhn D (1976) Rapid colorimetric assay of β-galactosidase and N-acetyl-β-glucosaminidase in human urine. Clin Chim Acta 73:453–461
13. Nakano M, Itoh G (1983) Elevation of urinary trehalase in mercuric chloride-induced nephrotic rabbits: urinary trehalase as a specific indicator of renal brush border damage. Chem Biol Interact 45:179–189
14. Nakano M, Aoshima K, Katoh T, Teranishi H, Kasuya M (1986) Urinary trehalase activity and renal brush border damage in inhabitants of a cadmium-polluted area (Jinzu river basin). Toxicol letters 34:159–166
15. Posey LE, Morgan LA, Levin S, Schwartz KM (1979) Measurement of arylsulfatse A in urine. Clin Chem 25:328–331
16. Rej R (1980) An immunochemical procedure for determination of mitochondrial aspartate aminotransferase in human serum. Clin Chem 26:1694–1700
17. Rej R, Horder M (1983) Aspartate aminotransferase, routine UV-method. In: Bergmeyer HU, Bergmeyer J, Grassl M (eds.) Methods of enzymatic analysis, 3rd edn. Vol. 3. Verlag Chemie, Weinheim, Deerfield Beach FL, Basel, pp 418–424
18. Schmidt E. Schmidt FW (1983) Glutamate dehydrogenase. In: Bergmeyer HU, Bergmeyer J, Grassl M (eds.) Methods of enzymatic analysis, 3rd edn. Vol. 3. Verlag Chemie, Weinheim, Deerfield Beach FL, Basel, pp 216–227
19. Wolf GB, Scherberich JE, Fischer P, Schoeppe W (1989) Isolation and characterization of dipeptidyl aminopeptidase IV from human kidney cortex. Clin Chim Acta 179:61–72

9 Basis of Reference

K Jung and D. Maruhn

Three essential and associated questions arise in the consideration of how to interpret urinary enzyme data:

1. What kind of urine (night urine, morning urine, 24-h urine, etc.) should be used for analysis?
2. Is a timed urine collection necessary as generally recommended in nephrology for the characterization of the excretion of analytes, or is it possible to use random urine samples?
3. What is the reference basis to express the excreted quantity of enzymes?

Different expressions to characterize the excretion of urinary enzymes are discussed in the literature [2, 19, 26]. The most frequently used modes for this purpose and their advantages and disadvantages are summarized in Table 1. To date, there is no general agreement for a standardized reference basis. Only a few studies have been performed to resolve this problem, and the arguments given for a particular recommendation have not always been conclusive [2].

In contrast to the release of enzymes into blood or other body fluids, the release into urine is characterized by an excretion into an open system [9]. Therefore, physiological factors like diuresis, physical activity, circadian rhythms, etc. which affect enzyme excretion are immediately reflected in the final urine (see Chap. 3). The result is a wide range of reference values of urinary enzymes with broad inter- and intraindividual variabilities which complicate the interpretation of data. In connection with the above-mentioned questions, decisive criteria for the selection of a suitable reference basis are (a) reduction of these variabilities, (b) convenient feasibility in routine praxis, and (c) best possibility to discriminate between healthy and diseased persons.

In addition to the "normal" physiological factors, the urinary enzyme concentration (enzyme volume activity; Table 1, point 1) is essentially influenced by urine volume [4, 10, 13]. The quantity of urine volume changes the enzyme concentration both directly and indirectly. The "direct" effect is attributed to the true influence of the urine flow rate on the excretion of enzymes. The cause of this effect is unknown, but the result is different for a particular enzyme. For example, if the urine flow increases, the excretion of brush border enzymes is higher than that of lysosomal enzymes [13]. In addition, there are differences of excretion rates not only between the various brush border enzymes [13] but also between enzymes of low molecular mass like lysozyme or ribonuclease [14]. The

Urinary Enzymes
K. Jung/H. Mattenheimer/U. Burchardt (Eds.)

Table 1. Different reference bases to express the excretion of urinary enzymes

Reference basis	Expression	Advantages	Disadvantages	Ref.
1. Enzyme concentration (volume activity)	U/l	No additional measurement or calculation	Dependent on diuresis	[2, 4, 10, 13, 15, 21]
2. Enzyme excretion per time (excretion rate)	U/t (e.g., U/24 h, U/h)		Accurate timed urine collection and volume measurement necessary Dependent on diuresis Inactivation of enzymes during long collection periods	[2, 19–21]
3. Ratio of enzyme concentration to urine creatinine concentration (enzyme activity/creatinine ratio)	U/mmol U/g	Use of random urine samples Partial compensation of diuresis effect	Additional measurement Creatinine excretion depends on muscle mass, sex, age, urine flow Analytical problems of creatinine determination	[2, 9, 19–22]
4. Ratio of enzyme concentration to specific gravity of urine		Same as in 3	Additional measurement Errors caused by increased excretions of protein, glucose, salt etc.	[10, 18]
5. Ratio of enzyme concentration to osmolality	U/mosmol	Same as in 3	Additional measurement Errors caused by changes of salt excretion (dietary and hormonal effects)	
6. Ratio of enzyme excretion rate to creatinine clearance (fractional excretion)	U/1	Indicator of the functioning kidney mass	Additional measurements and calculation Accurate timed urine collection necessary	[2, 3, 24]

"indirect" effect (concentration/dilution effect) results because the urine volume is variable over a wide range and therefore also the excreted enzyme concentration appears more or less concentrated or diluted. Diluted urine samples will give low values, whereas concentrated urine samples will show high values. Therefore, it is rather pointless to use "volume" as a reference basis.

The urinary excretion of substances is generally characterized in nephrology by a time-related excretion rate, especially by the rate of excretion per 24 h. This procedure is selected in order to reduce the variability of the excreted analyte caused by diurnal rhythms. Similar recommendations were also made for urinary enzymes; 8-h and 24-h samples were used [19]. As these long collection periods may cause inactivation of urinary enzymes either in vivo (in the bladder) or in vitro (when the samples are not cooled), shorter sampling periods (e.g., 3 h) have been proposed [19–21]. To avoid the influences of diurnal rhythm of enzyme excretion and the increasing problems of sampling errors when short collection periods are applied, the collection of the second morning urine sample after voiding the night urine is recommended [9, 19–22]. The collection period should cover between 2 and 4 h and it may be individually selected by the patient [19]. The time is noted by the patient, and therefore the excretion rate can be calculated by the following equation:

$$\text{Enzyme excretion rate (U/h)} = \frac{\text{Enzyme activity (U/l)} \times \text{Urine volume (l)}}{\text{Collection period (h)}}$$

As the accurate collection of timed urine samples is often unreliable and especially impracticable for screening purposes and for the investigation of outpatients, it is recommended to determine urinary analytes in spot urine samples [17]. This is a common practice in occupational medicine [1] and has also been suggested in nephrology [17, 23]. To reduce the variation of the respective indicator concentration due to the inconstant dilution/concentration of urine samples, adjustment to reference parameters has been proposed [6]. Specific gravity, osmolarity, and urinary creatinine level are the most commonly used parameters for such an adjustment procedure [1, 10, 18, 19]. An ideal reference parameter has to fulfill various preanalytical and analytical conditions: It should have slight interindividual variations and a constant day-to-day excretion, it should be little influenced by biological factors (urine volume, diet, physical effort, etc.), and it should be easily and specifically determined. Specific gravity is seldom used as reference basis in urinary enzyme analysis [10, 18] (Table 1), whereas the relation to urinary creatinine is now generally applied [19–22]. Studies on the use of the different reference bases, comparing the relation of urinary enzyme activities to volume, time, and creatinine, showed that the variabilities of enzyme excretion were least when the results were expressed as enzyme activity/creatinine ratio [21, 22]. These results are supported by clinical data. The ratio of enzyme activity to creatinine concentration exceeds the diagnostic validity of the volume- and time-related enzyme excretions [12]. However, it must be pointed out that urinary creatinine is also criticized as a reference parameter because of biological and analytical reasons

(Table 1). The variability of creatininuria in dependence on age or sex, for example, more substantially influences the enzyme activity/creatinine ratio than the variability of the truly excreted enzyme quantity [15]. This effect must be taken into account when reference values have to be established. Moreover, urine contains "non-creatinine chromogens" that may cause an overestimation of the creatinine concentration; the extent of the deviation from the true value is method dependent [25].

The relation of enzyme excretion to the glomerular filtration rate, i.e., the fractional excretion, is a further approach to characterize urinary enzyme excretion (Table 1, point 6). It could be demonstrated that the kidney enzyme content decreased in the case of patients with renal disease [3, 7, 24]. Thus, the fractional excretion has been suggested as a well-founded reference basis to express enzymuria, taking into account the functioning renal tissue [2]. For the example of the 24-h excretion rate, the following equations show that this reference basis additionally considers the serum creatinine value:

$$\text{Fractional excretion} = \frac{\text{enzyme excretion per 24 h}}{\text{glomerular filtration rate}}$$

$$= \frac{\text{enzyme activity (U/l)} \times \text{urine volume per 24 h}}{\text{urine/serum creatinine concentration} \times \text{urine volume per 24 h}}$$

$$= \frac{\text{enzyme activity (U/l)} \times \text{serum creatinine concentration } (\mu\text{mol/l})}{\text{urine creatinine concentration (mmol/l)}}$$

Amodio et al. [2] compared the fractional excretion of urinary enzymes with the enzyme excretion per time and per creatinine and related all these expressions additionally to the body surface. They concluded from their results that for both "tubular" (characterized by low molecular mass proteins like lysozyme coming from serum) and "parenchymatous" enzymuria (characterized by enzymes coming from renal tissue) the fractional excretion was the best way to express urinary enzyme excretion. Similar conclusions were drawn by other authors [3, 24]. However, the inspection of the data given in these papers calls these statements into question. On the other hand, it is doubtful from the theoretical point of view that the consideration of the serum creatinine level in addition to that of urine creatinine is suited to improve the results essentially. Further investigations are necessary to confirm these results.

No more homogeneous data of enzyme excretions were obtained when the relation was made to the body surface [2, 5] or body mass [5], although other authors found that this method reduces the variability of urinary enzyme excretions [16]. However, in time-expanded experiments with fast-growing animals (e.g., rats), reference to body weight seems to be imperative [8].

In conclusion, the use of the second morning urine as the random urine sample and the relation of enzyme activity to urinary creatinine level seem to form a convenient compromise to characterize urinary enzyme excretions. It is a practical approach for clinical and outpatient situations. The calculation of enzyme excretion per time or other modes of expression should be restricted for special purposes (e.g., nephrotoxicity studies [19] or after nephrectomy [11]).

References

1. Alessio L, Berlin A, Dell'Orto A, Toffoletto F, Ghezzi I (1985) Reliability of urinary creatinine as a parameter used to adjust values of urinary biological indicators. Int Arch Occup Environ Health 55:99–106
2. Amodio P, Bazzerla G, Malatesta R, Gatta A (1985) Reference ranges and methodological aspects in the urinary measuring of lysozyme, malate-dehydrogenase, γ-glutamyltransferase and α-glucosidase. Enzyme 33:216–225
3. Beck PR (1980) The significance and value of urinary gamma-glutamyltransferase assay. In: Goldberg DM, Werner M (eds) Progress in clinical enzymology. Masson, Washington, pp 109–115
4. Burchardt U, Hempel A, Jöpfner JB, Hempel RD (1976) Alaninaminopeptidase- und Kreatininausscheidung unter verschiedenen Diuresezuständen beim Menschen. Z Inn Med 31:550–552
5. Dietz AA, Hodges LK (1967) Correlation of activities of the phosphatases of the urine and serum of normal and cirrhotic persons. Clin Chim Acta 15:393–402
6. Elkins HB, Pagnotto LD, Smith HL (1974) Concentration adjustments in urinalysis. Am Ind Hyg Assoc J 3:559–565
7. Gatta A, Amodio P, Frigo A, Merkel C, Milani L, Zuin R, Ruol A (1981) Evaluation of renal tubular damage in liver cirrhosis by urinary enzymes and beta-2-microglobulin excretions. Eur J Clin Invest 11:239–243
8. Grötsch H, Hropot M, Klaus E, Malerczyk V, Mattenheimer H (1985) Enzymuria of the rat: biorhythms and sex differences. J Clin Chem Clin Biochem 23:343–347
9. Guder W, Heidland A (1986) Urine analysis. Report on the workshop conference of the German society for clinical chemistry and the society of nephrology in Würzburg, 25–26 Oct 1985. J Clin Chem Clin Biochem 24:611–620
10. Houser MT (1986) The effects of age and urine concentration on lysozyme and *N*-acetyl-β-D-glucosaminidase (NAG) content in urine. Ann Clin Biochem 23:297–302
11. Jung K (1990) Urinary enzyme excretion after donor nephrectomy — how should we express and compare excretion rates of the remaining kidney after donor nephrectomy? Nephron 54:196
12. Jung K, Diego J, Strobelt V (1985) Diagnostic significance of urinary enzymes in detecting acute rejection crises in renal transplant recipients depending on expression of results illustrated through the example of alanine aminopeptidase. Clin Biochem 18:257–260
13. Jung K, Schulze G, Reinholdt C (1986) Different diuresis-dependent excretions of urinary enzymes: *N*-acetyl-β-D-glucosaminidase, alanine aminopeptidase, alkaline phosphatase, and γ-glutamyltransferase. Clin Chem 32:529–532
14. Jung K, Pergande M, Porstmann B, Porstmann T (1988) Diuresis-dependent excretions of low-molecular mass proteins in urine: β_2-microglobulin, lysozyme, and ribonuclease. Scand J Clin Lab Invest 48:33–37
15. Jung K, Hempel A, Grützmann KD, Hempel RD, Schreiber G (1990) Age-dependent excretion of alanine aminopeptidase, alkaline phosphatase, γ-glutamyltransferase and *N*-acetyl-β-D-glucosaminidase in human urine. Enzyme 43:10–16
16. Klingmüller V, Bergmann E, Claaßen P (1984) Alaninaminopeptidase-Ausscheidung bei Kindern: Referenzbereiche und der Einfluß nierengängiger Kontrastmittel. J Clin Chem Clin Biochem 22:787–789
17. Krieg M, Gunßer KJ, Steinhagen-Thiessen E, Becker H (1986) Vergleichende quantitative Analytik klinisch-chemischer Kenngrößen im 24-Stunden-Urin und Morgenurin. J Clin Chem Clin Biochem 24:863–869
18. Leonard CD (1974) Urinary beta-glucuronidase in man: relation to urinary specific gravity and creatinine concentration. Clin Nephrol 2:106–112
19. Maruhn D (1983) Methodische Aspekte der Harnenzymologie. Z Gesamte Inn Med 38:557–564

20. Maruhn D, Fuchs I, Mues G, Bock KD (1976) Normal limits of urinary excretion of eleven enzymes. Clin Chem 22:1567–1574
21. Maruhn D, Strozyk K, Gielow L, Bock KD (1977) Diurnal variations of urinary enzyme excretion. Clin Chim Acta 75:427:433
22. Schulze G, Goebel H (1988) Harnenzymdiagnostik—ein Vorschlag zur Definition der präanalytischen Bedingungen. Z Med Lab Diagn 29:381–386
23. Shaw AB, Risdon P, Lewis-Jackson JD (1983) Protein creatinine index in assessment of proteinuria. Br Med J 287:929–932
24. Thiele KG (1973) γ-Glutamyltranspeptidase-Aktivität im Urin bei Gesunden und Nierenkranken. Klin Wochenschr 51:339–345
25. Tietz NW (1986) Textbook of clinical chemistry. Saunders, Philadelphia, pp 1276–1277
26. Werner M, Heilbron DC, Maruhn D, Atoba M (1970) Patterns of urinary enzyme excretion in healthy subjects. Clin Chim Acta 29:437–449

10 Reference Intervals

K. Jung

The survey on reference ranges given in this chapter are presented in addition to the values reported in the sections on the respective enzymes (see Chap. 8). They should be regarded as guidelines for the clinician and clinical chemist. It must be considered that these values may vary significantly due to methodological differences as especially observed in all cases of enzyme activity measurements. Thus, the methods are briefly characterized by the substrate and temperature in the reaction mixture although other reaction conditions also influence the catalytic activities.

The reference intervals either extracted or calculated from the literature sources are generally given as 0.95-interfractile intervals [13] or upper reference limits. The physiological factors (e.g. age and sex) influencing urinary enzyme excretion (see Chap. 5) are indicated as far as possible. Since the distribution of enzyme values could often not be inferred from the literature data, numerous calculations of intervals were performed under the not always well-founded assumption of a normal distribution of values. This emphasizes the use of the reported intervals as guiding data. To obtain more detailed information, the reader should refer to the original literature listed at the end of the chapter.

Age group Sex (years)	0.95-Interfractile reference interval	Sources
N-Acetyl-β-D-glucosaminidase (EC 3.2.1.30)		
2-chloro-4-nitrophenyl-N-acetyl-β-D-glucosaminide; 37°C		
Newborns		[32]
Children	0.18–1.70 U/mmol creatinine	
Adults		
3-cresolsulphonphthaleinyl-N-acetyl-β-D-glucosaminide; 37°C		
2–3	0.19–1.47 U/mmol creatinine	[15]
4–5	0.14–1.10	
6–8	0.20–1.00	
9–11	0.07–0.90	
Adults	0.56	[18]
4-nitrophenyl-N-acetyl-β-D-glucosaminide; 37°C		
0–0.5	1.13–19.1 U/mmol creatinine	[26]
0.6–2	0.50–16.7	
3–6	0.32–2.50	
7–12	0.22–1.55	

Urinary Enzymes
K. Jung/H. Mattenheimer/U. Burchardt (Eds.)

Age group (years)	Sex	0.95-Interfractile reference interval	Sources
13–18	F	0.23–1.20	
	M	0.21–2.50	
Adults	F	0.05–0.93	
	M	0.06–0.66	
Adults		0.07–0.65	[19]
Adults	F	0.28–1.18 U/mmol creatinine; 0.31–1.58 U/3 h	[25]
	M	0.19–1.06 U/mmol creatinine; 0.26–1.83 U/3 h	
20–30	F	0.01–0.33 U/mmol creatinine	[24]
	M	0.01–0.17	
31–40	F	0 –0.36	
	M	0 –0.28	
41–50	F	0 –0.45	
	M	0 –0.41	
51–60	F	0 –0.48	
	M	0 –0.44	
61–70	F	0.04–0.40	
	M	0 –0.52	
71–80	F	0.02–0.46	
	M	0 –0.49	
2-methoxy-4-(2′-nitrovinyl)phenyl-N-acetyl-β-D-glucosaminide; 37°C			
Adults		0 –0.51 U/mmol creatinine	[46]
4-methylumbelliferyl-N-acetyl-β-D-glucosaminide; 37°C			
Adults		0.03–0.19 U/mmol creatinine	[46]
Adults		0.02–0.65 U/mmol creatinine	[45]
Alanine aminopeptidase (EC 3.4.11.2) alanine hydrazide; 37°C			
1.5–14		0–1.99 U/mmol creatinine	[4]
Adults		0.09–0.98	
alanine-β-naphthylamide; 25°C			
3–15		0.29–0.90 U/mmol creatinine	[38]
18–30	F	0.14–0.39	
	M	0.17–0.46	
31–45	F	0.19–0.43	
	M	0.15–0.48	
46–65	F	0.19–0.70	
	M	0.21–0.61	
alanine-4-nitroanilide; 25°C			
0–2	M	0.46–1.55 U/d; 1.67–5.65 U/dxm^2	[29]
2–5	F	0.47–1.24 U/d; 0.71–1.89 U/dxm^2	
	M	0 –2.20 U/d; 0.01–3.12 U/dxm^2	
6–11	F	0.62–1.90 U/d; 0.54–1.70 U/dxm^2	
	M	0.48–2.68 U/d; 0.43–2.39 U/dxm^2	
12–16	F	0.62–2.08 U/d; 0.42–1.40 U/dxm^2	
	M	0.59–4.51 U/d; 0.22–3.20 U/dxm^2	
alanine-4-nitroanilide; 30°C			
Adults	F	0.24–0.55 U/mmol creatinine	[11]
	M	0.32–0.75	

Age group (years)	Sex	0.95-Interfractile reference interval	Sources
alanine-4-nitroanilide; 37°C			
see Methods I, II, in Sect. 8.1.1			
2–3		0.82–3.30 U/mmol creatinine	[15]
4–5		0.60–2.38	
6–8		0.52–2.12	
9–11		0.50–2.60	
leucine-4-nitroanilide; 25°C			
Adults	F	0.10–0.43 U/mmol creatinine; 0.31–1.58 U/3 h	[33]
	M	0.10–0.61 U/mmol creatinine; 0.26–1.83 U/3h	
Alanine aminotransferase (EC 2.6.1.2) optical test (alanine, 2-oxoglutarate); 37°C			
Adults		0–0.25 U/12 h	[14]
Alkaline phosphatase (EC 3.1.3.1) 4-nitrophenyl phosphate; 25°C			
Adults	F	0.03–0.32 U/mmol creatinine; 0.04–0.42 U/3 h	[33]
	M	0.03–0.23 U/mmol creatinine; 0.05–0.41 U/3 h	
Adults		0–0.71 U/mmol creatinine	[41]
4-nitrophenyl phosphate; 30°C			
Adults		0.13–1.88 U/mmol creatinine	[39]
4-nitrophenyl phosphate; 37°C			
see in Sect. 8.1.2			
0–4		0–2.27 U/mmol creatinine	[34]
5–10		0–1.40	
11–15		0–0.90	
Adults		0–0.07 U/h	[5]
Angiotensin I converting enzyme (EC 2.4.15.1) glycyl-L-histidyl-glycine; 37°C			
Adults		0.57–1.96 U/mmol creatinine	[3]
hippuryl-histidyl-leucine; 37°C			
Adults		0.27–0.85 U/mmol creatinine	[20]
Adults		0.21–0.55 U/mmol creatinine	[30]
Adenylate kinase (EC 2.7.4.3) optical test (ATP, AMP, phosphoenolpyruvate); 37°C			
Adults		0–5 U/l	[6]
Arylsulfatase A (EC 3.1.6.1) 4-methylumbelliferyl sulphate; 37°C			
Adults		3–22 U/mol creatinine	[23]
4-nitrocatechol sulphate; 37°C			
Adults		0.03–0.76 U/mmol creatinine	[37]
Adults		0.70–4.30 U/24 h	[40]
Aspartate aminotransferase (EC 2.6.1.1) optical test (aspartate, 2-oxoglutarate); 37°C			
Adults	F	0.05–0.32 U/mmol creatinine	[12]
	M	0.02–0.21	

Age group (years)	Sex	0.95-Interfractile reference interval	Sources
Dipeptidyl aminopeptidase IV (EC 3.4.14.5)			
glycyl-proline-4-nitroanilide; 37°C			
Adults		0.15–1.24 U/mmol creatinine	[9]
7-glycylproline-4-methylcoumarinamide; 37°C			
Adults		0.11–0.87 U/mmol creatinine; 0–4.5 U/l	[16]
***β*-Galactosidase** (EC 3.2.1.23)			
4-methoxy-4-(2′-nitrovinyl)phenyl-*β*-D-galactopyranoside; 37°C			
Adults		0–0.17 U/mmol creatinine	[46]
4-methylumbelliferyl-*β*-D-galactopyranoside; 37°C			
4–14		0–0.34 U/mmol creatinine	[36]
15–20		0.01–0.18	
21–45		0.01–0.19	
> 46		0–0.29	
Adults		0.06–0.44 U/mmol creatinine	[46]
Adults		0–0.13 U/h	[5]
4-nitrophenyl-*β*-D-galactopyranoside; 37°C			
Adults	F	0.05–0.74 U/mmol creatinine; 0.05–0.96 U/3 h	[33]
	M	0.11–0.81 U/mmol creatinine; 0.18–1.58 U/3 h	
Adults		0–0.36 U/mmol creatinine	[46]
***α*-Glucosidase** (EC 3.2.1.20)			
maltose; 37°C			
Children		0.02–7.52 U/10 h	[42]
Adults		0.24–0.52 U/h	[7]
Adults		0.78–10.8 U/24 h	[27]
Adults	F	0.35–1.38 U/mmol creatinine; 0.23–1.84 U/3 h	[33]
	M	0.29–1.03 U/mmol creatinine; 0.42–1.95 U/3 h	
4-nitrophenyl-D-glucopyranoside; 37°C			
Adults		0–22.1 U/mmol creatinine	[10]
Adults		0.03–0.20 U/mmol creatinine	[1]
***β*-Glucuronidase** (EC 3.2.1.31)			
4-methylumbelliferyl-*β*-D-glucuronide; 37°C			
4–14		0–0.32 U/mmol creatinine	[36]
15–20		0.02–0.18	
21–45		0.03–0.17	
> 46		0.01–0.23	
4-nitrophenyl-*β*-D-glucuronide; 37°C			
Adults	F	0.04–0.29 U/mmol creatinine; 40–358 mU/3 h	[33]
	M	0.05–0.25 U/mmol creatinine; 68–437 mU/3 h	
phenolphthalein mono-*β*-D-glucuronide; 56°C			
Adults		0–0.74 U/mmol creatinine	[17]
***γ*-Glutamyltransferase** (EC 2.3.2.2)			
γ-glutamyl-4-nitroanilide; 25°C			
Adults	F	0.80–4.01 U/mmol creatinine; 1.09–5.97 U/3 h	[33]
	M	1.02–3.35 U/mmol creatinine; 1.38–6.51 U/3 h	

Age group (years)	Sex	0.95-Interfractile reference interval	Sources
Adults	F	2.13–12.8 U/8 h	[43]
	M	3.88–17.9 U/8 h	
Adults		11–71 U/24 h	[28]
γ-glutamyl-4-nitroanilide; 37°C			
0–0.5		5.18–43.3 U/mmol creatinine	[26]
0.6–2		5.26–45.4	
3–6		3.32–15.8	
7–12		3.01–9.3	
13–18		1.88–11.3	
Adults	F	1.61–7.0	
	M	0.96–6.2	
γ-glutamyl-3-carboxy-4-nitroanilide; 25°C			
Adults		0.52–2.53 U/mmol creatinine	[41]
γ-glutamyl-3-carboxy-4-nitroanilide; 37°C			
2–3		1.97–9.11 U/mmol creatinine	[15]
4–5		2.10–8.11	
6–8		1.03–9.30	
9–11		3.52–9.97	
0–4		0.82–6.32	[34]
5–10		0.72–4.48	
11–15		0.63–3.12	
Adults	F	1.93–5.11 U/mmol creatinine; 0.84–2.52 U/h	[1]
	M	1.36–4.09 U/mmol creatinine; 0.84–2.52 U/h	
Kallikrein (EC 3.4.21.35) D-valyl-L-leucyl-L-arginine-4-nitroanilide; 37°C			
see in Sect. 8.4			
Adults		0.23–1.97 U/24 h	[31]
thiobenzyl benzyloxycarbomyl-L-lysinate; 37°C			
> 15		0–4.7 U/12 h	[2]
Lactate dehydrogenase (EC 1.1.1.27) pyruvate; 25°C			
Adults	F	0.04–1.56 U/mmol creatinine; 0.06–1.59 U/3 h	[33]
	M	0.07–0.68 U/mmol creatinine; 0.14–1.41 U/3h	
Adults		0–0.74 U/mmol creatinine	[41]
pyruvate; 30°C			
Adults		0.03–0.53 U/h	[5]
Adults		0.17–4.53 U/mmol creatinine	[39]
pyruvate; 37°C			
0–4		0–4.34 U/mmol creatinine	[34]
5–10		0–2.88	
11–15		0.15–1.62	
Lysozyme (EC 3.2.1.17) turbidimetric (*Micrococcus lysodeikticus*); 37°C			
see in Sect. 8.5.1			
Children		0–210 μg/l	[22]

Age group (years)	Sex	0.95-Interfractile reference interval	Sources
Adolescents		0–210 μg/l	[22]
Children		0–65 mg/mol creatinine	[21]
Adolescents			
Adults			
enzyme immunoassay			
Adults	F	13.2 mg/mol creatinine	[25]
	M	2.4	
Adults		1.7–123 μg/l	[8]
radioimmunoassay			
Adults		200–810 μg/l	[44]
Malate dehydrogenase (EC 1.1.1.37) optical test (oxalacetate); 25°C			
Adults		0–20 U/24 h	[47]
Trehalase (EC 3.2.1.29) Trehalose; 37°C			
Adults	F	15.9–97.7 U/mol creatinine; 16–132 mU/3 h	[33]
	M	1.0–79.5 U/mol creatinine; 12–148 mU/3h	
Adults		3.28–9.11 U/mol creatinine	[35]

References

1. Amodio P, Bazzerla G, Malatesta R, Gatta A (1985) Reference ranges and methodological aspects in the urinary measuring of lysozyme, malate dehydrogenase, γ-glutamyltransferase and α-glucosidase. Enzyme 33:216–225
2. Ash KO, Smith JB, Lynch M, Dadone M, Tolman KG, Williams RR (1985) Urinary kallikrein: assay validation and physiological variability. Clin Chim Acta 151:133–140
3. Baggio B, Favaro S, Cantaro S, Bertazzo L, Frunzio A, Borsatti A (1981) Increased urine angiotensin I converting enzyme activity in patients with upper urinary tract infection. Clin Chim Acta 109:211–218
4. Balogh A, Peiker G, Finke G, Traeger A (1984) Harnenzymausscheidung während einer Therapie mit Cephalotin bei Erwachsenen und Kindern. Z Urol Nephrol 78:425–435
5. Bernard A, Buchet JP, Roels H, Masson P, Lauwerys R (1979) Renal excretion of proteins and enzymes in workers exposed to cadmium. Eur J Clin Invest 9:11–22
6. Bernstein LH, Russell PJ, Foster WD (1982) Preliminary use of linear discriminant function to predict success or failure of kidney transplant. Ann Clin Lab Sci 12:60–67
7. Bonini PA, Ceriotti G, Franzini C (1970) Human urinary α-glucosidase as an index of kidney tubular damage. Clin Chim Acta 27:415–419
8. Brouwer J, Van Leeuwen-Herberts T, Otting-Van de Ruit M (1984) Determination of lysozyme in serum, urine, cerebrospinal fluid and feces by enzyme immunoassay. Clin Chim Acta 142:21–30
9. Burchardt U, Franke M, Krauss J, Barth A (1986) Renale Dipeptidylpeptidase-IV-Ausscheidung bei medikamenteninduzierten Nierenalterationen. Z Urol Nephrol 79:587–593
10. Ceriotti F, Barenghi L, Pontillo M, Murone M, Bonini PA (1985) Optimization of the urinary acid α-glucosidase determination. J Clin Chem Clin Biochem 23:845–850

11. Diener U, Knoll E, Langer B, Rautenstrauch H, Ratge D, Wisser H (1981) Urinary excretion of N-acetyl-β-D-glucosaminidase and alanine aminopeptidase in patients receiving amikacin or cis-platinum. Clin Chim Acta 112:149–157
12. Dietrich B, Haschen RJ, Hütter HJ, Pantscheva-Haschen R (1988) Die mitochondriale Aspartataminotransferase in der Differenzierung von Hyperenzymurien. Z Med Lab Diagn 29:245–253
13. Dybkaer R, Solberg HE (1987) Approved recommendation (1987) on the theory of reference values, 6. Presentation of observed values related to reference values. J Clin Chem Clin Biochem 25:657–662
14. Emanuelli G, Anfossi G, Calcamuggi G, Marcarino C, Lanzio M (1988) Urinary enzyme release following aminoglycoside administration in single low dose. Enzyme 39:119–122
15. Feldmann D, Flandrois C, Jardel A, Phan TM, Aymard P (1989) Circadian variations and reference intervals for some enzymes in urine of healthy children. Clin Chem 35:864–867
16. Hama T, Nagatsu T, Kobayashi S, Azumo S, Miyachi T, Kumagai Y, Kato T (1981) Changes in form of dipeptidyl-aminopeptidase IV in urine from patients with renal disease. Clin Chim Acta 113:217–221
17. Ho KJ, Ho LHC (1980) Determination of urinary β-glucuronidase activity. Single-point versus enzyme kinetic measuring system. Enzyme 25:361–370
18. Hofmann W, Guder WG (1989) A diagnostic programme for quantitative analysis of proteinuria. J Clin Chem Clin Biochem 27:589–600
19. Horak E, Hopfer SM, Sunderman FW (1981) Spectrophotometric assay for urinary N-acetyl-β-D-glucosaminidase. Clin Chem 27:1180–1185
20. Hosojima H, Miyauchi E, Morimto S (1989) Urinary excretion of angiotensin-converting enzyme in NIDDM patients with nephropathy. Diabetes Care 12:580–582
21. Houser MT (1983) Improved turbidimetric assay for lysozyme in urine. Clin Chem 29:1488–1493
22. Houser MT (1986) The effects of age and urine concentration on lysozyme and N-acetyl-β-D-glucosaminidase (NAG) content in urine. Ann Clin Biochem 23:297–302
23. Hultberg B (1979) Fluorimetric assay of the arylsulphatases in human urine. J Clin Chem Clin Biochem 17:795–797
24. Hultberg B, Isaksson A, Berg B, Tryding N, Ekman S, Nilsson JE (1988) The effect of age and sex on β-hexosaminidase in urine. Clin Chim Acta 177:271–273
25. Jung K, Schulze BD (1988) Comparative determinations of different urinary enzymes and low-molecular mass proteins in urine and serum of patients suffering from chronic renal diseases. In: Lubec G, Bartosch B (eds) Non-invasive diagnosis of kidney disease. Libbey, London, pp 65–80
26. Jung K, Hempel A, Grützmann KD, Hempel RD, Schreiber G (1990) Age-dependent excretion of alanine aminopeptidase, alkaline phosphatase, γ-glutamyltransferase and N-acetyl-β-D-glucosaminidase in human urine. Enzyme 43:10–16
27. Kai M, Kamada T, Baba Y, Shitomoto M, Setoyama S, Otsuji S (1980) Elevated levels of urinary acid maltase and maltose: riboflavin glucosyltransferase activities in diabetic patients. Clin Chim Acta 108:259–266
28. Kley R, Bahre G, Holzhüter H (1973) Die Bestimmung der γ-Glutamyltranspeptidase im Urin von nierengesunden und nierenkranken Personen mit einer automatisierten kinetischen Methode. J Clin Chem Clin Biochem 11:537–542
29. Klingmüller V, Bergmann E, Claaßen P (1984) Alaninaminopeptidase-Ausscheidung bei Kindern: Referenzbereiche und der Einfluß nierengängiger Kontrastmittel. J Clin Chem Clin Biochem 22:787–789
30. Kokubu T, Kato I, Nishimura K, Huwada K, Ueda E (1978) Angiotensin I-converting enzyme in human urine. Clin Chim Acta 89:375–379
31. Koolen MI, Daha MR, Frölich M, Van Es LA, Van Brummelen P (1984) Direct and indirect measurement of urinary kallikrein excretion in patients with essential hypertension and normotensives: relation to age and plasma renin and aldosterone levels. Eur J Clin Invest 14:171–174
32. Makise J, Saito E, Obuchi M, Kanayama M, Harakawa K, Yoshida K (1990) Improved kinetic assay of urinary N-acetyl-β-D-glucosaminidase with 2-chloro-4-nitrophenyl-N-acetyl-β-D-glucosaminide as substrate. Clin Chem 36:319–322
33. Maruhn D, Fuchs I, Mues G, Bock KD (1976) Normal limits of urinary excretion of eleven enzymes. Clin Chem 22:1567–1574
34. Matsukura H, Suzuki Y, Takai R, Okada T, Naiki S, Sakuragawa N (1987) Chromatographic profiles of urinary isoenzymes in healthy children. Clin Chim Acta 169:209–216

35. Nakano M, Aoshima K, Katoh T, Teranishi H, Kasuya M, Katoh T (1987) Severity of tubular brush border damage in cadmium-polluted area (Jinzu River Basin): clinical role of urinary trehalase. Environ Res 44:161–168
36. Paigen K, Peterson J (1978) Coordinacy of lysosomal enzyme excretion in human urine. J Clin Invest 61:751–762
37. Paigen B, Yarfitz S, Tabron D (1984) Urinary glucuronidase and arylsulfatases in identical twins of bladder cancer patients. Cancer Res 44:3624–3626
38. Peters JE, Mampel E, Schneider I, Burchardt U, Fukala E, Ahrens I, Haschen RJ (1972) Alaninaminopeptidase (L-Alanyl-Peptidhydrolase) im Harn bei Nierenerkrankungen. Clin Chim Acta 37:213–224
39. Pinto N, Eidelman A, Zaidman JL (1980) The diagnostic value of urinary alkaline phosphatase and lactate dehydrogenase enzyme and isoenzyme excretion in renal and urologic diseases and reference individuals. In: Goldberg DM, Werner M (eds) Progress in clinical enzymology. Masson, Washington, pp 122–123
40. Posey LE, Morgan LR, Levin S, Schwartz MK (1979) Measurement of arylsulfatase A activity in urine. Clin Chem 25:328–331
41. Prinsen JH, Günther H, Breuer J (1986) Bestimmung von Enzymaktivitäten in Urinen von Calciumoxalatstein-Patienten. J Clin Chem Clin Biochem 24:1001–1007
42. Straus B, Bajakic D, Cepelak I, Oberiter V, Kunovic B (1985) Changes in enzyme activities in urine in children with renal diseases. Z Med Lab Diagn 26:138–145
43. Szasz G (1970) γ-Glutamyltranspeptidase-Aktivität im Urin. J Clin Chem Clin Biochem 8:1–8
44. Thomas MJ, Russo A, Craswell P, Ward M, Steinhardt I (1981) Radioimmunoassay for serum and urinary lysozyme. Clin Chem 27:1223–1226
45. Whiting PH, Ross IS, Borthwick L (1979) Serum and urine N-acetyl-β-D-glucosaminidase in diabetics on diagnosis and subsequent treatment, and stable insulin dependent diabetes. Clin Chim Acta 92:459–463
46. Yuen CT, Price RG, Chattagoon L, Richardson AC, Praill PFG (1982) Colorimetric assays for N-acetyl-β-D-glucosaminidase and β-D-galactosidase in human urine using newly-developed ω-nitrostyryl substrates. Clin Chim Acta 124:195–204
47. Zekert F, Mautner Markhof C (1976) Untersuchungen über die normale Enzymausscheidung im Harn des Menschen und der Ratte. Wien Klin Wochenschr 88:467–469

Part 3

Urinary Enzymes in Clinical Medicine

11 Clinical Interpretation of Urinary Enzyme Excretion*

J.E. Scherberich

It is the purpose of the present chapter to summarize the general clinical and pathophysiological background which underlies the release of enzymes of renal origin into urine under pathological conditions. These aspects are compiled in Table 1 and Fig. 1. The data discussed may help as general guidelines to

Table 1. Interpretation of enzymuria under clinical conditions

Excretion of nephron marker enzymes in urine is:	Interpretation/comments
Enhanced	
1. (Transitorily) glomerulotubular damage, shedding of cell material/enzymes	
2. High urinary flow ("washout"), e.g., during recovery phase after acute tubular necrosis	
3. Physico-osmotic alterations of renal cells: combined osmotic damage/high urine flow (contrast media, HAES, mannitol, other osmocolloidal solutions)	
4. Increased protein turnover, compensatorily enhanced protein synthesis due to cellular reactions/adaptation towards: –toxins (lysosomotropic drugs, aminoglycoside); immunomediated injury –growth factors, hormones, cytotoxic lymphokines: T3, T4, growth hormone, angiotensin II (protooncogen activation, cell hypertrophy), epidermal growth factor (= enhances angiotensin II effect), platelet-derived growth factor, insulin-like growth factors, transforming growth factor-alpha/beta, interleukin-2, interferon-α; tumor necrosis factor α, etc.	
5. Progressive tubulointerstitial damage in primary glomerulonephritis; acute/chronic interstitial nephropathy, systemic disease (SLE, Wegener's granulomatosis, Churg-Strauss syndrome, multiple myeloma, diabetes mellitus, Goodpasture's syndrome), arterial hypertension; chronic renal graft rejection	
6. Presence of enzyme activators, decrease in enzyme inhibitors, drug interference	
7. Nonrenal enzymuria: serum-related low molecular weight enzymes, enzymes from infiltrating/resident blood cells; leucocytes and macrophages (lysozyme, elastase, cathepsin G, myeloperoxidase, proteinase III, glucosidases, sulfatases, ribonucleases, etc.)	
Decreased	
1. Tubular/glomerular atrophy (chronic cell damage with impaired recovery or repair of proteins)	
2. Intrarenal obstruction (luminal tubular aggregates of cellular debris, fragments, cast nephropathy, obstructing blebs)	
3. Overall decrease in renal plasma flow and glomerular ultrafiltration, "clusters of nonfiltering nephrons"; low washout	
4. Presence of enzyme inhibitors, loss of activating agents, drug interference; intrarenal protein digestion/inactivàtion; protein internalization; enzyme denaturation (loss of catalytic activity)	
5. Inborn error of synthesis of a renal marker enzyme	
6. Metabolic reasons (hypothyreosis; pH changes, "storage diseases")	

* Original work was supported by grant of the Dr. Cilli Weil foundation, Frankfurt/Main, and a grant (no. III-005/89) from the BGA/BMFT Bonn, Berlin.

Urinary Enzymes
K. Jung/H. Mattenheimer/U. Burchardt (Eds.)

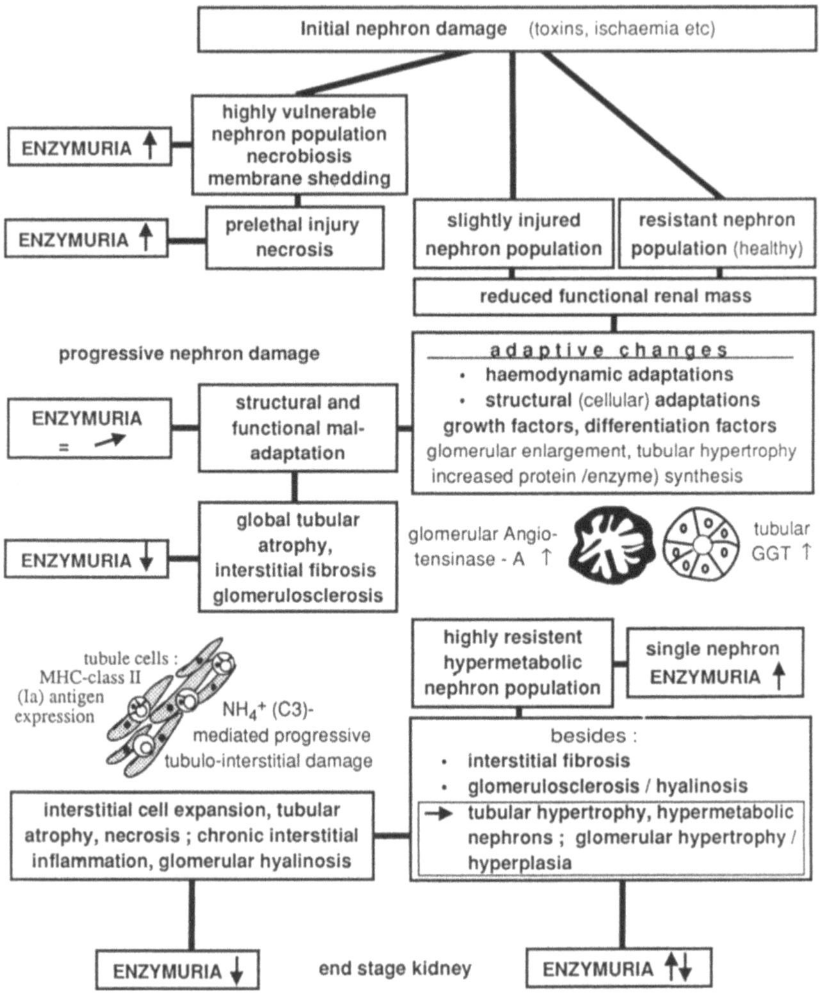

Fig. 1. Interpretation of enzymuria at various stages of progressive renal failure

interpret "enzymuria" in terms of a diagnostic tool in clinical medicine, although several aspects still require further validation [4, 6, 10, 13, 18]. Details on changes in enzymuria taking into account the characteristics of the respective diseases are described in the following Chaps. 12–15.

To interpret changes in enzymuria under various clinically defined conditions, light must be shed on the energy metabolism of kidney cells (tubular, glomerular epithelia), as well as on their ability to face acute and chronic injury by adaptive responses and protein repair mechanisms. Details on this subject are discussed in Chaps. 2–5.

As pointed out in Chap. 3, the amount of enzyme released into urine may vary depending on the different vulnerability of renal cells, the severity, progression, and possible circadian variations of the underlying renal disease, the residual renal functional reserve, the metabolic status, and the intrinsic energy potential of damaged cells to recover from injury as well as from the degree of interstitial proliferative activity and extracellular matrix deposition [4, 7, 15, 22, 24].

Increased urinary enzyme activity is observed in patients without apparent renal involvement (see Chap. 5), e.g., in those suffering from metabolic diseases (hyperthyreodism, active acromegaly; see Sect. 15.6) and during pregnancy (see Sects. 5.6 and 12.7). Under these conditions, various growth factors such as epidermal growth factor, endothelial-derived growth factor, transforming growth factors, insulin-like growth factors, etc. may influence the size, metabolism, protein (enzyme) synthesis, mitotic activity, repair mechanisms, and proliferative activity of renal epithelia (see Chap. 3) [7, 8, 19, 22, 24]. Angiotensin II is involved, partially through gene activation (c-*fos*, c-*myc*) in tubular hypertrophy and cooperates with other growth hormones [22–24]. Clusters of hypertrophic remnant nephrons in patients suffering from chronic progressive kidney disease are hypermetabolic and disclose increased enzyme synthesis, e.g., of membrane γ-glutamyltransferase (GGT) [17, 19]. However, due to tubulointerstitial fibrosis and atrophy, the overall excretion of the marker enzyme is lower when compared with normal controls (Fig. 1; see Sect. 12.4).

As compared with healthy controls, enzymuria is significantly higher in patients suffering from glomerulonephritis, interstitial nephritis (see Sects. 12.3 and 12.4), essential arterial hypertension (see Sect. 15.2), and lupus nephritis (see Sect. 15.7). However, the excretions of alanine aminopeptidase (AAP), GGT, alkaline phosphatase (AP), and *N*-acetyl-β-D-glucosaminidase (NAG) are usually lower than after acute ischaemic or toxic renal cell damage (see Chaps. 12–15) [11, 17, 25]. Some aspects dealing with the interpretation of enzymuria at different stages of renal damage are summarized in Fig. 1.

Renal cells are main targets for most of the potential cytotoxic drugs (see Chap. 14). Enzymuria following injection of, e.g., aminoglycosides is dose-dependent [9, 14]. A chronobiologic rhythmicity in renal drug tolerance (e.g., gentamicin) may help to explain repeating sequences in enzyme excretion profiles after administration of aminoglycosides [2, 5, 12]. Enzymuria is significantly lower after injection of low osmolar X-ray contrast media than after high osmolar contrast media [17, 20], indicating that physicochemical properties of contrast media may play a crucial role in contrast media-mediated renal toxicity.

Ca^{2+}-antagonists are capable of preventing, to some degree, tubular injury from acute ischaemia or cyclosporin-A-induced nephrotoxicity [3, 16], possibly by stabilizing cytoskeletal components. Thus, after administration of potentially nephrotoxic drugs, enhanced enzymuria may be decreased by Ca^{2+}-channel blockers [3, 16]. However, after renal or coronary angiography, the increased enzyme output was not reduced in patients receiving Ca^{2+}-antagonists [17, 20].

Enzymuria may also be influenced by angiotensin converting enzyme inhibitors (e.g., captopril, enalapril, ramipril), since angiotensin II enhances tubular protein (collagen) synthesis, stimulates protooncogen expression, cell proliferation, and tubular epithelial hypertrophy [23, 24], which is commonly associated with an adaptive increase of membrane-bound enzyme activity in remnant nephrons under conditions of reduced functional renal mass [17, 19]. In this light, decreased enzymuria in these patients apparently reflects possible antiproliferative ("renoprotective") properties of these enzyme inhibitors.

Enzymuria in patients with infectious diseases including malaria tropica, hepatitis B, and AIDS is elevated and can be explained by more or less specific prelethal lesions of glomerular and tubular cells under these conditions [1, 21].

For interpretation of enzymuria in patients with systemic diseases like multiple myeloma, amyloidosis, impaired glucose tolerance, diabetes mellitus (types I, II), and arterial hypertension, see Chap. 15.

References

1. Bourgoignie JJ (1989) Acquired immunodeficiency syndrome (AIDS)-related renal disease. Klin Wochenschr 67:889–894
2. Cal JC, Dorian C, Catroux P, Cambar J (1989) Nephrotoxicity of heavy metals and antibiotics. In: Lemmer B (ed) Chronopharmacology. Marcel Dekker, New York, pp 655–681
3. Chan L, Schrier RW (1990) Effects of calcium channel blockers on renal function. Annu Rev Med 41:289–302
4. Davies M. Coles GA, Thomas GJ, Martin J, Lovett DH (1990) Proteinases and the glomerulus: their role in glomerular diseases. Klin Wochenschr 68:1145–1149
5. Diener U, Knoll E, Langer B, Rautenstrauch H, Ratge D, Wisser H (1981) Urinary excretion of *N*-acetyl-β-D-glucosaminidase and alanine aminopeptidase in patients receiving amikacin or *cis*-platinum. Clin Chim Acta 112:149–157
6. Dubach UC, Schmidt U (eds) (1979) Diagnostic significance of enzymes and proteins in urine. Hans Huber, Bern (Curr Probl Clin Biochem 9)
7. Fine L (1986) The biology of renal hypertrophy. Kidney Int 29:619–634
8. Habenicht A (ed) (1990) Growth factors, differentiation factors and cytokines. Springer, Berlin Heidelberg New York
9. Heinert G, Wyrobnik J, Scherberich JE (1980) Quantitative computed histophotometry of membrane-integrated and lysosomal enzymes indicating inductive and alternative effects of aminoglycosides. Curr Chemother Immunother 11:252–254
10. Heinert G, Scherberich JE, Mondorf W, Weber W (1983) Quantitative enzymatic and immunologic computer-assisted histophotometry of human kidney tissue following neoplastic and other clinically significant alterations. Eur Urol 9:235–241
11. Jung K, Schulze BD, Sydow K (1987) Diagnostic significance of different urinary enzymes in patients suffering from chronic renal diseases. Clin Chim Acta 168:287–293
12. Maruhn D, Paar D (1982) Urinary enzymes in therapy monitoring and recognition of side effects of drugs. Clin Enzymol Symp 3:137–152
13. Mattenheimer H (1977) Enzymes in renal diseases. Ann Clin Lab Sci 7:422–432
14. Mondorf W, Scherberich JE, Falkenberg FW (1984) Brush border enzymes and drug nephrotoxicity. In: Solez K, Whelton A (eds) Acute renal failure. Marcel Dekker, New York, pp 281–292
15. Paddock JK, Taddeo AR, Paddock JP, Loewenstein LM (1985) Glycoprotein reutilisation in regenerating microvilli after renal ischaemia in rats. Kidney Int 28:498–503
16. Rodicio JL, Morales JM, Ruilope LM (1990) Calcium antagonists and the kidney. Nephrol Dial Transplant 5:81–86

7. Scherberich JE (1990) Urinary proteins of tubular origin: basic immunochemical and clinical aspects. Am J Nephrol 10 [Suppl 1]:43–51
8. Scherberich JE, Gauhl C, Heinert G, Mondorf W, Schoeppe W (1984) Characterization and clinical significance of membrane bound proteases from human kidney cortex. Adv Exp Med Biol 167:179–190
9. Scherberich JE, Wolf G, Albers C, Nowack A, Stuckhardt C, Schoeppe W (1989) Glomerular and tubular membrane antigens reflecting cellular adaptation in human renal failure. Kidney Int 36 [Suppl 27]:S38–S51
0. Scherberich JE, Rautschka E, Fischer A, Kollath J, Riemann HE (1990) Tubular histuria: clinical evaluation of the different nephrotoxic potential of x-ray contrast media. Contr Nephrol 83:229–236
1. Scherberich JE, Schneider A, Nockher A, Staszewski S, Schoeppe W (1991) Pattern of serum- and kidney-specific proteinuria in patients with HIV-infection. Kidney Int 39:1312
2. Weinberg JM (1991) The cell biology of ischaemic renal injury. Kidney Int 39: 476–500
3. Wolf G, Neilson EG (1991) Angiotensin II induces cellular hypertrophy in cultured murine proximal tubular cells. Am J Physiol 259:F768–F777
4. Wolf G, Neilson EG (1991) Molecular mechanisms of tubulointerstitial hypertrophy and hyperplasia. Kidney Int 39:401–420
5. Wolf G, Scherberich JE, Nowack A, Stein O, Schoeppe W (1990) Urinary excretion of dipeptidyl aminopeptidase IV in patients with renal diseases. Clin Nephrol 35:136–142

12 Urinary Enzymes in Renal, Renal-Associated and Urological Disorders

K. Jung

12.1 Introduction

In addition to clinical aspects and information from the case history, the diagnosis of renal disease is made through an evaluation of the functional and/or structural integrity of the kidney. Clinicians demand cost-efficient tests which sensitively and specifically reflect renal function and do not annoy patients. Highly specialized techniques like biopsy, arteriography and pyelography, although revealing specific and often unique information, require specially trained clinicians and are invasive. Non-invasive approaches are needed which can be used as parameters for screening and follow-up studies.

Documenting pathological processes by their effect on urinary composition is the oldest available non-invasive procedure in medicine. In this respect, measurement of "enzymuria" could be considered an advanced form although its clinical value in renal and urological diseases has not yet been fully realized. In the following sections we discuss the diagnostic potential of urinary enzyme measurements in renal, renal-associated and urological disorders.

12.2 Enzymuria in Acute Renal Failure
(together with U. Burchardt)

Acute renal failure is marked by a rapid deterioration of renal function, indicated by the increase of substances such as urea and creatinine. A reduction of urine volume may (oliguric/anuric type) or may not (non-oliguric type) occur. The causes of such renal damage can be classified by their pre-renal, post-renal or renal origin. The main aetiologies leading to acute renal failure defined in a more strict sense are exposure to nephrotoxic agents or reduced renal perfusion leading to hypoxia or ischaemia [18]. The progress of acute renal failure has traditionally been separated into four distinct phases: the *initial phase* as the time between the injurious event and the second phase, the *oliguric phase*; the *diuretic phase* with subsequent increases in flow rates over a period of hours or days and the *phase of functional recovery* after several days of diuresis. Thus, different processes of both a destructive and regenerative nature simultaneously take place in the tubular cells during these various phases and change the

Urinary Enzymes
K. Jung/H. Mattenheimer/U. Burchardt (Eds.)

excretion pattern of urinary enzymes. Enzyme measurements in microdissected cortical nephrons from human biopsies showed that the high *N*-acetyl-*β*-D-glucosaminidase (NAG) output during the maintenance and recovery phase of acute renal failure could be explained by its enhanced release from tubular segments and the increased cellular enzyme activity caused by regenerative processes [81]. The behaviour of urinary enzyme excretion in the form of peaks and troughs could result from these different processes [16].

Whereas there are numerous reports on experimental studies of urinary enzyme excretion during the course of acute renal failure, systematic studies in humans are rare [13, 16, 27, 80, 92]. A typical picture of urinary enzyme excretion during the clinical course of acute renal failure after intoxication by carbon tetrachloride is shown in Fig. 1. Besides a high output of the brush border enzyme alanine aminopeptidase (AAP), large increase of the lysosomal enzyme glucuronidase was evident [13]. Similar results of high excretion rates of NAG were observed in patients suffering from acute renal failure after hypotensive episodes [92]. The maximum excretion rate was 1200 times over the normal limit [92] occurring especially with or just after the onset of diuresis [16]. The excretion rates in acute renal failure are essentially higher than in other kidney diseases [27]. Enzyme excretion returned towards normal as renal function

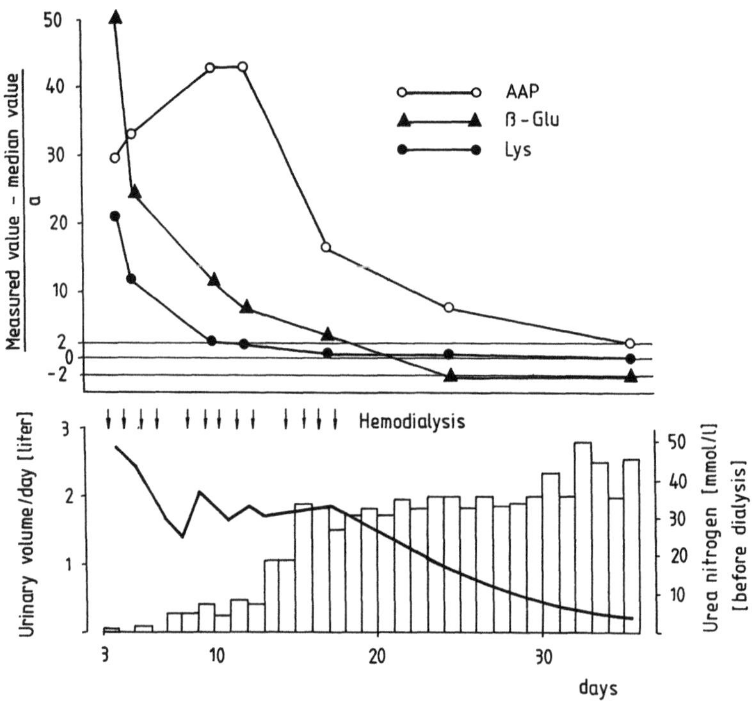

Fig. 1. Excretion of alanine aminopeptidase (*AAP*), glucuronidase (*β-Glu*) and lysozyme (*Lys*) in a case of acute renal failure (according to [13])

indicated by serum creatinine levels recovered. However, Wellwood et al. [92] described patients after hypotensive periods presenting with high urinary enzyme activities without other indices of renal dysfunction such as elevated serum creatinine concentration. In patients undergoing surgery with controlled arterial hypotension and an uncomplicated postoperative course, the serum creatinine level was unchanged whereas urinary NAG excretion increased for the first 7 postoperative days and later fell towards normal [34].

These results demonstrate that urinary enzyme excretions very sensitively reflect changes in the kidney. Unfortunately, no systematic follow-up studies have been performed yet to use the time course and the level of excretion rate of a pattern of different enzymes as predictive indicators. For example, patients with severe acute renal failure who needed haemodialysis treatment generally showed a higher urinary AAP activity than patients undergoing a lighter course who did not require haemodialysis [80]. In general, all the results presented in the literature are exclusively descriptive and do not give any evidence whether the serial determination of urinary enzymes in patients at risk (e.g. after abdominal aorta or open heart surgery, crush syndrome) could be helpful as diagnostic and prognostic parameters. Similar comparative studies as performed in evaluating nephrotoxicity [26] are necessary in order to come to practical recommendations.

12.3 Enzymuria in Acute Glomerulonephritis, Acute Pyelonephritis and Urinary Tract Infections

Acute glomerulonephrits and pyelonephritis were found to lead to increased excretion of the urinary enzymes AAP, aminotransferases, alkaline phosphatase (AP), galactosidase, gamma-glutamyltransferase (GGT), glucuronidase, lactate dehydrogenase (LDH) and NAG [21, 39, 84, 86, 92]. The levels of enzyme excretion correlated well with the activity and intensity of the renal disease (Fig. 2). Increased enzymuria returned to the normal range with the remission of the disease [84, 86, 92]. When clinical data and conventional nephrological parameters such as urinary protein and serum creatinine levels suggested that the acute renal disease was over, the persistence of raised urinary NAG excretion identified patients at risks of relapse [92]. For example, NAG determination has been recommended as a valuable test for the identification of active renal injury in patients with vesico-ureteric reflux [39]. It can be concluded that the determination of enzymuria in acute nephritis could be clinically relevant as a monitoring parameter for follow-ups rather than a diagnostic parameter. However, systematic studies taking this aspect into account have not yet been performed.

Urinary enzyme determinations have been recommended for diagnostic purposes in urinary tract infections [60]. These infections are traditionally

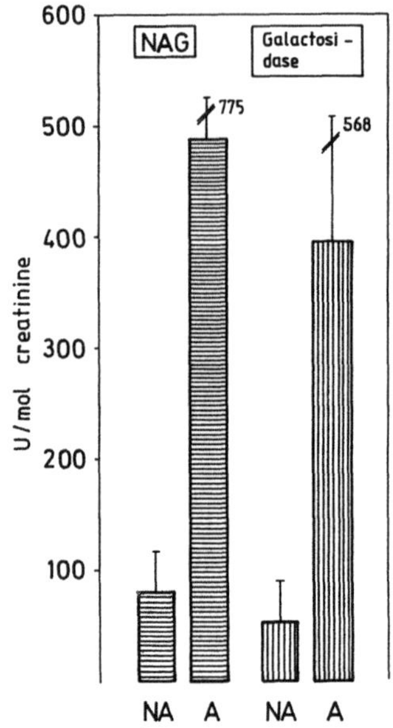

Fig. 2. Excretion of *N*-acetyl-β-D-glucosaminidase (*NAG*) and β-galactosidase during active (*A*) and non-active (*NA*) stages of renal disease. Values ($\bar{x} \pm$ SD) were calculated from data of Sheth and Good [84]

separated into upper and lower tract infections. The accurate identification of the infection site and whether the kidney or the bladder is affected are mandatory data in the diagnostic work-up as the treatment and prognosis are obviously different for these two clinical entities. The tubular involvement in upper urinary tract infection causes increased excretions of urinary enzymes, and therefore enzymuria estimation has been suggested to be of value for discrimination between the two forms. Several enzymes (angiotensin I converting enzyme, glucuronidasc, NAG and its isoforms, LDH and its isoenzymes, lysozyme) were tested, and contradictory results of their diagnostic validity were reported [3, 7, 10, 17, 30, 60, 68, 78, 79, 87, 88, 90]. It was concluded in comparative determinations of AAP, glucuronidase, glucosidase and NAG in patients with upper urinary tract infections and with fever of non-renal origin that the excretions were too variable and unpredictable to be useful as indicators [78, 79]. The separation of NAG in its multiple forms A and B may improve that discrimination [90]. Whereas some patients with different forms of urinary tract infection had overlapping values of total NAG excretion, all patients with upper urinary tract infection showed an increased excretion of isoenzyme B [90]. However, the determination of urinary LDH V seems to be the most reliable procedure to locate the site of infection [60]. Carvajal et al. [17] introduced this test in 1975 and showed that 94% of patients with pyelonephritis had an elevated excretion of LDH V whereas in all patients with cystitis the excretion of LDH V was absent. In human kidney tissue of healthy people LDH I and II predominate, and LDH V is present in relatively small quantities [88]. The usual isoenzyme pattern in urine is also predominately (about 90%) LDH I and

II [53]. In the renal cortex from patients with pyelonephritis isoenzyme V prevails, indicating that infection alters the isoenzyme composition and thus may cause the changed pattern seen in the urine of patients with upper urinary tract infections [17]. Meanwhile, the sensitivity and specificity of this test have been well established by several clinical studies in children and adults [60, 87]. It has been concluded that urinary LDH isoenzyme analysis is a reliable and practicable procedure to differentiate between upper and lower urinary tract infection when clinical and microbiological data indicate infection [87].

12.4 Enzymuria in Chronic Glomerulonephritis, Interstitial Nephritis and Chronic Renal Insufficiency

12.4.1 Enzymuria as Early Marker of Chronic Renal Diseases and Its Relation to Traditional Nephrological Parameters

(together with J.E. Scherberich)

Most chronic renal diseases progress through chronic renal insufficiency, leading more or less rapidly to terminal renal failure. It is the aim of nephrologists to detect these changes at an early stage when medical intervention can possibly re-establish the normal function or retard further progress of disease. Subjective symptoms and abnormality in conventional laboratory tests of renal dysfunction, such as serum creatinine level, urinary sediment (dip-stick) and quantification of protein excretion, appear relatively late. The measurement of glomerular filtration rate is the standard procedure to detect reduced renal function in the early stages. Since creatinine clearance is unreliable and precise methods such as isotope clearance techniques are expensive, time-consuming and not always available, serum creatinine concentration is the widely accepted routine parameter to assess renal function. However, the glomerular filtration rate may decrease by about 50% without detectable changes in serum creatinine level. Therefore, non-invasive, practical methods for the early detection of kidney diseases are needed.

The determination of urinary enzyme excretion seems to meet these requirements [73, 89]. Although urinary enzymes have been widely investigated in patients suffering from chronic renal diseases since the beginning of the "era" of urinary enzymes, scientists differ in their views of the clinical validity of such estimations [89]. The reasons for this discrepancy are almost the same as discussed in enzymuria of allograft recipients (Chap. 13). Very often, only one enzyme was investigated, and with the result obtained a generalized view of urinary enzyme excretion was established. A straightforward comparison of the diagnostic validity between levels of different enzymes and standard nephrological parameters is necessary to decide on the diagnostic usefulness. Unfortunately, little attention has been given to this comparison [44, 56, 66]. Measurements of urinary enzymes are reasonable only if they provide some better diagnostic information than the above-mentioned, routine, cost-efficient parameters.

Elevated excretion of urinary enzymes originating from the proximal tubular cells has been found in patients with chronic glomerular and interstitial kidney diseases [9, 15, 20, 49, 57, 58, 66, 67, 69, 77, 83]. The relationships between enzyme excretion and conventional nephrological parameters in light of the early marker concept can be summarized by the following conclusions:

1. The relationship between enzyme excretion and residual kidney function (serum creatinine or isotope clearance) is generally poor (Table 1). Correlation coefficients depend on the particular enzyme estimated and on the type of kidney disease (Table 1). Varying and partly contradictory correlations were reported because the authors based their conclusions on different enzymes and on measurements in patients with a wide variety of residual kidney function including those with end-stage renal insufficiency. However, patients with an advanced reduced glomerular filtration rate generally show a subnormal excretion of AAP [36], GGT [9], NAG [37] and glucuronidase [58] due to a smaller renal cellular mass [37]. On the other hand, enzyme excretions often increase without changes of serum creatinine [49, 83].

   An unexpected high percentage of patients with primary chronic glomerulonephritis and significantly increased urinary NAG activity were observed [95]. This elevated NAG excretion may reflect progressive tubulointerstitial damage and active renal disease leading to endstage renal failure (see Fig. 1 in Chap. 11). As shown by Bohle et al. [93] in patients with various forms of glomerulonephritis, the serum creatinine concentrations and the relative renal mass occupied by interstitial tissue (tubular atrophy, interstitial fibrosis), but not the glomerular changes per se, were closely correlated. Thus, the expansion and remodeling of the renal interstitium, associated with proliferation of fibroblasts, angioblasts and resident blood macrophages, the neosynthesis of type IV/III collagen, tubular damage, repair and atrophy, constructs a secondary form of ischaemic glomerulopathy which apparently governs the fate of the disease and contributes to its progression and

Table 1. Correlation coefficients between urinary enzyme excretion and [^{99m}Tc]-diethylenetriaminopentaacetate (DTPA) clearance, serum creatinine level and urinary protein in patients suffering from chronic glomerulonephritis (GN) and chronic pyelonephritis (PN)

Enzyme	DTPA		Serum creatinine		Protein excretion	
	GN	PN	GN	PN	GN	PN
Alanine aminopeptidase	−0.05	−0.59**	−0.03	0.27	0.64***	0.20
Alkaline phosphatase	−0.13	−0.43*	0.06	0.39*	0.61***	0.14
Gamma-glutamyltransferase	0.06	−0.16	0.10	0.08	0.62***	0.11
N-acetyl-β-D-glucosaminidase	−0.41*	−0.62***	0.24	0.38*	0.82***	0.56**
RNase	−0.56**	−0.55**	0.79***	0.84***	0.58**	0.75

Statistically significant correlation: $*P < 0.05$; $**P < 0.01$; $***P < 0.001$ (according to [44])

functional derangement [93]. The amount of brush border enzymes excreted in urine depends on the degree of functional impairment (glomerular filtration rate) in chronic renal disease (Table 1; [94]). By contrast to GGT and AAP, the activity of NAG and angiotensinase A in urine is negatively correlated with the decrease of glomerular filtration rate (Table 1; [94, 96]). In more severe cases of renal impairment (serum creatinine concentrations > 270 μmol/l) angiotensinase A is the only known enzyme whose activity in urine progresses with the increasing serum creatinine level, possibly due to adaptive mechanisms of remaining nephrons (see Sect. 8.1.4; [96]).

Therefore, the intensity of enzymuria is generally more likely to indicate the activity of the process in the kidney rather than the kidney function [83]. Urinary enzyme levels were higher during active or progressive disease than when the disease was stable [38, 83, 86]. Enzymuria returns with the regression of the disease [84]. Thus, a strong correlation between the variables of kidney function and enzymuria cannot be expected.

2. Significant correlations exist between proteinuria and enzyme excretion in cases of glomerulopathic diseases but not for all enzymes in cases of interstitial diseases (Table 1). The increased enzymuria in patients with tubulointerstitial diseases but without proteinuria indicates that proteinuria does not account for the increased enzymuria [44]. The excretion rates differ between various histological types of glomerulonephritis [38, 56]. Urinary enzymes of the proximal tubule found in patients suffering from glomerulopathies could not be explained as the corresponding serum enzyme freely passing through a damaged filtering barrier [49, 83]. For example, the NAG isoenzyme pattern corresponded with that of the renal tissue [37]. Increased enzymuria in glomerulopathies suggests that there is a tubulo-interstitial component to renal diseases primarily considered to be of glomerular origin. Elevated excretion of NAG may be achieved by its release from tubular lysosomes by exocystosis or degeneration of proximal cells caused by excessive protein reabsorption and catabolism [49, 58]. A characteristic example is the damage to tubular cells by the enormous protein load of Bence-Jones proteinuria [20].
3. Taking into account the mentioned concept of an early marker it is meaningful to compare the enzyme excretion pattern in patients with normal and abnormal serum creatinine levels in relation to that in healthy subjects (Fig. 3). When values are related to a cut-off point of 106 μmol/l, different behaviours of enzymes become evident (Fig. 3). Whereas AAP, AP and NAG were already significantly higher in patients with normal serum creatinine values than in healthy people, GGT and RNase increased only in patients with increased serum creatinine values [44]. Similar results were described when comparative studies of various enzymes were made [57]. These results emphasize the varying ability of different enzymes to reflect renal dysfunction. A comparison of the diagnostic sensitivity of enzymuria, urinary protein, serum creatinine value and isotope clearance showed that the percentage of increased excretion of NAG exceeded that of all other enzymes and

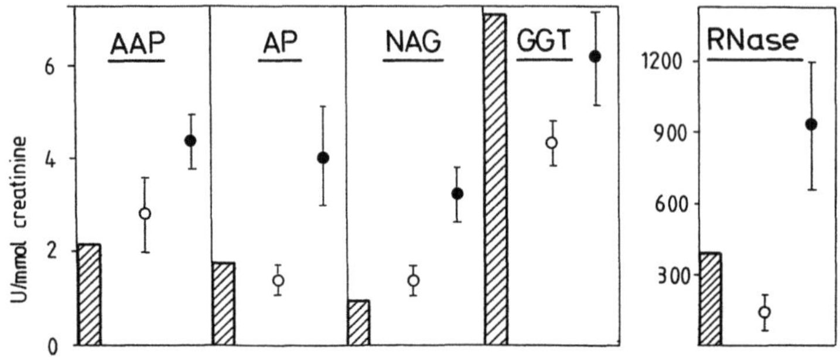

Fig. 3. Urinary enzyme excretion in relation to serum creatinine level. The 52 patients investigated were divided into two groups according to their serum creatinine values: ○, 35 patients with values < 106 μmol/l; ●, 17 patients with values > 106 μmol/l. Enzyme values are arithmetic means ± SEM, and the 97.5th percentile limits of a control group are indicated (hatched bar) at the side (according to [44]). *AAP*, alanine aminopeptidase; AP, alkaline phosphatase; *NAG*, *N*-acetyl-β-D-glucosaminidase; *GGT*, gamma-glutamyltransferase

Table 2. Percentage increase in values of urinary enzymes and standard nephrological parameters in patients with chronic renal diseases. For the calculation the 97.5th percentile limits of healthy persons were used (according to [44])

Parameter	Percentage increase in values	
	Total patients ($n = 52$)	Patients with serum creatinine level < 106 μmol/l ($n = 35$)
Alanine aminopeptidase	54	42
Alkaline phosphatase	44	31
Gamma-glutamyltransferase	23	23
N-acetyl-β-D-glucosaminidase	67	49
RNase	17	11
Creatinine	33	0
Urine protein	38	29

was found to be better than urinary protein and indicators of the glomerular filtration rate (Table 2).

The usefulness of NAG determination has been pointed out by several authors [21, 23, 49, 66, 72, 73, 83]. Maruhn and Bock [57] compared excretions of 10 enzymes (AP, arylsulfatase A, β-galactosidase, GGT, α-glucosidase, glucuronidase, LDH, leucine arylamidase, NAG and trehalase) with creatinine clearance and described the highest incidence of pathological results for NAG. About 84% of 122 patients with chronic renal disease presented an increased output of NAG. Whereas brush border enzymes such as AAP very sensitively reflect acute damage to the kidney (e.g. acute rejection crises in kidney transplantation, see Chap. 13; nephrotoxicity, see Chap. 14), they seem to be less sensitive than lysosomal enzymes like NAG for the detection of patients suffering from chronic renal diseases [44, 57].

Palmieri et al. [66] concluded from their study that assay of urinary enzymes was more reliable than the determination of urinary protein to uncover renal alterations. They recommended the combined determination of NAG, AAP and lysozyme, whereas Burchardt et al. [15] proposed the combination AAP, glucuronidase and lysozyme. In contrast to NAG, the glucuronidase or galactosidase as similar lysosomal enzymes did not appear to be identically suited for the detection of chronic renal diseases [58, 72]. For example, in comparative determinations NAG was increased in 98% of patients whereas elevated glucuronidase was observed in only 50% [58]. Glucuronidase resembles more the brush border enzymes in this respect [58].

Lysozyme or RNase as low molecular mass proteins are less often increased than tubular enzymes like NAG or AAP [42]. Tubular enzymuria and low molecular mass proteinuria probably reflect two different mechanisms. Damage to a small number of tubular cells may result in an increased enzymuria without affecting the large reabsorptive-catabolizing function of this nephron part [37]. Only if there is a substantial loss of tubular cells by permanent destruction do low molecular mass proteins appear in the urine, whereas tubular-specific enzymes may be reduced because of the diminished brush border membranes as mentioned before.

Patients with latent nephropathies showed increased AAP excretion after intravenous application of hypertonic solutions like diatrizoate, whereas it changed insignificantly in healthy subjects (Fig. 4). It was concluded that this elevation, caused by osmotic nephrosis after application of the solution, developed more intensively in people with pre-damaged than with healthy kidneys

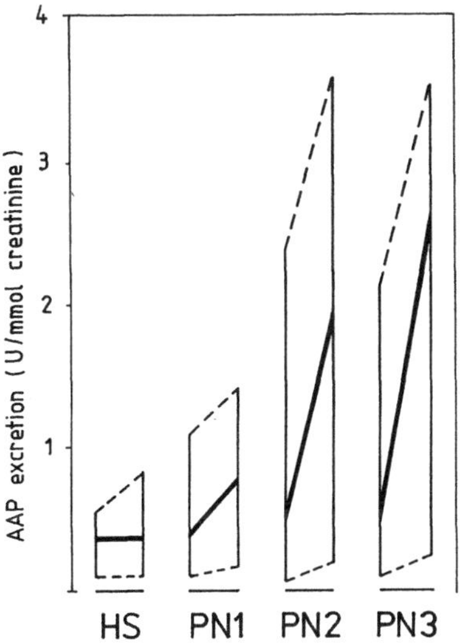

Fig. 4. Increase of urinary alanine aminopeptidase (*AAP*) excretion after intravenous application of diatrizoate (according to [15]). *HS*, 55 healthy subjects; *PN1*, 34 patients suffering from latent chronic pyelonephritis without retention and without bacteriuria; *PN2*, 40 patients suffering from active chronic pyelonephritis without retention and with bacteriuria; *PN3*, 35 patients suffering from active chronic pyelonephritis with retention and with bacteriuria. Values are given as median values plus ranges

[15]. This "load test" has been recommended to detect patients suffering from latent pyelonephritis [15].

Multiple forms of urinary AAP, AP and GGT (particulate and soluble fractions) were separated in patients with chronic renal disease [43, 69]. There was a tendency for the soluble fraction to increase in these patients. However, the separation into multiple forms led to no improvement in diagnostic reliability than the determination of total excretion activity of the respective enzymes [43]. Isoenzyme analysis of NAG by ion-exchange chromatography demonstrated elevation of the B-form [23]. Systematic investigations of the diagnostic validity of NAG isoenzyme separation compared with total activity have not yet been performed in patients suffering from chronic renal disease.

In summary, the determination of NAG excretion in urine can be considered to be a suitable additional indicator for the detection of chronic renal diseases. At present, NAG seems to be the best urinary enzyme for this purpose. As elevated levels are also observed in pathological conditions other than kidney diseases and even under physiological conditions, it should not replace the use of the conventional parameters serum creatinine level and urinary protein. Nevertheless, the measurement of NAG excretion can be recommended as an additional tool to improve diagnostic facilities in nephrology [35].

12.4.2 Enzyme Pattern as a Tool for the Differential Diagnosis of Chronic Renal Diseases

Different excretion rates of 10 urinary enzymes were observed in chronic glomerulonephitis and chronic pyelonephritis [56, 57]. Glomerular kidney diseases showed higher average enzyme excretions and more frequent pathological results than interstitial diseases [56, 83]. These qualitative and quantitative variations of the enzyme pattern suggest the value of a multivariate analysis of these data. Using the method of discriminant analysis, a pattern of 10 enzymes resulted in a correct classification of about 90% of patients with glomerular and interstitial renal diseases (Table 3). The diagnostic classification was more effective in patients with normal than with reduced kidney function because an assimilation of the enzyme pattern of glomerular and interstitial diseases was observed with advanced deterioration of kidney function [56]. However, it was concluded that so many enzymes in combination could not meet the requirement of cost efficiency, and it is the aim of further studies to reduce the necessary parameters for differential diagnosis [56]. Laterza et al. [50] studied four enzymes (NAG, galactosidase, glucosidase, glucuronidase) in patients suffering from chronic renal disease and applied the method of partial and multiple correlations. They suggested that an accurate estimate might be achieved with just galactosidase and glucuronidase. It is obvious that further investigations are necessary in this respect.

Table 3. Discrimination between glomerular and interstitial renal diseases with normal kidney function. Results from 36 patients with chronic glomerulonephritis and 15 patients with interstitial nephritis by discriminant analysis. At each step the enzyme giving the greatest increase in the discrimination entered the analysis (according to [57])

Step	Parameter	Correct diagnosis (%)		
		Glomerular disease	Interstitial disease	Overall
1	Glucuronidase	53	93	65
2	Step 1 + GGT	58	100	71
3	Step 2 + NAG	58	100	71
4	Step 3 + GAL	64	100	75
5	Step 4 + TRE	75	100	82
6	Step 5 + NAG/GGT	75	100	82
7	Step 6 + GLUC	81	100	86
8	Step 7 + LDH	83	100	88
9	Step 8 + ARS	86	100	90
10	Step 9 + LAP	89	100	92
11	Step 10 + AP	89	100	92
12	Step 11 + protein	94	100	96

GAL, galactosidase; TRE, trehalase; GLUC, glucosidase; ARS, arylsulfatase; LAP, leucine arylamidase; GGT, gamma-glutamyltransferase; NAG, *N*-acetyl-β-D-glucosaminidase; LDH, lactate dehydrogenase; AP, alkaline phosphatase

12.5 Enzymuria in Renal and Urinary Tract Tumours

Changed excretion rates have been described for AAP [76], AP [25, 51, 70], arylsulfatase [4, 55, 65, 71], collagenase [33], GGT [29, 31, 32], β-glucuronidase [25, 27, 33, 65], LDH [12, 25, 31, 51, 61, 62, 70, 75], malate dehydrogenase [31] and NAG [33, 76] in the urine of patients with renal and bladder tumours. Both increased and decreased activities were observed.

In the 1960s, especially AP and LDH were investigated, and increases in cancer patients to about five- to tenfold normal values were described. A review [25] of diagnostic results obtained with these two enzymes in various studies of patients with neoplasms of the kidney and bladder are summarized in Table 4. It has been suggested that the increase of LDH is caused by the LDH_5 fraction

Table 4. Urinary enzyme activity values in patients with neoplasm of the kidney and bladder (according to data given in [25])

Carcinoma	Lactate dehydrogenase		Alkaline phosphatase	
	n	Increased values	*n*	Increased values
Kidney	138	104 (75%)	42	26 (62%)
Bladder	257	162 (63%)	35	15 (43%)

[52] and that of AP by the kidney fraction [70]. Similar LDH isoenzyme changes were observed in patients with bladder carcinoma [12, 62]. These isoenzyme characteristics are probably caused by the corresponding changes in the tissue. In contrast to the renal tissue, malignant tumour cells are typically enriched with LDH_4 and LDH_5 [25]. However, these LDH changes could not be verified in all cases [70].

The limited diagnostic sensitivity of both enzymes to detect a tumour is evident (Table 4). Similar results are valid for AAP, ß-glucuronidase and NAG, although few studies were performed [27, 33, 62, 76]. As the authors often did not mention the stage of the tumour, the enzyme results cannot be evaluated in relation to the "early" marker concept. Russo et al. [76] suggested that increased AAP and NAG were the expression of infiltration into the surrounding renal parenchyma whereas a normal enzyme value showed that the neoplastic process was still confined to the excretory pathways. Small localized tumours could not be detected [25]. In addition to the limited sensitivity, the diagnostic specificity is also low because any non-specific damage to the kidney results in urinary enzyme changes, as in the case of LDH [25], or these changes more closely reflect cellular contamination of the urine rather than the histological findings [61]. Different results have been obtained using urinary enzyme excretion in order to discriminate between malignant and benign tumours. Whereas some authors found normal excretion levels of LDH and β-glucuronidase in patients with benign tumours, other authors described increased values [61, 65, 75]. It has been concluded that the LDH test lacks diagnostic specificity and sensitivity for the detection of neoplastic processes [51, 61, 75].

Hautmann et al. [31, 32] investigated 12 enzymes in urine collected by bilateral ureteric catherisation and found as a typical malignant pattern increased LDH and malate dehydrogenase (MDH) combined with decreased GGT excretion. The urinary GGT level is decreased because renal carcinoma tissue contains less GGT than does normal kidney [31]. In comparison with normal renal GGT, a novel variant could be detected [29]. Using the above-mentioned pattern a sensitivity of 91% and a specificity (patients with benign tumours and inflammatory renal diseases) of 92% were obtained in 141 patients with a malignant tumour. That parameter combination gave better results than the decision criterion made only on the basis of decreased GGT activity [31]. Until now, these positive results have not been confirmed. However, it stresses that multivariate analysis by combining data from various enzymes may improve diagnostic results.

Patients with bladder cancer showed an increased urinary arylsulphatase activity [4, 55, 65, 71]. This enzyme's level is more consistently elevated than those of LDH and AP [71]. Arylsulfatase A increased in all patients investigated whereas isoenzyme B did in 83%. It is hypothesized that increased arylsulfatase excretion is correlated with the ability of a bladder tumour to spread [71].

In summary, it must be concluded that all urinary enzymes investigated at present cannot be considered as reliable tumour markers because of their low diagnostic sensitivity and specificity. The numerous studies performed in the past have not fulfilled expectations of a sensitive marker. Meanwhile, sophisticated methods such as special roentgenographic techniques, ultrasound and organ scanning using computerised tomography give a high level of confidence in diagnosis.

12.6 Enzymuria in Nephrolithiasis and After Extracorporeal Shock-Wave Lithotripsy

Increased AAP, AP, GGT, LDH and NAG excretion and decreased aminotransferase excretion have been found in patients suffering from nephrolithiasis [6, 8, 46, 74, 82]. These observations are important both from pathobiochemical and diagnostic viewpoints. The changed enzyme excretion pattern is explained by:

1. A consequence of a primary metabolic disorder in the kidney which is responsible for or at least promotes stone formation [6]. It is well-known that L-glutamic acid inhibits kidney stone formation. Whereas healthy people have sufficient renal aminotransferase to form this L-glutamic acid and thus to retard stone formation, stone formers have a decreased aminotransferase activity and the disadvantage of low L-glutamic acid formation. The reduced urinary aminotransferase activity follows from the decreased renal enzyme activity. There are similar conclusions of changed metabolic processes in the kidney of stone-forming patients and their association with increased LDH and MDH excretion [82].
2. Tubular damage. It has been suggested that stone-forming patients suffer from injured proximal tubules because crystal deposition in the kidneys alters the cellular membrane [22]. Experimental studies demonstrated that the increase of various enzymes was evident before calcium oxalate crystalluria occurred and crystals in the cortex or papillary tubes appeared [46]. Thus, the damage to the proximal tubules is not primarily induced by the mechanical damage by the crystals but by hyperoxaluria. A highly significant positive correlation between urinary oxalate level and NAG excretion could be demonstrated [46].

It has been reported that AAP, AP, GGT, LDH and NAG excretion can be used to discriminate between healthy subjects and stone-forming patients [74]. Even the simple ratio GGT/AAP has a diagnostic specificity of 76% and a sensitivity of 80%. The diagnostic efficiency could be improved by using multidimensional discrimination methods with which a specificity of 100% was obtained [74].

Extracorporeal shock-wave lithotripsy has become the treatment of choice for most urinary tract stones. Although it has been accepted as a routine method, its short-term and long-term side effects on kidney function are not yet known exactly [1]. The time-dependent behaviour of the excretion of various urinary enzymes (AAP, AP, ß-galactosidase, GGT, LDH, lysozyme, NAG) were measured after extracorporeal shock-wave lithotripsy to assess possible renal injury by this treatment method [1, 5, 45, 47, 48, 54]. Both increased and unchanged excretion levels were found [5, 45, 47, 48, 54]. Post-lithotripsy enzyme increase was interpreted to be due to a transient proximal tubule injury although other signs of renal dysfunction were not established [5]. The number of shock waves used did not correlate with enzyme output [1, 45] and did not differ substantially between studies with contradictory results [1, 45, 48]. Consequently, it cannot be excluded that the various instruments used for lithotripsy (e.g. Lithostar from Siemans or Lithotripter from Dornier) produce a different injurious effect on the renal parenchyma.

12.7 Enzymuria in Pregnancy-Associated Nephropathy

Normal pregnancy is accompanied by gestional-dependent changes of urinary enzyme excretion. Kidney involvement in pregnancy occurs in three forms:

1. First appearance of a primary kidney disease; acute pyelonephritis occurs in 1%–2% of pregnant women
2. Effect of pregnancy on pre-existing renal disease (e.g. chronic glomerulonephritis, diabetic nephropathy)
3. Pre-eclampsia or eclampsia

Here we shall discuss only enzymuria in pre-eclampsia/eclampsia because the other involvements are described in Sects. 12.3, 12.4, 15.6.

Pre-eclampsia/eclampsia is characterized by hypertension and oedema and leads to a non-selective proteinuria through glomerular lesions and a vasospastic effect. The vasospastic renal ischaemia may induce renal tubular damage which intensifies the decrease of tubular reabsorption of proteins already observed in normal pregnancy [19] and the release of enzymes from proximal tubular cells into the urine [11, 28, 40]. Median values of AAP [28] and NAG [28, 40] excretion in women with mild and severe pre-eclampsia were significantly higher than the corresponding values in healthy pregnant women. The increased NAG activity was accompanied by a change of isoenzyme pattern in favour of the B form [24]. However, many women suffering from light and moderate pre-eclamptic symptoms (and also from severe pre-eclampsia) had low levels of enzyme excretion [11, 28]. This phenomenon is explained by the intermittent effect of vasospasm [28]. The overlap of enzymuria values of pregnant women with pre-eclamptic symptoms and those of healthy pregnant women does not permit us to recommend enzymuria as a sensitive test for detecting pre-eclampsia [28] and predicting foetal outcome [63]. Severe kidney damage is indicated earlier by enzymuria than by conventional kidney function tests [11]. As a negative test of enzymuria does not preclude pre-eclampsia, the clinical value of monitoring enzymuria in this case is limited and can be useful only if the other causes of changed enzyme values in pregnancy are considered [63, 64, 85, 91].

12.8 Enzymuria in Renal and Renovascular Hypertension, Renal Artery Stenosis and Essential Hypertension

Changes of the urinary excretion pattern in patients with hypertension and renal artery stenosis are described in Sect. 15.2. As hypertension can be both cause and effect of renal diseases, measurement of urinary enzymes has been suggested to be a valuable adjunctive test for differentiating and monitoring hypertension [2, 41, 59].

12.9 Enzymuria in Systemic Diseases with Secondary Renal Involvement and in Inborn Errors of Metabolism

Urinary enzyme excretion patterns in diabetes mellitus (diabetic nephropathy) and connective tissue diseases are described in Sects. 15.6 and 15.7. Chapter 15 deals with further extrarenal disorders and hereditary diseases affecting enzyme excretion.

References

1. Ackaert KSJW, Schröder FH (1989) Effects of extracorporeal shock wave lithotripsy (ESWL) on renal tissue. Urol Res 17:3–7
2. Alderman MH, Melcher L, Drayer DE, Reidenberg MM (1983) Increased excretion of urinary N-acetyl-ß-glucosaminidase in essential hypertension and its decline with antihypertensive therapy. N Engl J Med 309:1213–1217
3. Appelmelk BJ, Maclaren DM (1981) Localization of urinary-tract infection with urinary lactic dehydrogenase isoenzyme 5. Lancet i:1417–1418
4. Arenas J, Mayor J, Gella FJ, Fraile B, Bornstein B, Jarillo MD, Martinez A, Santos I (1988) Arylsulfatase A in urine of patients with urothelial tumors. Clin Biochem 21:73–77
5. Assimos DG, Boyce WH, Furr EG, Espeland MA, Holmes RP, Harrison LH,Kroovand RL, McCullough DL (1989) Selective elevation of urinary enzyme levels after extracorporeal shock wave lithotripsy. J Urol 142:687–690
6. Azoury R, Garti N, Perlberg S, Sarig S (1982) May enzyme activity in urine play a role in kidney stone formation? Urol Res 10:185–189
7. Baggio B, Favaro S, Cantaro S, Bertazzo L, Frunzio A, Borsatti A (1981) Increased urine

angiotensin I converting enzyme in patients with upper urinary tract infection. Clin Chim Acta 109:211–218
8. Baggio B, Gambaro G, Ossi E, Favaro S, Borsatti A (1983) Increased urinary excretion of renal enzymes in idiopathic calcium oxalate nephrolithiasis. J Urol 129:1161–1162
9. Beck PR (1980) The significance and value of urinary gamma-glutamyltransferase assays. In: Goldberg DM, Werner M (eds) Progress in clinical enzymology. Masson, Washington, pp 109–115
10. Bonadio M, Donadio C, Catania B (1979) Lysozymuria and upper-urinary-tract infection. N Engl J Med 301:1065–1066
11. Borenstein R, Rozenman D, Lancet M, Bar-Khayim Y, Cohen N Gertler A (1980) Significance of urinary enzymes in gestosis. Gynecol Obstet Invest 10:193–203
12. Bredin CH, Daly JJ, Prout GR (1975) Lactic dehydrogenase isoenzymes in human bladder cancer. J Urol 113:487–490
13. Burchardt U, Mampel E, Peters JE, Thulin H (1975) Pathophysiologische und klinische Aspekte der Harnenzymausscheidung bei akuten Nierenversagen. Dtsch Gesundheitswes 30:1067–1070
15. Burchardt U, Haschen RJ, Krosch H (1979) Clinical usefulness of enzyme determinations in urine. Curr Probl Clin Biochem 9:106–112
16. Butterworth PJ, Moss DW, Pitkanen E, Pringle A (1965) Patterns of urinary excretion of alkaline phosphatase in acute renal failure. Clin Chim Acta 11:212–219
17. Carvajal HF, Passey RB, Berger M, Travi LB, Lorentz WB (1975) Urinary lactic dehydrogenase isoenzyme 5 in the differential diagnosis of kidney and bladder infections. Kidney Int 8:176–184
18. Chapman A, Legrain M (1979) Acute tubular necrosis and interstitial nephritis. In: Hamburger J, Crosnier J, Grünfeld JP (eds) Nephrology. Wiley, New York, pp 383–410
19. Cheung CK, Lao T, Swaminathan R (1989) Urinary excretion of some proteins and enzymes during normal pregnancy. Clin Chem 35:1978–1980
20. Coward RA, DeLamore IW, Mallick NP, Robinson LE (1984) Urinary N-acetyl-ß-D-glucosaminidase as an indicator of tubular damage in multiple myeloma. Clin Chim Acta 138:293–298
21. Dance N, Price RG, Cattell WR, Lansdell J, Richards B (1970) The excretion of N-acetyl-ß-glucosaminidase and ß-galactosidase by patients with renal disease. Clin Chim Acta 27:87–92
22. Dykstra M, Hackett RL (1979) Ultrastructural events in early calcium oxalate crystal formation in rats. Kidney Int 15:640–650
23. Ellis BG, Tucker SM, Thompson AE, Price RG (1975) Presence of serum and tissue forms of N-acetyl-ß-glucosaminidase in urine from patients with renal disease. Clin Chim Acta 64:195–202
24. Furiya K, Yoshida M, Yoshida S, Isono S, Nagano T, Kakuta Y, Shiozaki M, Takeuchi M, Morikawa M (1985) Urinary N-acetyl-ß-D-glucosaminidase (EC 3.2.1.30) activity during pregnancy. J Tokyo Woman's Med Coll 55:55–63
25. Gault HM, Geggie PHS (1969) Clinical significance of urinary LDH, alkaline phosphatase and other enzymes. Can Med Assoc J 101:208–215
26. Gibey R, Dupond JL, Alber D, Leconte des Floris R, Henry JC (1981) Predictive value of urinary N-acetyl-beta-D-glucosaminidase (NAG), alanine aminopeptidase (AAP) and beta-2-microglobulin ($ß_2$M) in evaluating nephrotoxicity of gentamicin. Clin Chim Acta 116:25–34
27. Gonick HC, Kramer HJ, Schapiro AE (1973) Urinary ß-glucuronidase in renal disease. Arch Intern Med 132:63–69
28. Goren MP, Sibai BM, El-Nazar A (1987) Increased tubular excretion in preeclampsia. Am J Obstet Gynecol 157:906–908
29. Hada T, Higashino K, Yamamoto H, Yamamura Y, Matsuda M, Osafune M, Kotake T, Sonoda T (1978) A novel γ-glutamyl transpeptidase in renal carcinoma with normal kidney enzyme. Clin Chim Acta 85:267–277
30. Harris RE, Thomas VL, Underwood JM, Gaudot F (1978) A comparison of two indirect methods for localizing the site of urinary tract infection: beta glucuronidase levels and the presence of antibody-coated bacteria. Am J Obstet Gynecol 131:647–649
31. Hautmann R (1979) Diagnosis of renal disorders: comparison of urinary patterns with corresponding tissue patterns. Curr Probl Clin Biochem 9:58–70
32. Hautmann R, Kisters R, Lutzeyer W (1976) Diagnosis of renal tumor by γ-glutamyltranspeptidase (EC 2.3.2.2). Urology 7:12–15
33. Heller W, Harzmann R, Bichler KH, Gericke D, Grötsch H, Schmidt K (1978) ß-Glucuronidase, N-acetyl-glucosaminidase and collagenase in cancer of the urinary bladder: experimental and clinical aspects. J Clin Chem Clin Biochem 16:203–204

. Hofer E, Keiler A, Kotanko P (1988) Enzymuria in controlled hypotension in humans—preliminary results. In: Lubec G, Bartosch B (eds) Non-invasive diagnosis of kidney disease. Libbey, London, pp 99–106
. Hofmann W, Guder WG (1989) Moderne Methoden zur Proteindifferenzierung im Urin. Lab Med 13:336–344
. Horpacsy G (1983) Diagnostic significance of standardized determination of urinary enzymes in the practice of human kidney transplantation and renal toxicology. Contrib Nephrol 35:105–127
. Hultberg B (1980) Urinary excretion of ß-hexosaminidase in different forms of proteinuria. Clin Chim Acta 108:195–199
. Hultberg B, Ravnskov U (1981) The excretion of N-acetyl-ß-glucosaminidase in glomerulonephritis. Clin Nephrol 15:33–38
. Hultberg B, Wieslander J (1982) Urinary excretion of ß-hexosaminidase in patients with vesicoureteric reflux. Acta Med Scand 211:257–259
. Hultberg B, Isakson A, Krutzen E, Nilsson-Ehle P (1989) Urinary excretion of N-acetyl-ß-D-glucosaminidase in normal and complicated pregnancy. J Clin Chem Clin Biochem 27:487–489
. Johnston IDA, Jones NF, Scoble JE, Yuen CT, Price RG (1983) The diagnostic value of urinary enzyme measurements in hypertension. Clin Chim Acta 133:317–325
. Jung K, Schulze BD (1988) Comparative determinations of different urinary enzymes and low-molecular mass proteins in urine and serum of patients suffering from chronic renal diseases. In: Lubec G, Bartosch B (eds) Non-invasive diagnosis of kidney disease. Libbey, London, pp 65–80
. Jung K, Pergande M, Reinholdt C, Schulze B, Scholz D, Strobelt V (1984) Multiple forms of alanine aminopeptidase, alkaline phosphatase and gamma-glutamyltransferase in urine of healthy persons, patients suffering from kidney diseases and patients with kidney transplants. J Clin Chem Clin Biochem 22:523–525
. Jung K, Schulze BD, Sydow K (1987) Diagnostic significance of different urinary enzymes in patients suffering from chronic renal diseases. Clin Chim Acta 168:287–295
. Jung K, Brien G, Kirschner P, Schröder K, Schmiedel T (1989) Excretion of urinary enzymes after extracorporeal shock-wave lithotripsy. Clin Chem 35:2000–2001
. Khan SR, Shevock PN, Hackett RL (1989) Urinary enzymes and calcium oxalate urolithiasis. J Urol 142:846–849
. Kishimoto T, Yamamoto K, Sugimoto T, Yoshihara H, Maekawa M (1986) Side effects of extracorporeal shock-wave exposure in patients treated by extracorporeal shock-wave lithotripsy for upper urinary tract stone. Eur Urol 12:308–313
. Kitada S, Kuramoto H, Kumazawa H, Yamaguchi A, Nakasu H, Hara S (1989) Effects of extracorporeal shock wave lithotripsy on urinary excretion of N-acetyl-beta-D-glucosaminidase. Urol Int 44:35–37
. Kunin CM, Chesney RW, Craig WA, England AC, DeAngelis C (1978) Enzymuria as a marker of renal injury and disease: studies of N-acetyl-beta-glucosaminidase in the general population and in patients with renal disease. Pediatrics 62:751–760
. Laterza G, Santini P, Russo G, Scibinetti F, Gentile V, Liberti M, De Berardinis E, Clemenzia G (1985) Partial and multiple correlation between four enzymes in two groups of chronic nephritis sufferers. Minerva Med 76:99–104
. Lumberopoulos S, Lutzeyer W (1966) Praktische Bedeutung der Fermentdiagnostik in der Urologie. Urologe 5:181–189
. Macalalag EV, Prout GR (1964) Confirmation of the source of elevated urinary lactic dehydrogenase in patients with renal tumor. J Urol 92:416–423
. Maekawa M (1988) Lactate dehydrogenase isoenzymes. J Chromatogr 429:373–398
. Marcellan FJR, Servio LI (1986) Evaluation of renal damage in extracorporeal lithotripsy by shock waves. Eur Urol 12:73–75
. Maru A, Ishibashi T, Imai Y, Makita A, Tsuji I (1980) Arylsulfatase A activity of urine in patients with various genitourinary tract disorder. Clin Chim Acta 107:155–161
. Maruhn D (1979) Evaluation of urinary enzyme patterns in patients with kidney diseases and primary hypertension. Curr Probl Clin Biochem 9:135–145
. Maruhn D, Bock KD (1979) Urinary enzyme excretion in renal disease. In: Burlina A, Galzigna L (eds) Clinical enzymology symposia. Piccin, Padua, pp 491–499
. Maruhn D, Hartmann H, Heredia D, Singchoovong L, Bock KD (1978) Urinary excretion of trehalase, N-acetyl-ß-glucosaminidase and ß-glucuronidase in patients with glomerulopathies. In: Goldberg DM (ed) Enzymes in health and disease. Karger, Basel, pp 201–205
. Maruhn D, Paar D, Bock KD (1979) Lysosomal and brush border membrane enzymes in urine

of patients with renal artery stenosis and with essential hypertension. Clin Biochem 12:228–230
60. Mengoli C, Lechi A, Arosio E, Rizzotti P, Lechi C, Corgnati A, Micciolo R, Pancera P (1982) Contribution of four markers of tubular proteinuria in detecting upper urinary tract infections. A multivariate analysis. Nephron 32:234–238
61. Mirabile CS, Bower GN, Berlin BB (1966) Urinary lactic dehydrogenase: a report based on 250 hospitalized patients. J Urol 95:79–82
62. Mokomiya Y, Yamada K, Mitsushima S, Ijuim M, Inya K, Okajima E (1975) Studies on urinary isoenzyme of lactic dehydrogenase and ß-glucuronidase in patients with bladder tumors. Urol Res 3:41–48
63. Mullan DP (1967) Urinary excretion of leucine aminopeptidase in pregnancy. J Clin Pathol 20:660–664
64. Noble MCB, London MJ, Davison JM (1977) The excretion of γ-glutamyl transferase in pregnancy. Br J Obstet Gynaecol 84:522–527
65. Paigen B, Yarfitz S, Tabron D (1984) Urinary glucuronidase and arylsulfatases in identical twins of bladder cancer. Cancer Res 44:3624–3626
66. Palmieri L, Ronca G, Cioni L, Puccini R (1984) Enzymuria as a marker of renal injury and disease: studies of N-acetyl-ß-glucosaminidase, alanine aminopeptidase and lysozyme in patients with renal disease. Contrib Nephrol 42:123–129
67. Peters JE, Mampel E, Schneider I, Burchardt U, Fukala, Ahrens I, Haschen RJ (1972) Alaninaminopeptidase (L-Alanyl-Peptidhydrolase) im Harn bei Nierenerkrankungen. Clin Chim Acta 37:213–224
68. Pfau A, Ashkenazi A, Sacks TG (1968) The value of urinary ß-glucuronidase activity in the assessment of urinary tract infections. Isr J Med Sci 4:1249–1253
69. Pfleiderer G, Baier M, Mondorf AW, Stefanescu T, Scherberich JE, Müller H (1980) Change in alkaline phosphatase isoenzyme pattern in urine as possible marker for renal disease. Kidney Int 17:242–249
70. Pinto N, Eidelman A, Zaidman JL (1980) The diagnostic value of urinary alkaline phosphatase and lactate dehydrogenase enzyme and isoenzyme excretion in renal and urologic diseases and in reference individuals. In: Goldberg DM, Werner M (eds) Progress in clinical enzymology. Masson, Washington, pp 122–123
71. Posey LE, Morgan LR (1977) Urine enzyme activities in patients with transitional cell carcinoma of the bladder. Clin Chim Acta 74:7–10
72. Pott G, Zumkley H, Intorp HW (1976) Enzymaktivitätsveränderungen der ß-Hexosaminidase im Urin bei Nierenkranken. Klin Wochenschr 54:1001–1003
73. Price RG (1982) Urinary enzymes, nephrotoxicity and renal diseases. Toxicology 23:99–134
74. Prinsen JH, Günther H, Breuer J (1986) Bestimmung von Enzymaktivitäten in Urinen von Calciumoxalatstein-Patienten. J Clin Chem Clin Biochem 24:1001–1007
75. Riggins RS, Kiser WS (1963) A study of lactic dehydrogenase in urine and serum of patients with urinary tract disease. J Urol 90:594–601
76. Russo G, Gentile V, Scibinetti F, Liberti M, De Berardinis E, Perugia G, Gallo G, Clemenzia G (1986) Study of urinary enzymes in kidney cancer. Panminerva Med 28:35–38
77. Salgo L, Szabo A (1982) γ-Glutamyl transpeptidase activity in human urine. Clin Chim Acta 126:9–16
78. Sandberg T, Cooper EH, Lidin-Janson G, Yu H (1985) Fever and proximal tubular function in acute pyelonephritis. Nephron 41:39–44
79. Sandberg T, Bergmark J, Hultberg B, Jagenburg R, Trollfors B (1986) Diagnostic potential of urinary enzymes and β_2-microglobulin in acute urinary tract infection. Acta Med Scand 219:489–495
80. Scheler F, Bergmann H (1963) Die Leucin-Aminopeptidase-Ausscheidung beim akuten Nierenversagen. Verh Dtsch Ges Inn Med 69:732–735
81. Schmid H, Mall A, Bockhorn H (1986) Catalytic activities of alkaline phosphatase and N-acetyl-ß-D-glucosaminidase in human cortical nephron segments: heterogenous changes in acute renal failure and acute rejection following kidney allotransplantatio. J Clin Chem Clin Biochem 24:961–970
82. Schwille P, Bäurle H, Bäurle-Uhländer G, Sigel A (1973) Untersuchungen zum Verhalten von Enzymen im Serum und Urin bei Nephrolithiasis: Lactatdehydrogenase, Leucinaminopeptidase, Malatdehydrogenase und α-Amylase. Urol Int 28:405–416
83. Sherman RL, Drayer DE, Leyland-Jones BR, Reidenberg MM (1983) N-Acetyl-ß-glucosaminidase and β_2-microglobulin. Their urinary excretion in patients with renal parenchymal disease. Arch Intern Med 143:1183–1185

84. Sheth KJ, Good TA (1978) Urinary acidic hydrolases in renal diseases in children. Clin Nephrol 10:151–157
85. Skrha J, Perusicova J, Sperl M, Bendl J, Stolba P (1989) N-Acetyl-ß-glucosaminidase and albuminuria in normal and diabetic pregnancies. Clin Chim Acta 182:281–288
86. Straus B, Bajakic D, Cepelak I, Oberiter V, Kunovic B (1985) Changes in enzyme activities in urine in children with renal diseases. Z Med Lab Diagn 26:138–145
87. Sun T, Chow C, Mcvicar M, Mailloux L (1985) Urinary lactate dehydrogenase isoenzyme analysis in adult population. Ann Clin Lab Sci 15:32–38
88. Thiele KG, Mattenheimer H (1968) The isoenzymes of lactate dehydrogenase in the nephron of the healthy human kidney. J Clin Chem Clin Biochem 6:132–138
89. Vanderlinde RE (1981) Urinary enzyme measurements in the diagnosis of renal disorders. Ann Clin Lab Sci 11:189–201
90. Vigano A, Assael BM, Villa AD, Gagliardi L, Principi N, Ghezzi P, Salmona M (1983) N-Acetyl-ß-D-glucosaminidase (NAG) and NAG isoenzymes in children with upper and lower urinary tract infections. Clin Chim Acta 130:297–304
91. Vinacur JC, Casellas JM, Rubi RA, Oneto EP (1974) Serum anti-Escherichia coli antibodies and ß-glucuronidase for the diagnosis and control of evolution of urinary infection during pregnancy. Am J Obstet Gynecol 120:812–816
92. Wellwood JM, Ellis BG, Price RG, Hammond K, Thompson AE, Jones NF (1975) Urinary N-acetyl-ß-D-glucosaminidase activities in patients with renal disease. Br Med J 3:408–411
93. Bohle A, Mackensen-Haen S, v Gise H (1987) Significance of tubulointerstitial changes in the renal cortex for the excretory function and concentration ability of the kidney; a morphometric contribution. Am J Nephrol 7:421–433
94. Scherberich JE (1990) Urinary proteins of tubular origin: basic immunochemical and clinical aspects. Am J Nephrol 10, Suppl 1:43–51
95. Scherberich JE, Mösbauer A, Meißner A, Bentele D, Schoeppe B, Schoeppe W (1989) Kidney- and serum-derived proteins in urine of patients suffering from renal diseases or arterial hypertension. Klin Wochenschr 67, Suppl 17:44–47
96. Scherberich JE, Wolf G, Albers C, Nowack A, Stuckhardt C, Schoeppe W (1989) Glomerular and tubular membrane antigens reflecting cellular adaptation in human renal failure. Kidney Int 36, Suppl 27:S38–S51

13 Urinary Enzymes in Kidney Transplantation

K. Jung and P. Kotanko

13.1 Introduction

Renal transplantation provides a well-established treatment for patients with end-stage renal disease. The postoperative period may be complicated by a broad variety of graft alterations, especially graft rejection, acute tubular necrosis, nephrotoxicity, infections and surgical problems.

In addition to the clinical examination and modern techniques of radiology, nuclear medicine, sonography, immunology and histology, clinical chemistry is necessary for the detection of deterioration of the graft [8]. Differential diagnosis of the underlying aetiological disorder provides a challenging clinical problem.

Since the early 1960s the post-transplant excretion pattern of urinary enzymes was studied in order to provide additional information on the condition of the graft. The rationale of this approach was that damage to the graft should lead to a subsequent release of cellular enzymes into the urine.

Animal experiments performed at the beginning of the era of renal transplantation demonstrated the usefulness of urinary enzymes to detect rejection episodes [6, 12, 43, 45]. Up to now, all the commonly measured urinary enzymes have been determined in human graft recipients [22]. By evaluating the output of urinary enzymes after transplantation, numerous working groups have outlined the behaviour of different urinary enzymes in renal transplant recipients and have especially investigated the clinical utility of the measurement of urinary enzyme activities for an early diagnosis of acute rejection episodes [3, 5, 11, 16, 25, 28–33, 35, 47, 49, 50].

Despite a remarkable amount of results published in this field, the clinical use of urinary enzyme estimation after transplantation has not generally been accepted because the findings of various groups have been contradictory with regard to the diagnostic validity. In the following chapter we try to give a critical evaluation of the present state-of-the-art. Earlier reviews on this topic have been published [16, 22, 44, 52]. For systematical reasons we describe the urinary excretion of enzymes in graft recipients with an uncomplicated transplantation, acute tubular necrosis or a functionless graft after transplantation and in acute graft rejection.

Urinary Enzymes
K. Jung/H. Mattenheimer/U. Burchardt (Eds.)

13.2 Enzymuria in Uncomplicated Transplantation

In an immediately functioning graft a high initial urinary enzyme output is uniformly observed. One must keep in mind that graft function and morphological integrity is influenced by pre-transplant variables, including the donor's condition (living donor, cadaveric grafting; for example, blood pressure and underlying diseases), preservation time and method, duration of cold and warm ischaemia times. Thus, the pre-transplant graft damage and the intraoperative alterations define in addition to other factors like rejection or infection the post-transplant enzymuria.

Pathophysiological reasons for increased postoperative enzyme excretions are:

1. Release of enzymes from the damaged and not fully functioning kidney cells
2. Disturbed glomerular filtration barrier
3. Insufficient reabsorptive-catabolizing function of the tubular cells

This elevated excretion rate concerns enzymes derived from the kidney and plasma (Fig. 1). Intracellular enzymes (e.g. AAP and NAG as in Fig. 1) are released, and brush border membranes with enzymatic properties are shed [23, 39] into the urine. The enzymes are then washed out by the revived diuresis and appear in the urine. Brush border enzymes occur as soluble and particulate forms in the urine [23]; the appearance of membranous forms reflects cellular necrosis and of soluble forms, cellular recovery [46]. The initial increased excretion declines as graft function improves (Fig. 1).

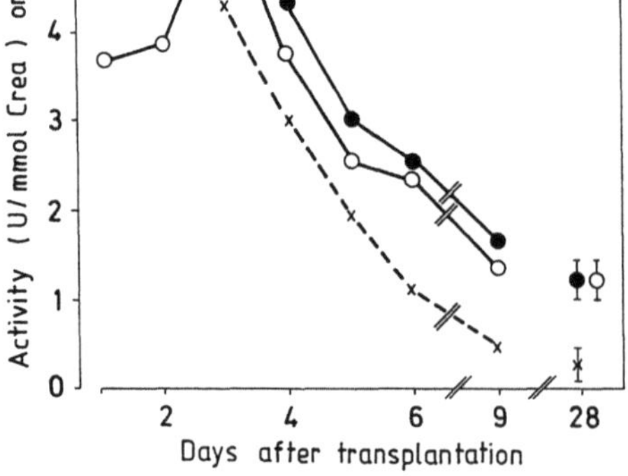

Fig. 1. Urinary activities of alanine aminopeptidase (*AAP*), N-acetyl-β-D-glucosaminidase (*NAG*) and lysozyme after kidney transplantation. Mean values of 23 (AAP, NAG) and 21 (lysozyme) renal transplant recipients (modified according to [22]). *Crea*, Creatinine

In addition, a disturbed glomerular filtration barrier results in a nonselective proteinuria leading to the appearance of serum enzymes in the urine. For example, the characteristic serum isoenzyme AAP is found in the urine when proteinuria exceeds 1 g/day [14]. Similar results could be demonstrated for NAG and AP [7, 37].

Deteriorated proximal tubular cells have a reduced capacity to reabsorb and catabolize filtered low-molecular mass proteins. Additionally, the filtered load of low molecular mass protein is increased because pre-transplant uraemia leads to an elevated serum concentration. For example, the pre-transplant serum concentration of lysozyme is increased 5–10-fold over the normal level, and its urinary excretion immediately after transplantation is 30–50-fold above normal (Fig. 1). Thus, a declining excretion of urinary lysozyme has been recommended as a useful functional indicator of tubular recovery [19].

About 5–10 days after transplantation a stable individual baseline of enzymuria is reached. However, even at discharge, patients with an uncomplicated postoperative course show higher urinary enzyme activities than healthy controls [47]. The long-term enzyme excretion rates after transplantation have been studied only occasionally [21, 32, 47]. Wellwood et al. [47] examined the excretion of NAG for 90 days after transplantation and found a decrease of enzyme output depending on the interval after transplantation. AAP, AP, GGT and NAG showed a similar behaviour over a period of 15 months after transplantation (Fig. 2) whereas the concentration of serum creatinine remained unchanged [21]. Post-transplant graft regeneration and steroid tapering have been suggested as causes [21]. During the early period after transplantation, regenerative processes take place in the transplant and can induce a higher enzyme excretion [36]. As there is a heterogenous change in the long-term

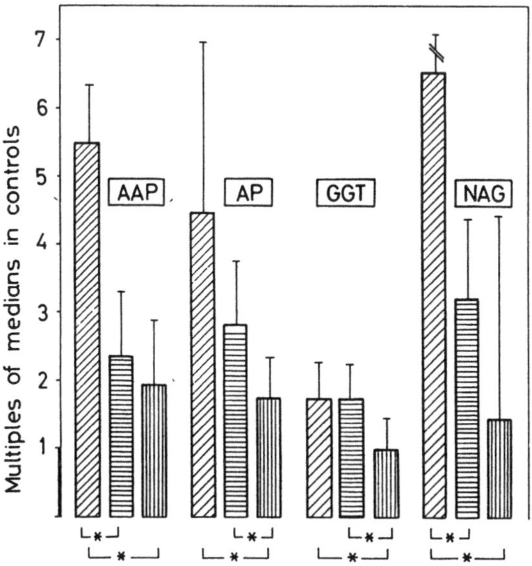

Fig. 2. Time-dependent behaviour of urinary enzyme excretion in six renal transplant recipients. Data (▨, 2 months; ▤, 9 months; ▥, 15 months after transplantation) are given as mean value ± SD of the median value of a control group. Time-dependent significance of difference ($p < 0.05$) calculated by nonparametric test according to Wilcoxon is indicated with asterisk (modified according to [21]). *AAP*, alanine aminopeptidase; *AP*, alkaline phosphatase; *GGT*, gamma-glutamyltransferase; *NAG*, *N*-acetyl-β-D-glucosaminidase

excretion of urinary enzymes (Fig. 2), the different intrarenal localizations and the time elapsed since grafting may explain these observations [40]. Until now, the exact mechanism is still a matter of debate.

13.3 Enzymuria in Acute Tubular Necrosis and Functionless Graft

Acute tubular necrosis is the consequence of the potentiation of pre-transplant damages and intraoperative alterations mentioned in Sect. 3.2. It leads to oliguria and is in general accompanied by a high urinary enzyme output. No decrease of enzymuria occurs as a function of the post-transplant interval. Although the urinary excretion of various enzymes in acute tubular necrosis is significantly higher than in acute rejecton episodes (Fig. 3), for the individual differential diagnosis urinary enzymes are insufficient as the only parameter. However, urinary LDH level related to glomerular filtration rate was recommended as a diagnostic tool to assess the extent of ischaemic injury [26]. In this respect it is worth mentioning that the measurement of perfusate NAG activity during preservation by continuous perfusion allowed the quantification of pre- and intra-perfusion injury and the prediction of postoperative performance of the kidney [2].

A combined evaluation of enzyme excretion and sodium concentration in the urine has been recommended as a useful indicator for the differentiation of acute tubular necrosis and acute graft rejection [18]. Acute tubular necrosis was characterized by high urinary enzyme activities and urinary sodium concentrations > 40 mmol/l, whereas a distinct decrease of sodium concentration in the

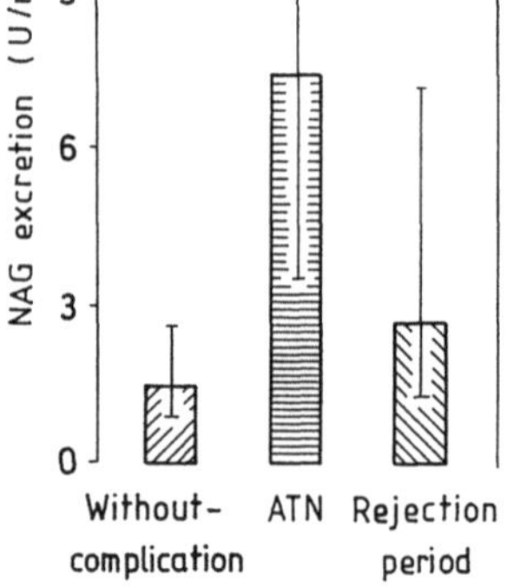

Fig. 3. Urinary *N*-acetyl-β-D-glucosaminidase (*NAG*) excretion on uncomplicated post-transplant days, in the "rejection period" and with acute tubular necrosis (*ATN*). Values (median, 25% and 75% percentiles) are given as U/mmol creatinine (according to P. Kotanko et al., manuscript in preparation)

presence of high urinary enzyme excretion could be observed in rejection episodes.

In the case that acute tubular necrosis causes a permanent loss of function (so-called primarily functionless transplant) and the number of perfused nephrons is small, low urinary enzyme output appears [16]. Urinary lysozyme activity may be high due to increased serum levels and the lack of tubular reabsorption. Interestingly, an abnormal low molecular mass variant of AAP was described [34] which resembled the fraction found in patients suffering from Balkan nephropathy and congenital renal Fanconi's syndrome. Hence, this AAP variant was discussed as a possible biochemical equivalent to tubular atrophy in patients with a functionless transplant [34].

When a graft is irreversibly damaged by rejection and then remains functionless (so-called secondarily functionless transplant), low brush border enzyme excretion is observed [16]. A decrease of urinary kallikrein to undetectable concentrations characterized recipients with a definite transplant failure after rejection [4].

13.4 Enzymuria in Rejection Episodes

Acute rejection episodes are characterized by an acute loss of function and occur in about 80% within the first post-transplant weeks. Rejection leads to a reduction of graft blood flow and thus to tissue hypoxia with a consecutive increase of membrane permeability and release of cellular enzymes into the urine (see Chap. 3).

Figure 4 displays the time course of urinary AAP and NAG activities over a period of 4 days prior to acute rejection. Both enzyme patterns are characterized by an increase of urinary output whereas serum creatinine shows a smaller elevation. Similar patterns were established for adenylate kinase [3], GGT [45], glucuronidase [13], glutathione S-transferase [1], LDH [52], lysozyme [41] and malate dehydrogenase [52]. In reversible rejection enzymuria decreases again [49]. Irreversible rejections often cause an extremely high increase of urinary enzymes [13] which may be followed by very low excretion rates [16]. However, the degree of enzymuria is not a reliable predictor of the outcome of rejection [51], a fact well-known from other fields of clinical enzymology, e.g. post-infarct behaviour of myocardial enzymes.

In addition to these quantitatively changed excretion rates, acute rejection is preceded by an altered isoenzyme pattern, which has been described for AP [36], GGT [20], NAG [5, 28, 51] and LDH [38]. The increased excretion of AP is essentially due to the intestinal-like isoenzyme fraction [36]. The liver-like form of urinary AP separated by ultracentrifugation or immunoprecipitation increases negligibly [36]. Changes of the NAG isoenzyme pattern after transplantation have been described by several groups [5, 28, 51]. Healthy people

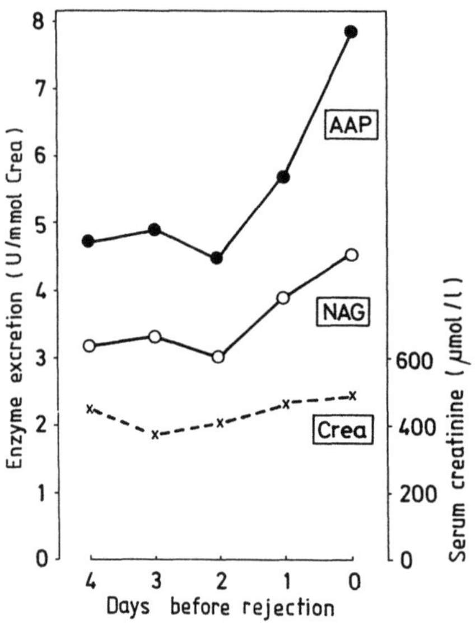

Fig. 4. Urinary alanine aminopeptidase (*AAP*) and *N*-acetyl-β-D-glucosaminidase (*NAG*) excretion in comparison with the behaviour of serum creatinine (*Crea*) during the 4 days preceding the clinical diagnosis of rejection. Data are given as mean values of 67 rejection episodes (modified and extended according to [25])

exhibit mainly fractions A and B (82% and 17%, respectively) and the intermediate forms I_1 and I_2 (about 1%) elucidated by diethylaminoethyl (DEAE)-cellulose fractionation. During acute rejection episodes the intermediate forms increased to above 20% at the expense of fraction A [28, 51]. In some cases a unique isoenzyme form located between I_2 and A could be observed [28, 37]. All rejection episodes studied by Yuen et al. [51] were associated with an increase of intermediate forms whereas the total NAG excretion was not uniformly increased. Successful treatment of a rejection crisis coincided with a return to the normal NAG isoenzyme distribution [51].

No systematic studies have been conducted to investigate the behaviour of urinary enzymes in chronic rejection. No doubt the performance of such a study is worthwhile, especially in the era of cyclosporine therapy, as the differential diagnosis of chronic rejection and cyclosporine nephrotoxicity provides a diagnostic challenge.

13.5 Diagnostic Significance of Urinary Enzymes in Renal Transplantation

Determination of urinary enzyme activity in patients with a kidney graft is motivated by the fact that graft damage leads to qualitative and quantitative changes in enzyme excretion. The crucial questions are whether or not the observed changes in the enzyme patterns described in Sect. 13.4 can be attributed to a specific clinical and pathological condition and whether or not a certain clinical condition (e.g. rejection) is heralded by a significant enzymuria.

From 1976 to 1978 a broad discussion on the diagnostic significance of urinary enzymes took place in clinical chemistry [10, 12, 27, 42]. In 1981 Vanderlinde [44] reviewed the opinions concerning the contradictory views on the diagnostic significance of urinary enzyme analysis. There are manifold clinical and methodological factors which account for divergent experiences (Table 1). Essential clinical reasons for these different results are the general uncertainty in the detection of rejection crises. Different groups of investigators used different methods to diagnose rejection episodes. The diagnostic spectrum ranged from clinical signs (e.g. fever, tenderness of the graft, increase of graft size, hypertension, oliguria) and basic clinical chemistry (e.g. rise of serum creatinine level) to sophisticated immunological monitoring, urinary cytology, intracapsule pressure measurements and biopsy. The "golden standard" of pathohistological examination of graft biopsies was not used uniformly by all groups. Additionally, numerous clinical conditions, including the use of nephrotoxic antibiotics, cytomegaloviral infection, urinary tract infection and urinary leaks may hamper the differential diagnosis of rejection.

At the beginning of the cyclosporine era high therapeutic doses were administered and produced frequently nephrotoxic side effects. For example, AAP and NAG excretion correlated significantly with blood levels of cyclosporine [9]. Even the histological diagnosis of nephrotoxicity versus rejection proved to be difficult. Today, lower therapeutic doses are applied, and acute cyclosporine nephrotoxicity is less frequent. No comparative study has been performed to assess the diagnostic power of urinary enzyme excretion for a rejection diagnosics in patients receiving different immunosuppressive regimens.

Moreover, the ability to compare the results achieved by different groups is limited as no generally accepted definition of the "rejection period" exists. The rejection period describes a specified time interval before the clinical diagnosis of

Table 1. Reasons for the differences of diagnostic potential of urinary enzyme measurements in detecting acute rejection episodes in renal transplant recipients, observed by various working groups (modified according to [24])

Clinical reasons
1. General uncertainty in detecting rejection crises
2. Use of different clinical and laboratory criteria for detecting rejection crises
3. Definition of a rejection period
4. Insufficient number of patients studied
5. Comparison of results from non-comparable groups of patients
6. Increases of urinary enzyme excretion by non-rejection factors

Methodological reasons
1. Determination of different enzymes
2. Determination of identical enzymes with different methods
3. Selection of different reference bases for the characterisation of enzyme excretion
4. Selection of different cut-off points for denoting a significant increase of enzyme excretion
5. Insufficient consideration of diagnostic criteria (sensitivity, specificity, efficiency) in evaluating the diagnostic significance of enzyme excretion

rejection, in which a certain enzyme change is valued as "positive". Such a well-defined period is mandatory for the comparison of results. Only two groups introduced a clear-cut time interval for data analysis [15, 24].

There are also numerous methodological reasons which account for the disagreement concerning the clinical use of urinary enzyme activities after kidney transplantation. Disappointing experience with some enzymes gave rise to the opinion that their use in general cannot be recommended as a diagnostic tool. Even with the same enzyme, different groups drew different conclusions. For example, some authors underlined the special usefulness of urinary NAG in detecting acute rejection [16, 29, 32, 37, 49]. Subsequently, a test paper and miniphotometer for self-control were developed [37]. On the other hand, other authors deny the usefulness of NAG determination [33] or concede it only confirmative value [27]. One possible explanation is that the colorimetric method is less sensitive than the fluorimetric one [37].

An additional confusing point is the basis of reference for urinary enzyme activities. Enzyme activities in the urine have been expressed as units/volume, units/time and units/urinary creatinine (see Chap. 9). Comparative investigation on this issue showed that the enzyme excretion related to the urinary creatinine concentration offered a better diagnostic performance than those given as volume activities or time-related output (Fig. 5). By relating enzyme excretion to urinary creatinine two effects improve the diagnostic value: firstly, enzyme excretion is adjusted to different urinary volumes, thus reducing the inter-individual variability (see Chap. 5); secondly, the signal, i.e. the activity/creatinine ratio, is amplified by the diminished excretion of urinary creatinine during acute rejection [24].

No generally accepted method exists for the selection of an "optimal" cut-off point to denote significantly increased enzyme excretion. For example, Wellwood et al. [47] considered an increase of 50% compared with the preceding day in NAG excretion as a positive test whereas Loertscher et al. [33]

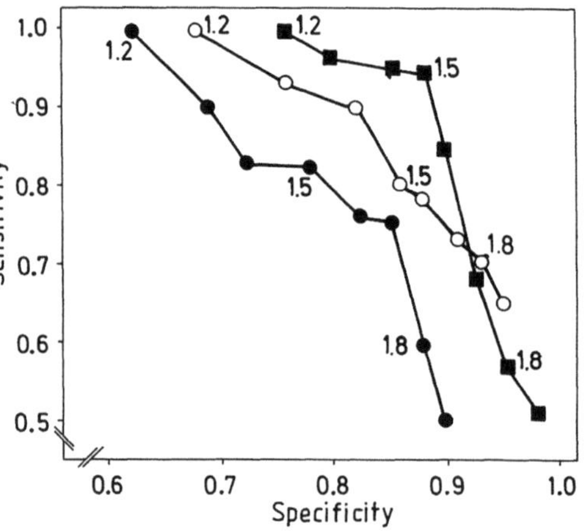

Fig. 5. Diagnostic sensitivity and specificity of urinary alanine aminopeptidase excretion in detecting rejection episodes, depending on the expression of enzyme excretion. ■——■, U/mmol creatinine; ●——●, U/l; ○——○, U/4 h. The numbers by the lines indicate the different activity ratios. The activity behaviour up to 4 days before the clinical diagnosis of rejection (altogether 47 rejection episodes) and the activity behaviour in 40 patients not suffering rejection during a period of an average of 46 days after transplantation were considered (according to [24]). Definition of sensitivity and specificity as in Table 2; definition of the "activity ratio" in the text

and Whiting et al. [49] valued a twofold increase as significant. The influence of a certain cut-off point as the determinator of the diagnostic efficiency cannot be overestimated. The different levels for accepting a result as "significant" or "positive" severely impair the comparability of findings between different groups.

Arbitrary selection of the cut-off point was the most popular approach to overcome to obvious problem of classification. Two groups used more objective methods to estimate optimal cut-off points. Jung et al. [24] performed an analysis of the receiver operating characteristic (ROC) curve of the "activity ratio". Because the intra- and inter-individual excretion rates of urinary enzymes in renal transplant recipients vary so greatly, this "activity ratio" procedure has been recommended as a simple method to employ urinary enzyme measurements in the diagnosis of rejection crises [24]. This activity ratio denotes the ratio of the current enzyme activity and its preceding value. It considers the baseline value for each patient as a decision criterion to compare the activity measured at a certain time with the activity determined at the preceding measurement. Another approach was the maximization of the information content of the cut-off points [15].

Table 2 demonstrates the influence of different cut-off points of NAG excretion on the performance of the diagnostic test. As can be seen, a certain cut-off point defines the test's sensitivity and specificity, and by increasing sensitivity the specificity decreases and vice versa. Thus, the selection of a particular cut-off point is influenced by the clinical demand of whether a screening test (i.e. a test with high sensitivity) or a confirmative test (i.e. a test with high specificity) is needed. This effect of cut-off points on the sensitivity and specificity of test results could also be clearly demonstrated in a comparative study with the five enzymes AAP, AP, GGT, lysozyme and NAG using the "activity ratio" procedure (Fig. 6). The different diagnostic validity is evident among these enzymes at the same cut-off point.

Table 2. Diagnostic sensitivity and specificity of *N*-acetyl-β-D-glucosaminidase excretion used to discriminate between uncomplicated and complicated post-transplant courses as a function of various cut-off points (according to P. Kotanko et al., manuscript in preparation).

Cut-off point (U/mmol creatinine)	Sensitivity (%)	Specificity (%)
0.5	97	7
1	91	30
2	77	63
3	66	83
4	59	90
5	52	94
10	31	99

Sensitivity = true positives/(true positives + false negatives); specificity = true negatives/(true negatives + false positives).

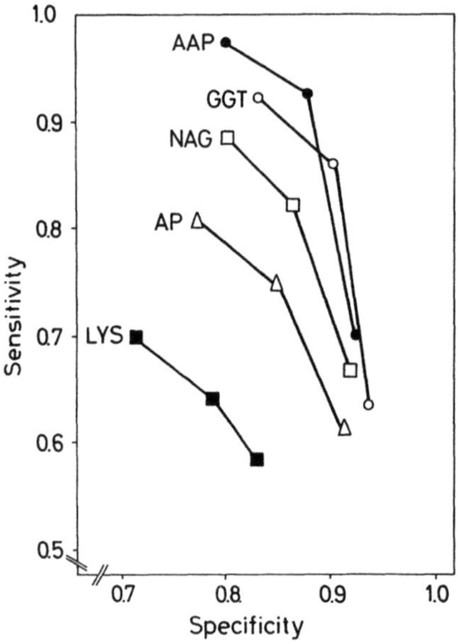

Fig. 6. Diagnostic sensitivity and specificity of various urinary enzymes in detecting rejection episodes depending on the selected cut-off point. Sensitivity and specificity were based on outcome for 82 patients using the activity ratios 1.2, 1.5 and 1.7 indicated by the three points in every line from left to right. Definition of sensitivity and specificity as in Table 2; definition of the "activity ratio" in the text. *AAP*, alanine aminopeptidase; *GGT*, gamma-glutamyltransferase; *NAG*, *N*-acetyl-β-D-glucosaminidase; *AP*, alkaline phosphatase; *LYS*, lysozyme (according to [25]).

Until now, detailed analysis of sensitivity and specificity has been performed for AAP, AP, fructose-1.6-bisphosphatase, GGT, glutathione S-transferase, lysozyme, NAG and pyruvate kinase [25, 30, P. Kotanko et al., manuscript in preparation]. As a consistent finding in agreement with other authors [16, 32, 35], urinary AAP and NAG activities proved to be of significant diagnostic value in terms of sensitivity and specificity. An activity ratio of 1.5 resulted in a sensitivity of 0.93 and a specificity of 0.88 in the case of AAP in the diagnosis of acute rejection episodes [25]. The sensitivity and specificity of NAG as a function of different cut-off points are displayed in Table 2. Simple and optimized methods are available for both enzymes (see Chap. 8) which offer the opportunity to perform reliable determinations even in non-specialised laboratories.

Meanwhile, two groups have applied multi-variate approaches, combining urinary enzyme evaluations with other parameters [15, 35]. Kotanko et al. [15, 31] used an information-theoretical model to discriminate non-rejection and rejection. In this model, a combined analysis of urinary enzymes, serum creatinine and urinary neopterin levels was performed. The combined analysis of parameters with a different pathophysiological background was superior to the use of either parameter alone. Pantscheva-Haschen et al. [35] calculated a discrimination function from 13 clinical and laboratory parameters. Urinary AAP activity was among the 5 most relevant parameters. Both approaches are indicative for further prospective investigations in this field because they permit an unbiased evaluation of clinical and laboratory data together.

Of clinical importance is the question of whether or not the analysis of urinary enzymes provides an earlier diagnosis of rejection than the other conventional tests used in this field. Increased AAP excretion [25] precedes the clinical diagnosis of rejection by about 1.9 days. Other authors demonstrated similar "fore warning" periods [17, 47]. For example, the increased enzyme output as early as 1–5 days before clinical diagnosis was described by Horpacsy et al. [17] for 35% of the investigated rejection crises in 235 transplant recipients. The increased excretion rate coincided with the clinical diagnosis in 57% and followed the clinical diagnosis in 15%. Wellwood et al. [47] showed an early NAG increase at a mean of 1.5 days prior to the clinical diagnosis of rejection. Loertscher et al. [33] could not confirm these positive results with NAG, especially in patients suffering from vascular rejection. The above-mentioned information-theoretical approach provided a positive test 2.4 days before the clinical diagnosis of rejection [15].

13.6 Conclusions and Future Aspects

On the basis of present evidence AAP and NAG activities are of documented significance in detecting graft rejection episodes. When determined on a daily basis, both provide sensitive indicators of renal damage, and a rise of urinary excretion heralds the clinical signs of rejection. Hence, both can be recommended as valuable additional markers of graft damage. Changes in both enzymes are not specific for a distinct clinical setting. Due to its sensitivity a rise of urinary enzyme output should be a stimulus to clinical suspicion [48] and trigger the use of additional diagnostic procedures.

The lack of specificity can be overcome by the combined analysis of parameters with different pathophysiological backgrounds, e.g. immunological, functional and cytological data. A fruitful line of research will be the development of mathematical procedures to analyse diagnostic data on a multi-variate basis.

Acknowledgements. The present work was supported by the Austrian "Fonds zur Förderung der wissenschaftlichen Forschung" (PK).

References

1. Bäckman L, Appelkvist EL, Ringden O, Dallner G (1988) Glutathione transferase in the urine: a marker for post-transplant tubular lesions. Kidney Int 33:571–577
2. Ballardie FW, Cohen GL, Johnson RWG (1976) N-acetylglucosaminidase release during cadaver kidney preservation. Br J Surg 64:299
3. Bernstein LH, Russell PJ, Horenstein JM (1985) Urinary adenylate kinase activity as a predictor of renal allograft crises. Clin Chem 31:1151–1154

4. Bönner G, Meider G, Evers J, Lent V (1985) Veränderungen der renalen Kallikreinausscheidung im Urin von Nierentransplantat-Empfängern. Med Welt 36:276–279
5. Bourbouze R, Gluckman JC, Frantz P, Paraire M, Baumann FC, Luciani J, Percheron F, Legrain M (1985) Early monitoring of human renal transplantation by N-acetyl-β-D-glucosaminidase isoenzyme activities in urines. Clin Chim Acta 149:185–195
6. Burlington H, Holloway S, Stewart WW, Nathan P (1967) Alterations in glomerular and tubular enzyme activity in dog kidney allografts. Transplantation 5:1248–1255
7. Butterworth PJ, Moss DW, Pitkanen E, Pringle A (1965) Some characteristics of alkaline phosphatase in human urine. Clin Chim Acta 11:220–226
8. Chatterjee SN (1980) Manual of renal transplantation. Springer, Berlin Heidelberg New York
9. Coratelli P, Giannattasio M, Schena A, Marzolla R, Buongiorno E (1988) Comparison of urinary enzyme pattern in renal transplant recipients receiving two different immunosuppressive regimens: azathioprine versus ciclosporin. Contrib Nephrol 68:203–206
10. Corbett CRR, Thompson AE, Price RG (1978) Urinary enzymes and renal transplantation: the early warning controversy. Clin Chem 24:724–725
11. Corbett R, Gardner GJ, Kind PRN, Thompson AE, Price RG (1983) Comparison of urinary N-acetyl-β-D-glucosaminidase and urinary fibrin degradation products for diagnosis of rejection after renal transplantation. Clin Chim Acta 128:141–150
12. Dietrich G, Egger E, Erdmann T (1968). Leuzin-p-nitranilid spaltende Enzyme im Harn nach Nierentransplantation. Acta Biol Med Ger 20:517–520
13. Dyck RF, Cardella CJ, Sacks MA (1979) Urinary lysosomal enzyme excretion after renal transplantation. Clin Chim Acta 91:111–116
14. Egger E, Jung K, Mallmann B (1973) Arylamidase in the urine of patients with kidney transplants. J Clin Chem Clin Biochem 11:565–566
15. Heiss H, Wild W, Margreiter R, Pfaller W, Kotanko P (1988) Noninvasive diagnosis of renal allograft rejections—application of an information-theoretical model. Klin Wochenschr 66:32–36
16. Horpacsy G (1983) Diagnostic significance of standardized determination of urinary enzymes in the practice of human kidney transplantation and renal toxicology. Contrib Nephrol 35:105–127
17. Horpacsy G, Zinsmeyer J, Schröder K, Mebel M (1977) Value of determining urinary enzymes after human kidney transplantation: early warning of rejection or not? Clin Chem 23:770–771
18. Horpacsy G, May G, Zinsmeyer J, Mebel M (1978) Biochemical parameters for determination of graft viability during early postoperative period. Urol Int 33:234–244
19. Horpacsy G, Zinsmeyer J, Schröder K, Mebel M (1978) Changes in serum and urine lysozyme activity after kidney transplantation: influence of graft function and therapy with azathioprine. Clin Chem 24:74–79
20. Jacyszyn K, Kemp E, Laursen T, Rasmussen F (1975) Investigations of the excretion of gamma-glutamyltranspeptidase into the urine. Int Urol Nephrol 7:205–214
21. Jung K, Diego J, Scholz D, Schröder K, Strobelt V (1982) Urinary enzyme excretion by renal-transplant recipients in relation to interval after transplantation. Clin Chem 28:1762–1764
22. Jung K, Scholz D, Diego J, May G (1983) Harnenzymausscheidung bei nierentransplantierten Patienten. Z Ges Inn Med 38:581–592
23. Jung K, Pergande M, Reinholdt C, Schulze B, Scholz D, Strobelt V (1984) Multiple forms of alanine aminopeptidase, alkaline phosphatase and γ-glutamyltransferase in urine of healthy persons, patients suffering from kidney diseases and patients with kidney transplants. J Clin Chem Clin Biochem 22:523–525
24. Jung K, Diego J, Strobelt V (1985) Diagnostic significance of urinary enzymes in detecting acute rejection crises in renal transplant recipients depending on expression of results illustrated through the example of alanine aminopeptidase. Clin Biochem 18:257–260
25. Jung K, Diego J, Strobelt V, Scholz D, Schreiber G (1986) Diagnostic significance of some urinary enzymes for detecting acute rejection crises in renal transplant recipients: alanine aminopeptidase, alkaline phosphatase, γ-glutamyltransferase, N-acetyl-β-D-glucosaminidase, and lysozyme. Clin Chem 32:1807–1811
26. Kehrer G, Blech M, Kallerhoff M, Bretschneider HJ (1989) Urinary LDH-release for evaluation of postischemic renal function. Klin Wochenschr 67:477–485
27. Keyser JW, Watkins GL, Salaman JR (1976) N-acetyl-β-D-glucosaminidase and γ-glutamyltransferase in urine of patients with renal transplants: not an early indicator of rejection. Clin Chem 22:925–926

28. Kind PRN (1982) N-acetyl-β-D-glucosaminidase in urine of patients with renal disease, and after renal transplants and surgery. Clin Chim Acta 119:98–97
29. Koivula T, Pitkänen E, Turto H, Tötterman T (1978) The excretion of urinary N-acetyl-β-D-glucosaminidase and β-glucuronidase as a sign of impending rejection of kidney transplants. Ann Clin Res 10:288–290
30. Kotanko P, Keiler R, Knabl L, Aulitzky W, Margreiter R, Gstrauthaler G, Pfaller W (1986) Urinary enzyme analysis in renal allograft transplantation. Clin Chim Acta 160:137–140
31. Kotanko P, Keiler R, Heiss H, Wild W, Gstrauthaler G, Margreiter R, Pfaller W (1988) Combined evaluation of urinary N-acetyl-β-D-glucosaminidase, urinary neopterin and plasma creatinine in renal allograft recipients. In: Lubec G, Bartosch B (eds) Noninvasive diagnosis of kidney disease. Libbey, London, pp 43–54
32. Krishna SK, Kirubakaran MG, Pandey AP, Kanagasabapathy AS (1985) Urinary N-acetyl-β-D-glucosaminidase and aminopeptidase N in the diagnosis of graft rejection after live donor transplantation. Clin Chim Acta 150:69–85
33. Loertscher R, Scholer A, Brunner F, Harder F, Thiel G (1982) N-Acetylglucosaminidase-Bestimmung im Urin bei Nierentransplantatempfängern mit und ohne Cyclosporin A. Schweiz Med Wochenschr 112:1658–1664
34. Neef L, Hütter HJ, Böhme I, Dotschev D, Danev ST, Haschen RJ (1982) A low molecular weight alanine aminopeptidase in urine. Biochemical equivalent of tubular atrophy? Enzyme 28:348–354
35. Pantscheva-Haschen R, Schulze R, Haerting J, Langkopf B (1988) Versuch zur Optimierung der Rejektionsdiagnostik nach allogener Nierentransplantation unter Heranziehung von Harnenzymen und multivariater Statistik. Z Med Lab Diagn 29:179–187
36. Pfleiderer G, Baier M, Mondorf AW, Stefanescu T, Scherberich JE, Müller H (1980) Change in alkaline phosphatase isoenzyme pattern in urine as possible marker for renal disease. Kidney Int 17:242–249
37. Price RG (1979) Urinary N-acetyl-β-D-glucosaminidase (NAG) as an indicator of renal disease. Curr Probl Clin Biochem 9:150–163
38. Prout GR, Macalalag EV, Hume DM (1964) Serum and urinary lactic dehydrogenase in patients with renal homotransplants. Surgery 56:283–295
39. Sachse HJ, Falkenberg FW, Scherberich JE, Stefanescu T, Fassbinder W, Mondorf AW, Schoeppe W (1981) Development of a radioimmunoassay for a high molecular mass tubular antigen in urine—its application for early detection of tubular damage. Clin Chim Acta 110:91–104
40. Schmid H, Mall A, Bockhorn H (1986) Catalytic activities of alkaline phosphatase and N-acetyl-β-D-glucosaminidase in human cortical nephron segments: heterogenous changes in acute renal failure and acute rejection following kidney allotransplantation. J Clin Chem Clin Biochem 24:961-970
41. Sennekamp J, Ricken D, Siedek M, Hildebrandt G (1973) Lysozym im Urin als Indikator der akuten und chronischen Abstoßungskrisen bei 17 Patienten mit Nierentransplantation. Klin Wochenschr 51:748–753
42. Thompson AE, Tucker SM, Price RG (1977) Assay of N-acetyl-β-D-glucosaminidase in urine of patients with renal transplants as an indicator of rejection. Clin Chem 23:301–302
43. Tyler HM, Williams MA, Kountz SL, Kapros C, Dempster WJ (1964) A biochemical approach to the study of rejection of canine renal homotransplantation. I. Some quantitative assessments of enzyme activity. Br J Exp Pathol 45:219–225
44. Vanderlinde RE (1981) Urinary enzyme measurements in the diagnosis of renal disorders. Ann Clin Lab Sci 11:189–201
45. Ward JP (1975) Gamma-glutamyltranspeptidase, a sensitive indicator of renal ischaemic injury in experimental animals and renal homograft rejection in man. Ann R Coll Surg Engl 57:248–261
46. Welbourne TC, Phifer T, Thomas M, Dass PD (1983) Gamma-glutamyltransferase release by the postischemic kidneye: multiple forms and cellular pH. Life Sci 33:1141–1147
47. Wellwood JM, Davies D, Leighton M, Thompson AE (1978) N-acetyl-ß-D-glucosaminidase assay in renal transplant recipients. Transplantation 26:396–400
48. Whiting PH (1988) Enzymuria: a stimulus to clinical suspicion. In: Lubec G, Bartosch B (eds) Non-invasive diagnosis of kidney disease. Libbey, London, pp 3–11
49. Whiting PH, Nicholls AJ, Catto GRD, Edward N, Engeset J (1980) Patterns of N-acetyl-β-D-glucosaminidase excretion after renal transplantation Clin Chim Acta 108:1–7

50. Whiting PH, Petersen J, Power DA, Stewart RDM, Catto GRD, Edward N (1983) Diagnostic value of urinary N-acetyl-β-D-glucosaminidase, its isoenzymes and the fractional excretion of sodium following renal transplantation. Clin Chim Acta 130:369–376
51. Yuen CT, Corbett CRC, Kind PRN, Thompson AE, Price RG (1987) Isoenzymes of urinary N-acetyl-β-D-glucosaminidase (NAG) in patients with renal transplants. Clin Chim Acta 164:339–350
52. Zekert F. Markhof CM. Pinggera WF, Piza F, Wagner O, Steinbereithner K, Krenn J (1976) Harnenzyme zur Frühdiagnose von Abstoßungsreaktionen nach Nierentransplantationen. Wien Klin Wochenschr 88 [Suppl 4]: 1–16

14 Urinary Enzymes and Nephrotoxicity in Humans

R.G. Price and P.H. Whiting

14.1 Introduction

Nephrotoxicity is a major problem, so how can the assay of urinary enzymes help in its detection and assessment? In most instances renal damage is irreversible, and commonly used renal function tests are insensitive; they therefore fail to identify early lesions, primarily because the reserve capacity of the kidney is more than adequate to compensate for damage in the early stages. Biopsy sampling is not acceptable practice except in the presence of well-established renal disease, and thus sensitive, early, non-invasive markers of renal injury are required. The need for this type of test is reinforced by the observation that around 30% of patients presenting with acute renal failure (ARF) have a normal renal histology. Essential prerequisites for general use are that the methods be rapid, cost-effective and readily automated. Kit procedures are now available which overcome the earlier problems of both inter-laboratory standardization and the requirement for expensive non-standard equipment [69, 94]. An elevated urinary enzyme level provides an early warning to the clinician which, since false-positives are rare, justifies further investigations. As experience has shown that the most reliable information is obtained from daily assays, collection of samples should be standardized, preferably with the first or second sample of the day because of difficulties encountered in the collection of 24-h samples. Spot sampling is acceptable provided activities are factored by creatinine, although abnormal values are diagnostic of pathological change whatever method of expression is used [13]. The test should be specific for the presence of renal disease, but since the classification of renal disease is based on morphological criteria, differential diagnosis of the type of disease will require further investigation. Only a limited number of urinary enzymes have been found to have potential usefulness in the assessment of nephrotoxicity, and these are listed in Table 1. The significance of the early detection of nephrotoxicity to health has been reviewed [3, 74].

Individuals may be exposed to nephrotoxins accidentally in the workplace (Table 2). Examples include agricultural workers using pesticides [53] and/or herbicides [54], contamination of stored foods [92] and a variety of chemicals used in industry, e.g. cadmium, lead and organic solvents [6].

Toxic chemicals can also enter the food chain [39, 68], and some mushrooms [67] are also toxic. Drug abuse can result in renal damage, heroin [84] and

Urinary Enzymes
K. Jung/H. Mattenheimer/U. Burchardt (Eds.)

Table 1. Urinary enzymes used in the detection of nephrotoxicity in humans

Source	Enzyme	Ref.
Renal lysosomes	NAG, NAG isoenzymes	[14]
		[90]
	β-Galactosidase	[46]
Renal brush border	Trehalase	[40]
	Alanine aminopeptidase	[47]
	Gamma-glutamyltransferase	[25]
	Alkaline phosphatase	[46]
Cytosol of renal cells	Ligandin, glutathione-S-transferase	[23]
	Lactate dehydrogenase	[85]
	Malate dehydrogenase	[49]
Distal/glomerular cells	Angiotensinase A	[79]
Distal tubular cells	Kallikrein	[29]
Filtered through glomerulus	Lysozyme (muramidase)	[82]

NAG, *N*-acetyl-β-D-glucosaminidase

Table 2. Nephrotoxins important for humans

Type of toxin	Example
1. Abuse of drugs	Analgesics, (NSAID), heroin, cannabis, morphine
2. Therapeutic drugs	Aminoglycoside antibiotics, cyclosporine, *cis*-platinum, diuretics, lithium
3. Diagnostic agents	Contrast media
4. Immuno-allergic agents	Penicillin, sulphonamides
5. Immunological side effects	Mercury, gold salts, D-penicillamine
6. Solvents	Chloroform, styrene, butadienes, organometallics, carbon tetrachloride, anti-freeze
7. Metals	Cadmium, lead, mercury
8. Agri-Chemicals	Paraquat, diquat
9. Fungi	Balkan nephropathy, toadstools, mushrooms
10. Contaminated foods	Spanish oil

NSAID, non-steroidal anti-inflammatory drugs

analgesics [22] are amongst the most important examples, with patients with analgesic nephropathy forming the largest group [81]. Some patients with chronic illness may be at risk, and the possibility that lithium treatment is toxic has not yet been disproved [35]. Drugs can be used to combat infection or control cancer, and provided the treatment with nephrotoxic drugs is carefully monitored, this approach to patient management has allowed their clinical use, e.g. *cis*-platinum [36]. Nephrotoxic antibiotics can also be useful providing precautions are taken to minimize their deleterious effect on the kidney [87]. Because of the kidney's high blood flow and ability to reabsorb, modify and transport chemicals [71], secondary effects of drugs or toxins on the kidney are important and should also be monitored. For this reason all human volunteers given new drugs should be screened for possible nephrotoxic effects. When a new drug becomes available, the susceptibility of groups such as children [31], pregnant women [84b] and the elderly should also be taken into account. The pre-disposition of individuals with certain types of renal disease is a further constraint to be considered when a new drug is evaluated [16].

14.2 Enzymes Useful for Screening Nephrotoxicity

When the assay of urinary *N*-acetyl-β-D-glucosaminidase (NAG) activity was carried out along with the estimation of proteinuria (albustix positive), 91% of the hypertensive patients with renal dysfunction could be detected [41]. Urinary NAG excretion was found to be associated with early diabetic nephropathy and poor long-term glycaemic control [88]. It is advisable to measure a single enzyme activity, or at most two, together with other simple parameters of renal function, e.g. urinary protein or sodium content when screening for potential nephrotoxicity (see Table 3). If NAG is the enzyme of choice, the second enzyme should be a brush border enzyme, alanine aminopeptidase [26, 58] or possibly kallikrein [29]. This would give preliminary information on the type and site of damage. The majority of nephrotoxins attack the proximal tubule, and lesions may occur in the S1, S2 or S3 segments. However, some toxins are known to damage the papilla, glomerulus or distal tubule, and because the function of the cells in these regions varies, their enzyme complement is different. It is therefore to be expected that the pattern of urinary enzyme excretion (both quantitative and qualitative) would vary with agents with different site specificities [20, 21]. In humans, damage ultimately involves more than one nephron region, and therefore the enzyme profile may not be diagnostic of particular intra-nephron sites. However, the ratios of different enzyme activities do vary at different stages of disease/damage. Urinary enzyme assays are of value in the early detection of nephrotoxicity and monitoring the development of the lesion—continuing high values indicate the presence of a developing lesion while falling activities indicate recovery—which should be confirmed by other tests [73]. Enzyme activities also provide a useful prognostic indicator [89]. It is possible that isoenzyme analysis will provide additional diagnostic information [34, 90, 95].

14.3 Relationship of Urinary Enzymes to Other Commonly Measured Parameters of Damage/Dysfunction

Recently, attempts have been made to demonstrate the relationship between enzymuria and other measurements of renal damage following the administration of a variety of nephrotoxic agents. Increased urine activities of lactate delydrogenase (LDH), alanine aminopeptidase (AAP), gamma-glutamyltransferase (GGT), NAG, alkaline phosphatase (AP) and lysozyme were all increased following administration of iohexol, a radiocontrast dye, but both glomerular filtration rate (GFR) and effective renal plasma flow (ERPF) were unaffected [17]. Similarly, Neumayer and co-workers [66] demonstrated that urinary GGT, AAP and NAG activities increased in parallel with urinary total protein concentrations. Other studies with aminoglycoside antibiotics [52, 60], interferon-2B [51] and high-dose folinic acid and 5-fluorouracil [61] found that urinary enzyme measurements provided a consistent indication of nephrotoxicity which predated alterations in both tubular proteinuria and creatinine clearance measurements in some patients. It has been recently shown that the hypomagnesaemia following *cis*-platinum therapy is associated with increased NAG enzymuria [59]. These authors also demonstrated that NAG excretion was reduced in patients receiving magnesium supplementation. Serum platin concentrations exhibited a strong positive correlation with the urinary excretion of the brush border enzymes AAP and GGT ($r = 0.82$), with a weaker associaton also noted for both NAG ($r = 0.48$) and AP ($r = 0.40$) following *cis*-platinum therapy [19]. Guder and Hoffman [33] assessed the sensitivity of several urine measurements to indicate renal abnormalities, including those for GFR (albumin/IgG), tubular reabsorption (α_1-microglobulin) and NAG in unselected clinical patients. They concluded that NAG enzymuria was more sensitive than α_1-microglobulin and that renal tubular dysfunction could be detected with quantitative measurements of urinary albumin and NAG enzymuria.

14.4 Examples of the Use of Enzyme Excretion to Monitor Drug-Induced Alterations in Patients

The clinical assessment of renal status involves mainly estimates of function including GFR, sodium excretion and both urinary flow rate and osmolality. Urine enzyme analysis is not an estimate of renal function (with the possible exception of lysozyme or muramidase) but of parenchymal integrity, and the presence of increased enzymuria provides additional useful information which at the very least should increase the index of clinical suspicion regarding the presence of potential renal insult (Fig. 1). Since they are non-invasive they can be

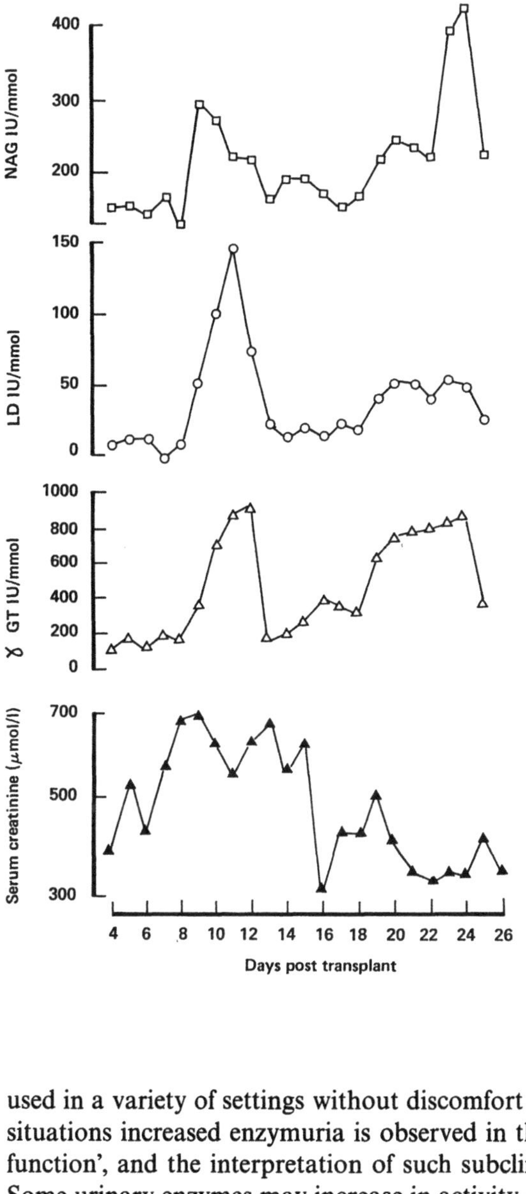

Fig. 1. Enzyme excretion and serum creatinine values in a recipient of a renal transplant. GT, glutamyltransferase; LD, lactate dehydrogenase; NAG, *N*-acetyl-*β*-D-glucosaminidase

used in a variety of settings without discomfort to the patient. In many clinical situations increased enzymuria is observed in the presence of 'normal or renal function', and the interpretation of such subclinical nephrotoxicity is difficult. Some urinary enzymes may increase in activity either in parallel with increased kidney levels due to enzyme induction (e.g. lysosomal hydrolases) or following cell leakage, damage or even diuresis (e.g. brush border enzymes). Furthermore, increased enzyme excretion may not be maintained during continued exposure to nephrotoxic agents (Fig. 2). For example, patients treated with aminoglycoside antibiotics or radiocontrast media may demonstrate only a transient enzymuria [17, 66]. However, the presence of enzymuria can predate alterations in other commonly measured parameters of renal dysfunction/damage. For

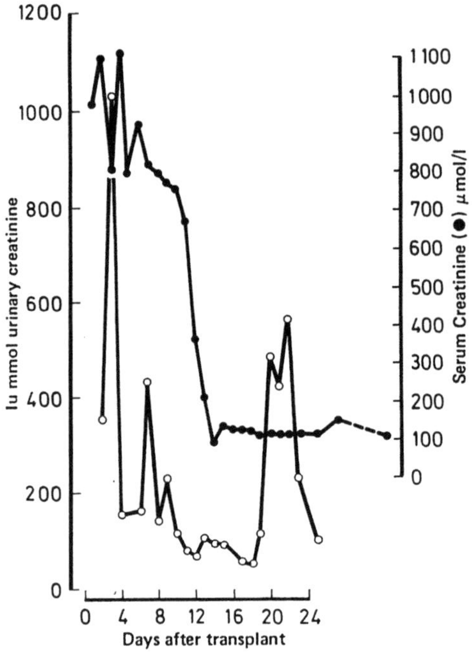

Fig. 2. Effect of gentamicin therapy on *N*-acetyl-β-D-glucosaminidase (NAG) excretion. The NAG peaks on days 3 and 7 are dialysis-related; gentamicin given from days 19–25 (used with permission).

example, in a cardiac allograft recipient treated with cyclosporin A, NAG enzymuria predated both increased glycosuria and fractional excretion of sodium values (Fig. 3) by day 4 and eventually a rise in serum creatinine concentration.

A further clinical example can be used to demonstrate the potential predictive and therapeutic value of urine enzyme analysis. The patient (female, 41 years old, body weight 44 kg) was in intensive care on intermittent positive pressure ventilation (IPPV). She had received chemotherapy for a Hodgkin's lymphoma and was currently receiving gentamicin therapy for septicaemia. Her blood chemistry revealed a hypernatraemia of 155 mmol/l and a serum creatinine of 75 μmol/l. Her urinary flow and creatinine clearance rate were 7.38 and 38.7 ml/min, respectively, and the fractional excretion of sodium was 5.22%. In addition, urinary enzyme activities of NAG, GGT and AAP were increased to 229-, 323- and 212-fold respectively, the upper limits of the reference ranges. This indicated an enormous renal tubular and brush border insult consistent with both the increased urine flow rate and fractional excretion of sodium. Indeed, lithium clearance studies not only indicated that the patient's kidneys were incapable of sodium reabsorption but also that there was a net movement of sodium into the glomerular filtrate. However, in order to treat the

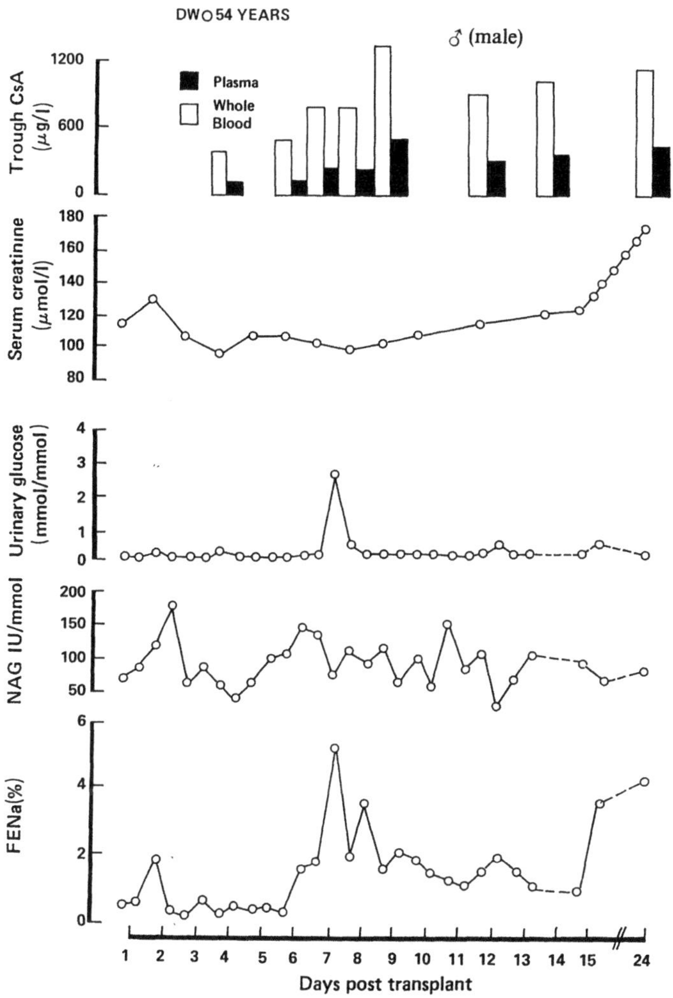

Fig. 3. Comparison of urinary sodium and glucose, *N*-acetyl-β-D-glucosaminidase (NAG) excretion and serum creatinine in a cardiac transplant recipient receiving cyclosporine. Trough cyclosporine A (trough CsA), fractional extraction of sodium (FENa%)

hypernatraemia it was decided to restrict the sodium intake (some 400 mmol/24 h). Consequently, the serum sodium fell from around 155 mmol/l to 135 mmol/l overnight. This approach was contrary to the advice given that the sodium intake be titrated against the serum sodium concentration, due to the poor renal tubular function, to avoid such dramatic swing in serum sodium values.

14.5 Coordination of Excretion

Although the mechanisms by which enzymes of renal origin are excreted into the urine are not clearly understood, there is evidence that enzymes from the same intracellular location are excreted together. In particular, both lysosomal hydrolases and brush border enzymes have been shown to exhibit an association under a variety of physiological and pathological conditions. For example, the four lysosomal hydrolases α- and β-galactosidase, β-glucuronidase and hexosaminidase, but not the cytosolic enzyme LDH, were excreted coordinately in normal individuals [70] as were the brush border enzymes AP, leucine aminopeptidase (LAP), GGT, β-glucosidase and trehalase following administration of radiocontrast media [35]. The former authors also suggested that although lysosomal enzyme activities in the urine demonstrated considerable variaton, both throughout the day and from day to day, the ratio of enzyme activities remains relatively constant. The observation that enzymes from the same intracellular location show different peak excretion rates although coordinately following either toxic insult [17] or during states of diuresis [44] or anti-diuresis [43] may be related to an altered permeability of tubular cells for various enzymes and a diuresis-dependent membrane turnover of enzymes [43], the different localization of the corresponding enzymes in the cell (e.g. GGT is situated on the lumen surface and AAP on the intracellular surface of the brush border membrane) or to the differing renal activities of the various enzymes [77].

There is still no clear consensus as to the relationship between the rates of enzyme excretion from different intracellular locations. Some studies demonstrate coordination between brush border, lysosomal and cytosolic enzymes [17, 19, 66] (Fig. 1), while in others no relationship is apparent [10, 38, 70]. It is clear that the excretion of enzymes into the urine is a multi-factorial process [80], and the resultant enzymuria must depend on the severity of the damage and on the nature and site-specificity of either the nephrotoxicant or renal insult.

14.6 Monitoring Patients and Adjusting Regimes, Volunteer Studies and Screening Populations at Risk

NAG enzymuria, in particular, has been used to screen both the general population and certain high-risk groups for the presence of early renal parenchymal damage. Urinary NAG activity was frequently elevated in patients with tubulointerstitial disease even in the absence of proteinuria [50] and also in workers exposed to nephrotoxic chemicals, e.g. lead, mercury or organic solvents, but without overt clinical signs of renal disease [62]. NAG enzymuria has been employed in evaluating hypertensive patients for underlying renal damage [1, 55, 56], a factor which may predispose them to chemical-or drug-induced renal insult. However, the diagnostic significance of single estimations of urine enzyme activity as a screen for patients and populations at risk may be questioned due to the daily variability in enzyme excretion. Consequently, particularly in the clinical situation, it is important to obtain baseline measurements prior to exposure of the patient/volunteer to potential nephrotoxicants and to consider the pattern of excretion both during and after therapy. This can only be achieved by daily measurement of enzyme excretion of, for example, NAG and AAP (Table 1). Population screening exercises also require baseline measurements for comparison and regular assessment of risk, and in this context, enzymuria is attractive due to its non-invasive nature.

It is difficult to assess critical excretion levels, but consistently increased urinary enzyme excretion in the absence of decreased renal function must be taken as a serious indication of renal insult. Single observations of elevated enzyme activity estimations must be confirmed on fresh samples. Conversely, it has been suggested that the observation of normal enzyme excretion may indicate either a non-nephrotoxic situation [17] or the absence of rejection in cyclosporin A-treated renal allograft recipients [2, 24, 42]. However, in the latter case the assay of glutathione-S-transferase excretion may be valuable [4].

14.7 Chronic Versus Acute

In the early days of investigating urinary enzymes as diagnostic aids, the commonly held view was that their primary value lay in the detection of acute insults. However, it has subsequently become apparent that they have potential utility in following chronic conditions. One obvious example is monitoring industrial personnel who are exposed to nephrotoxins over extended periods of time [6, 62]. Other examples include following up patients with analgesic nephropathy [32] and diabetes as well as estimating immunologically mediated damage resulting from chronic exposure to nephrotoxins [18]. Chronic injury can also result from long-term drug abuse and therapeutic drugs taken for chronic illness [28]. While a good marker of chronic injury needs to be defined, the statement that urinary enzyme assays are of no value in chronic disease should be challenged. In addition, urinary enzyme excretion will need to be interpreted against a background of declining renal function as well as modified morphology.

Much of the initial criticism of the use of enzymes to assess chronic damage was methodological in origin. Since the elevations of urinary enzymes in chronic injury are much lower than those seen in acute damage (e.g. with antibiotics), the establishment of accurate reference ranges and the elimination of variation due to age, sex, collection of samples, inhibitors, activators and pH are essential prerequisites. None of these parameters need be a problem if the enzyme of choice is NAG [26] or AAP [58], and gradually similar information is becoming available for other enzymes. Much of the current debate arises from the accepted use of insensitive methods by clinicians, and any changes in urinary enzymes may therefore often conflict with clinical judgement. With medicine adopting a more scientific approach, the biochemical evidence obtained from urinary enzyme assays should be checked and confirmed and, if abnormal, prompt further clinical investigation. This approach is certainly worth pursuing because morphological studies should be reserved for well-developed disease, and there is the possibility that the causative agent (drug or toxin) can be eliminated by changing the treatment or modifying the exposure.

Urinary enzymes have been described as "transient", and since it is unusual to obtain a false-negative result, these findings do reflect changes in the kidney. Fortunately, renal disease can also be transient, and the criticism of urinary enzyme data on the basis of the apparent reversibility of increased levels may reflect an embedded view that renal disease proceeds inevitably towards renal failure. The incorporation of urinary enzyme assay into protocols for monitoring chronic nephrotoxic damage is one of the major challenges facing investigators in this field. These studies should be approached with the knowledge that enzymes do have a valuable role to play in the controlled use of nephrotoxic drugs which can result in acute injury.

In acute toxic injury it is usual to take daily samples [84a] until values return to normal; however, this is impractical in chronic cases. Individuals usually attend outpatient clinics or see their general practitioner or industrial doctor at infrequent intervals. A number of studies with industrial workers [12, 48] and diabetic suffers [88] have been extended over long periods of time, and it is now possible to visualize trends. It should also be remembered that susceptibility to nephrotoxins may have a genetic basis or results from the presence of incipient disease such as hypertension. Rises in enzyme activities are usually smaller in chronic injury than those seen in acute injury, and when values return to normal some renal impairment may still be present. However, although this may not have any functional significance, exposure to the potential toxin must be stopped as soon as possible. Chronic exposure can also be exacerbated by a change in drug regimen or in toxic exposure levels [91].

Further studies should involve the re-evaluation of changes in excretion of isoenzymes [57, 63] and the use of groups of tests, e.g. a combination of urinary enzyme, specific protein and electrolyte assays (Table 3). Should elevations in urinary enzyme activity be maintained, then steps should be taken to investigate the individual thoroughly for the presence of renal damage. This approach is particularly important as in more than 50% of all cases of end-stage renal

Table 3. Non-invasive tests indicating renal damage

1. Renal tubular function:	
Concentrating ability	Volume (specific gravity, osmolality), creatinine
Reabsorption	Albumin, glucose
Distal, proximal function	pH, sodium, calcium
2. Celluria	Bacteria, blood cells
3. Enzymuria	*N*-Acetyl-β-D-glucosaminidase, alanine aminopeptidase, kallikrein, lactate dehydrogenase
4. Proteinuria:	
Specific tests	β_2-microglobulin, retinol binding protein, α_1-microglobulin, Tamm–Horsfall glycoprotein, neopterin
Molecular size separation	glomerular, tubular and mixed proteinuria
5. Tubular antigens	Adenosine deaminase binding protein, tubular antigens
6. Glomerular	Microalbuminuria

failure, both the cause and/or effect of the nephrotoxicity is unknown. Further, it is important that methods be developed which identify groups at risk of toxic nephropathy [16].

14.8 Heavy Metal Exposure

Although the nephrotoxicity of heavy metals has been well documented with the development of increasingly sensitive techniques, the levels of exposure which are considered to be safe have been steadily reduced. Acceptable levels of exposure vary widely throughout the world, and moves towards the enforcement of agreed international standards would be greatly aided by the availability of a group of easy-to-perform, non-invasive tests which could be used to screen the general population and factory workers. Apart from monitoring tissue levels of the heavy metals, the assay of specific proteins including β_2-microglobulin and retinol binding protein have been widely used. β_2-Microglobulin is unstable in acid urine, which can cause problems, while at present no commercial kit is available for the assay of retinol binding proteins. This has led a number of groups to consider the possibility of using urinary enzymes for the early detection of renal damage. However, to date no systematic attempt has been made to recognize and quantitate different forms of proteinuria (including enzymuria) in the way that Hofmann and Guder [37] have carried out with renal patients. Such an approach would greatly aid the development of appropriate screening programmes in the future.

One of the most widely studied groups of workers is those exposed to cadmium. In an early study by Bernard et al. [7], 42 workers exposed to cadmium were compared with 77 control workers. It had previously been established that low molecular weight proteinuria was characteristic of cadmium exposure [11, 27], indicating a tubular abnormality. High molecular weight proteinuria is characteristic of long-term exposure (up to 25 years) and was found in individuals whose blood concentration of cadmium exceeded 1 μg/100 ml and urine level, 10 μg/g creatinine. A number of urinary enzymes were also measured (β-galactosidase, LDH, AP, acid phosphatase and catalase), but only β-galactosidase was excreted in significantly higher amounts in exposed workers. The increased excretion of this enzyme indicated tubular cellular dysfunction and interference of cadmium with the lysosomal system. The same group extended their studies to include inorganic lead- and mercury vapour- as well as cadmium-exposed groups [9]. High or low molecular weight proteinuria and enzymuria (increased β-galactosidase excretion) was found in all three groups. Since cadmium and mercury can cause damage via an auto-immune mechanism, β-galactosidase may enter the urine via faulty glomerular filtration; however, it is probable that the additional enzyme excreted originates from the renal tubule.

Recent studies on the excretion of urinary enzymes following cadmium exposure have been extended to include NAG [45] and trehalase [12]. Tohyama et al. carried out an epidemiological investigation of the significance of the urinary excretion of metallothionein in people living in a cadmium-polluted area of Japan. Although the correlation between urinary metallothionein and NAG was not as high as that for metallothionein and α_1-microglobulin, there was a significant relationship which supports the idea that metallothionein is released after damage to tubular cells which contain high levels of NAG activity. The potential of NAG assays was further emphasized in a study of cadmium pigment workers [45], for whom it was found that the excretion of NAG had a closer correlation to urinary cadmium concentration than β_2-microglobulin. Since methodological difficulties had been encountered with the assay of urinary β_2-microglobulin, the assay of urinary NAG was recommended for the assessment of renal damage due to exposure to cadmium. A constant elevation in NAG activity was observed at an exposure level of between 3 and 10 μg/l of cadmium, suggesting that this parameter may be of value in establishing the biological limit of cadmium, which is probably too high at 5 μg cd/g creatinine. Trehalase is located in the brush border of renal proximal tubular cells that has recently been assayed in individuals living in a cadmium-polluted area [40]. No significant correlation was found between the excretion of this enzyme and urinary cadmium. Seventeen of the subjects studied showed elevated trehalase with normal β_2-microglobulin, suggesting that the assay of this enzyme may provide an earlier indicator of nephrotoxicity. In individuals with high levels of exposure, the excretion of trehalase fell suggesting a decline in tubular cell mass. Though the assay of trehalase appears to be promising, a longitudinal study in which the excretion of β-galactosidase, NAG and trehalase are compared with the excretion of specific protein markers, e.g. β_2-microglobulin and retinol binding protein, is required to establish the relative sensitivities of the various assay procedures.

The potential of urinary enzyme assays in screening for nephrotoxicity in the work force has been reviewed [62]. In addition to heavy metals, solvents also pose a danger. There is strong evidence which suggests that NAG is the enzyme of choice for screening, but it will be interesting to determine its relative sensitivity. The potential of NAG was confirmed by Yoshida et al. [93] who found that it was chronically elevated in older workers exposed to aromatic nitro-amino compounds. Urinary enzyme assays can therefore play a role in the determination of standard exposure levels to, e.g. lead [8, 76]. Any further studies should include NAG [72] and trehalase.

14.9 Practical Application of Urinary Enzyme Assays

Elevated serum creatinine is used as an indicator of renal damage/disease; however, it is not an indicator of tubular function. The assay of a urinary enzyme such as NAG will provide information on tubular integrity, and both creatinine and NAG have been used to monitor nephrotoxic drugs including *cis*-platin [15], cyclosporine [83] and gentamicin [89]. The advantage of this approach is that the treatment can be modified on a day-to-day basis, which allows the use of therapeutically beneficial but nephrotoxic drugs.

The potential nephrotoxicity of newly developed drugs should be identified early in trials, and urinary enzyme assays have been valuable here. Adverse drug interactions can be assessed in volunteer studies, which may also be useful in the detection of contra-indications. The value of new drugs can be determined and the lowest effective dose levels established. The nephrotoxicity of some drugs are enhanced in certain diseases while in other situations there may be a protective effect [78]. The presence of hypertension [76] may have an adverse effect, and it is known that some individuals are high responders while others are low responders. This is particularly important when monitoring industrial exposure to toxins.

Urinary enzymes should be incorporated into any screening protocol of populations at risk, e.g. hypertensive or diabetic sufferers, patients with urogenital tract infection and individuals taking non-steroidal anti-inflammatory drugs (NSAID), lithium and certain antibiotics. An interesting problem arises in the case of patients with rheumatoid arthritis, as to whether the renal damage that develops is due to the disease or to exposure to nephrotoxic drugs [75]. Discontinuing the drug may not be an option, and careful monitoring and recognition of the nephrotoxic effect are therefore of paramount importance. Similarly, diabetes may afford some protection against nephrotoxicity [86].

When assessing urinary enzyme output in normal individuals or participants in drug trials, the lifestyle of the subjects should be considered. Intensive exercise, including marathon running, squash, etc., does cause some elevations in proteinuria and urinary enzymes, and some time should be allowed after exercise before taking samples.

14.10 Conclusions

The release of enzymes from the kidney is due to multiple effects on tubular cells. Changes in urinary enzyme excretion represent a sensitive indication of renal damage, which contrasts with the more commonly used light microscopy and renal function tests, which are relatively insensitive. Serum urea and creatinine concentrations and their respective clearance rates are poor indicators of GFR. Macroproteinuria also appears late in the damage process, so the commonly accepted parameters are insensitive. However, the use of radioimmunoassay for the detection of microalbuminuria has modified views on sensitivity, and the accusation that urinary enzymes are "too sensitive" is heard less often now. Essentially, urinary enzymes represent a different measure and provide an alternative set of information on biochemical integrity.

The requirement for sensitive indicators of renal damage has lead to the development of new methods of assay [30, 69, 94] and the availability of kits which can be readily used in the clinical chemistry laboratory. The presence of known inhibitors [64] can be eliminated by pre-treatment of the urine or dilution in the assay mixture. Interference by the nephrotoxic themselves or their metabolites can therefore be avoided. Reliable methods are available for the assay of NAG, GGT and AAP in urine, and so these enzymes can be used with confidence.

Technology has now advanced so that the assay of urinary enzymes presents no more difficulty than of those in serum. Enzyme standards are available now which allow standardization of assays between different instruments and laboratories, and the recently developed colorimetric procedures can be carried out manually or on any of the autoanalysers commonly found in clinical chemistry laboratories. The problems previously encountered with interlaboratory standardization when the fluorimetric methods were used can now be avoided. However, it is essential that the method of assay for any enzyme in urine instead of serum be modified as it is unusual to find one that can be directly applied, unchanged, as the kinetics of the assay are not usually the same. Variation in urine pH rarely causes problems as only occasionally does it rise above 8.0. With the methods now available, high blank values are not usually a problem, and in certain circumstances blanks can be omitted.

There is enough clinical information to confirm that the assay of urinary enzymes is valuable in the detection and monitoring of the effects of nephrotoxins. They can act as positive and negative discriminators as well as providing prognostic information. Both acute and chronic insults can be followed as can the effect of withdrawal or treatment. Simplicity, sensitivity and non-invasive nature make the assay of urinary enzymes a powerful additon to the clinician's diagnostic armoury.

Acknowledgements. This work was supported by the Commission of the European Commities (CEC) and Praill, Price, Richardson (PPR) Diagnostics (RGP) and The Grampian Health Board (PHW).

References

1. Alderman MH, Melcher L, Drayer DE, Reidenberg MM (1983) Increased excretion of urinary N-acetyl-β-glucosaminidase in essential hypertension and its decline with antihypertensive therapy. New Eng J Med 309:1213–1217
2. Avolio AW, Nanni GC, Herio F, Pozetto SC, Maggalini B, Zappacosta B, Tacchino RM, Zuppu C, Castagneto M (1987) Increase in N-acetyl-β-glucosaminidase excretion as a marker of kidney damage in cyclosporine-treated renal allografts. Transplant Proc 19:2086–2087
3. Bach P, Berlin A, Heseltine E, Krug E, Lauwerys R, Smith E, van der Venne MT (eds) (1989) Health significance and early detection of nephrotoxicity. Toxicol Lett 46:1–249
4. Bäckman L, Appelkvist EL, Ringderi O, Dallner G (1989) Urinary levels of basic glutathione transferase as an indicator of proximal tubular damage in renal transplant recipients. Transplant Proc 21:1514–1516
5. Bendz H (1983) Kidney function in lithium-treated patients. Acta Psychiatr Scand 68:303–324
6. Bernard A, Lauwerys R (1989) Epidemiological application of early markers of nephrotoxicity. Toxicol Lett 46:293–306
7. Bernard A, Bucket JP, Roels H, Masson P, Lauwerys R (1979) Renal excretion of proteins and enzymes in workers exposed to cadmium. Eur J Clin Invest 9:11–22
8. Bernard BP, Becker CE (1988) Environmental lead exposure and the kidney. Clin Toxicol 26:1–34
9. Buchet JP, Roels H, Bernard A, Lauwerys R (1980) Assessment of renal function of workers exposed to inorganic lead, cadmium or mercury vapour. J Occup Med 22:741–750
10. Burchardt U, Haschen RJ, Krosch H (1979) Clinical usefulness of enzyme determinations in urine. In: Dubach UC, Schmidt U (eds) Diagnostic significance of enzymes and proteins in urine, Hans Huber Berne, pp 106–112
11. Butler EA, Flynn FV, Harris H, Robson EB (1962) A study of urine proteins by two-dimensional electrophoresis with special reference to the proteinuria of renal tubular disorders. Clin Chim Acta 7:34–41
12. Chia KS, Ong CN, Ong HY Endo G (1989) Renal tubular function of workers exposed to low levels of cadmium. Brit J Ind Med 46:165–170
13. Corbett CRR (1980) Urinary enzyme excretion after renal transplantation and other surgical operations. M.S., Cambridge University
14. Corbett CRR, Gardner G, Kind PRN, Thompson AE, Price RG (1983) Comparison of urinary N-acetyl-β-glucosaminidae and urinary fibrin degradation products for the diagnosis of rejection after renal transplantation. Clin Chim Acta 128:141–150
15. Daugaard G, Abildgaard U, Holstein-Rathlou NH, Bruunshuus I, Bucher D, Leyssac PP (1988) Renal tubular function in patients treated with high-dose cisplatin. Clin Pharmacol Ther 44:164–167
16. Dieperink HH (1989) Identification of groups at risk for renal disease (including nephrotoxicity). Toxicol Lett 46:257–268
17. Donadio C, Tramonti G, Giordani R, Lucchetti A, Calderazzi A, Bianchi C (1989) Glomerular and tubular effects of the contrast medium iohexol. In: Proc IVth Int Symp Nephrotoxicity, Guildford 1989, in press

18. Druet P (1989) Contribution of immunological reactions to nephrotoxicity. Toxicol Lett 46:55–64
19. Ehninger G, Kaufman T, Vossler G, Wernet P, Pfleiderer G, Risler T (1985) Nephrotoxicity of cis-diamminedichloroplatinum measured by β_2-microglobulin, albumin, enzymes and magnesium in urine. In: Bach PH, Lock EA (eds) Renal heterogeneity and target Cell toxicity, John Wiley & Sons, pp 415–418
20. Ellis BG, Price RG, Topham JC (1973) The effect of tubular damage by mercuric chloride on kidney function and some urinary enzymes in the dog. Chem Biol Interact 7:101–113
21. Ellis BG, Price RG, Topham JC (1973b) The effect of papillary damage by ethyleneimine on kidney function and some urinary enzymes in the dog. Chem Biol Interact 7:131–142
22. Eyanagi R, Shigematsu H, Yoshida K, Yoshimata H (1985) Metabolism and nephrotoxicity of phenacetin and sulfanilamide. J Pharmacobrio-Dyn 8:95–105
23. Feinfeld DA, Fleigchner GM, Goldstein EJ, Levine RD, Levine SD, Avram MM, Arias IM (1979) Ligandinuria: an indication of tubular cell necrosis. Curr Probl Clin Biochem 9:273–279
24. Finn WF, Gitelman HJ (1985) N-acetyl-β-D-glucosaminidase excretion during cyclosporin A treatment. In: Bach PH, Lock EA (eds) Renal heterogeneity and target cell toxicity, John Wiley & Sons, pp 429–432
25. Flandrois C, Flandrois JP, Coullioud D, Maire I (1989) Valeur de l'enzymurie dans le suivi de la thérapeutique antibactérienne. Path Biol 37:657–663
26. Flandrois G, Lahet C, Feldmann D, Maire I, Aymard P, Mathieu M (1989) Optimisation of the determination in urine of alanine aminopeptidase, gamma-glutamyltransferase and N-acetyl-β-D-glucosaminidase. Ann Biol Clin 47:252–260
27. Friberg L (1948) Proteinuria and kidney injury among workmen exposed to cadmium and nickel dust. J Ind Hyg Toxicol 30:32–36
28. Garvey MJ, Furlong C, Schafter CB, Adelman R (1988) Elevation of renal enzyme N-acetyl-β-D-glucosaminidase in affectively disordered patients. Acta Psychiatr Scand 77:247–252
29. Girolami JP, Bascards JL, Pecher C, Cabos G, Moatti JP, Mercier JF, Haguenoer JM, Manuel Y (1989) Renal kallikrein excretion as a distal nephrotoxicity marker during cadmium exposure in the rat. Toxicology 55:117–129
30. Goren MP, Wright RK, Osborne S (1986) Two automated procedures for N-acetyl-β-D-glucosaminidase determination evaluated for the detection of drug-induced tubular nephrotoxicity. Clin Chem 32:2052–2055
31. Goren MP, Baker DK, Shenep JL (1989) Vancomycin does not enhance amikacin-induced tubular nephrotoxicity in children. Pediatr Infect Dis J 8:278–288
32. Gregg NJ, Elseviers MM, De Broe ME, Bach PH (1989) Epidemiology and mechanistic basis of analgesic-associated nephropathy. Toxicol Lett 46:141–151
33. Guder WG, Hoffman W (1991) Future markers for the diagnosis of renal lesions. In: Nephrotoxicity, mechanisms, early diagnosis and therapeutic management (Eds: Bach PU, Gregg NJ, Wilkes MF, and Delacruz L), Marcel Dekker pp 515–580
34. Halman J, Price RG, Fowler JSL (1984) Urinary enzymes and isoenzymes of N-acetyl-β-D-glucosaminidase in the assessment of nephrotoxicity. In: Goldberg DM, Werner M (eds) Selected topics in clinical enzymology, Vol. 2, Berlin, Walter de Gruyter & Co., pp 435–444
35. Hartman HG, Braedel HE, Jutzler GA (1985) Detection of renal tubular lesions after abdominal aortography and selective renal arteriography by quantitative measurements of brushborder enzymes in the urine. Nephron 39:95–101
36. Hirogawa A, Niitani H, Hayashibara K, Tsuboi E (1989) Effects of sodium thiosulfate on combination therapy of cisdichlorodiammine platinum and vindesine. Cancer Chemother Pharmacol 23:255–258
37. Hofman W, Guder WG (1989) A diagnostic programme for quantitative analysis of proteinuria. J Clin Chem Clin Biochem 27:589–600
38. Houser MT (1986) The effect of hydropenia and oral water loading on renal lysozyme handling and N-acetyl-β-D-glucosaminidase (NAG) content in urine. Ann Clin Biochem 23:297–302
39. Ikegwuonu FI, Egbunike GN, Emerole GO, Aire TA (1980) The effects of aflatoxin B_1 on some testicular and kidney enzyme activity in rat. Toxicology 17:9–16
40. Iwata K, Katch T, Morikawa Y, Aoshima K, Nishijo M, Teranishi H, Kasuya M (1988) Urinary trehalase activity as an indicator of kidney injury due to environmental cadmium exposure. Arch Toxicol 62:435–439
41. Johnston IDA, Jones NF, Scoble JF, Yuen CT, Price RG (1983) The diagnostic value of urinary enzyme measurements in hypertension. Clin Chim Acta 133:317–325

42. Jones MC, Whiting PH, Innes A, Propper DJ, Edwards N, Catto GRD (1989) Biochemical abnormalities in serum and urine from renal transplant patients receiving cyclosporine. Transplant Proc 21:1487–1489
43. Jösch W, Dubach UC (1967) Einfluβ der Diurese auf die Enzymurie. Clin Chem 32:529–532
44. Jung K, Schulze G, Reinholdt C (1986) Different diuresis-dependent excretions of urinary enzymes: N-acetyl-β-D-glucosaminidase, alanine aminopeptidase, alkaline phosphatase and gamma-glutamyltransferase. Clin Chem 32:529–532
45. Kawada T, Koyama H, Suzuki S (1989) Cadmium, NAG activity, and β_2-microglobulin in the urine of cadmium pigment workers. Brit J Ind Med 46:52–55
46. Khan SR, Shevoc PN, Hackett RL (1989) Urinary enzymes and calcium oxalate urolithiasis. J Urol 142:846–849
47. Klingmüller V, Bergmann E, Claaβen P (1984) Alaninaminopeptidase-Ausscheidung bei Kindern: Referenzberiche und der Einfluss nierengängiger Kontrastmittel. J Clin Chem Clin Biochem 22:787–789
48. Koskinen H, Järvisalo J, Pitkänen E, Mutanen P, Zitting A (1984) Serum β-N-acetylglucosaminidase and β-glucuronidase activities in silicosis patients and workers exposed to silica dust. Brit J Dis Chest 78:217–224
49. Kotanko P, Gstraunthaler G, Pfaller W (1984) Urinary enzymes in non-invasive diagnosis of kidney epithelial lesions in acute renal failure. Wien- Klin Wochenschr 96:625–629
50. Kunin CM, Chesney RW, Craig WC, England AC, de Angelis C (1978) Enzymuria as a marker of renal injury and disease: studies of N-acetyl-β-D-glucosaminidase in the general population and in patients with renal disease. Pediatrics 62:751–760
51. Kurschel E, Metz-Kurschel U, Graben N (1989) Investigations on the subclinical and clinical nephrotoxicity of interferon Alfa-2B in patients with myeloproliferative syndromes. In: Proc Int Symp Nephrotoxicity, Guildford 1989, 17
52. Langhendries JP, Battisti D, Bertrand JM (1988) Aminoglycoside nephrotoxicity and urinary excretion of N-acetyl-β-D-glucosaminidase. Biol Neonate 53:253–259
53. Lock EA (1989) Mechanism of nephrotoxic action due to organohalogenated compounds. Toxicol Lett 46:93–106
54. Lock EA, Ismael J (1979) The acute toxic effects of hexachloro-1,3-butadiene on the rat kidney. Arch Toxicol 43:47–57
55. Mansell MA, Jones NF, Zirogynnis PN, Tucker SM, Marston WS (1978) NAG: a new approach to the screening of hypertensive patients for renal disease. Lancet 2:803–805
56. Maruhn D, Paar D, Bock KD (1979) Lysosomal and brush border membrane enzymes in urine of patients with renal artery stenosis and essential hypertension. Clin Biochem 12:228–230
57. Matsukura H, Suzuki Y, Takai R, Okada T, Naiki S, Sakuragawa N (1987) Chromatographic profiles of urinary isoenzymes in healthy children. Clin Chim Acta 169:209–216
58. Mattenheimer H, Frölke W, Grötsch H, Maruhn D, Simane Z (1988) Recommendation for the measurement of "alanine aminopeptidase" in urine. J Clin Chem Clin Biochem 26:635–644
59. McAllister EJ, Willcox JC, Cathcart S, Calman KC (1985) Reduction of cis-platinum nephrotoxicity following supplementation with magnesium. In: Bach P, Lock EA (eds) Renal cell heterogeneity and target cell toxicity, John Wiley, Chichester, pp 411–414
60. Merle L, Valette JP, Charmes JP, Nouaille Y, Lachatre G, Nicot G (1985) Identification of glomerular and tubular renal damage in patients treated with aminoglycosides. In: Bach PH, Lock EA (eds) Renal hetrogeneity and target cell toxicity, John Wiley, Chichester, pp 373–376
61. Metz-Kurschel U, Kurschel E, Graber N (1989) Folate nephropathy occurring during cytotoxic chemotherapy with high-dose folinic acid and 5-fluorouracil. In: Proc IVth Int Symp Nephrotoxicity, Guildford 1989, 18
62. Meyer BR, Fischbain A, Rosenman K, Lerman Y, Drayer DE, Reidenberg MM (1984) Increased urinary enzyme excretion in workers exposed to nephrotoxic chemicals. Amer J Med 76:989–998
63. Mozar JL, Gibey R, Dupond JL, Henry JC (1988) Influence des aminoglycosides sur l'activité de la N-acetyl-bêta-D-glucosaminidase (NAG) urinaire et de ses isoenzymes. Path Biol 36:230–234
64. Mueller PW, MacNeill ML, Steinberg KK (1989) N-acetyl-β-D-glucosaminidase assay in urine: urea inhibition. J Analyt Toxicol 13:188–190
65. Mutti A (1987) The study of urinary excretion of kidney antigens to reveal early effects of exposure to exogenous chemicals. In: Foa V, Emmett EA, Maroni M, Colombi A (eds) Occupational and environmental chemical hazards. Cellular and biochemical indices for monitoring toxicity. Ellis Horwood, Chichester, pp 315–322

66. Neumayer HH, Junge W, Küfner A, Wenning A (1989) Prevention of radiocontrast-media-induced nephrotoxicity by the calcium antagonist nitrendipin: a prospective randomized clinical trial. In: Proc IVth Int Symp Nephrotoxicity, Guildford 1989, in press
67. Nieminen L, Möttönen M, Tirri R, Okonen S (1975) Nephrotoxicity of *Cortinarius speiosissimus*: a histological and enzyme histochemical study. Exp Path Biol 11:239–246
68. Nogawa K, Yamada Y, Honda R, Tsuritani I, Ishizaki M, Sakamoto M (1983) Urinary N-acetyl-β-D-glucosaminidase and β_2-microglobulin in 'itai-itai' disease. Toxicol Lett 16:317–322
69. Noto A, Ogawa Y, Mori S, Yoshioka M, Kitakaze T, Hori T, Nakamura M, Miyake T (1983) Simple, rapid spectrophotometry of urinary N-acetyl-β-D-glucosaminidase, with use of new colorimetric substrate. Clin Chem 29:1713–1716
70. Paigen K, Peterson J (1978) Coordinancy of lysosomal enzyme excretion in human urine. J Clin Invest 61:751–762
71. Porter G (1989) Risk factors for toxic nephropathies. Toxicol Lett 46:269–279
72. Pospischil E, Meisinger V, Cseike L, Jahn O (1987) Urinary N-acetyl-β-D-glucosaminidase excretion as an indicator of early renal effects: an investigation of workers exposed to lead and solvents. WHO, Occupational Commission on Health in the Chemical Industry 22:19–202
73. Price RG (1979) Urinary N-acetyl-β-D-glucosaminidase (NAG) as an indicator of renal disease. In: Dubach UC, Schmidt U (eds) Diagnostic significance of enzymes and proteins in urine, Hans Huber, Berne, pp 150–163
74. Price RG (1982) Urinary enzymes, nephrotoxicity and renal disease. Toxicology 23:99–134
75. Redondo FL, Bergon E, Tinture T, Miravalles EM, Pascual T (1987) Urinary enzyme activities in patients treated with gold and other rheumatic drugs. Clin Biochem 20:343–347
76. Rempel D (1989) The lead exposed worker. J Amer Med Assoc 262: 532–534
77. Ross BD, Guder WG (1982) Heterogeneity and compartmentation in the kidney. In: Sies H (Ed) Metabolic compartmentation, Academic Press, pp 363–409
78. Saito M, Masaki T, Kato M, Numasaka K (1988) A clinical evaluation of the protective effect of fosfomycin (FOM) against cis-diamminedichloroplatinum (CDDP)-induced nephrotixicity. Hinyokika Kiyo 34:782–789
79. Scherberich JE, Wolf G, Stuckhardt C, Kugler P, Schoeppe W (1987) Characterization and clinical significance of angiotensinase A of human kidney. Nephrol Dial Transplant 2:387
80. Schmidt E, Schmidt FW (1979) Enzyme diagnosis in diseases of the liver and biliary system. In: Schmidt E, Schmidt FW, Trautschold I, Friedel R (eds) Advances in clinical enzymology, S Karger, pp 239–292
81. Schwartz A (1987) Analgesic-associated nephropathy. Klin Wochenschr 65:1–16
82. Severini G, Aliberti LG (1987) Variation of urinary enzymes N-acetyl-β-glucosaminidase, alanine aminopeptidase and lysozyme in patients receiving radio contrast agents. Clin Biochem 20:339–341
83. Simpson JG, Saunders NJ, Thompson KJ, Whiting PH (1988) Chronic renal damage caused by cyclosporine. Transplant Proc 20:792–799
84. Sreepada Rao TK, Nicastri AD, Friedma EA (1977) Renal consequences of narcotic abuse. Adv Nephrol 7:261–290
84a. Stolarek I, Howey JEA, Fraser CG (1989) Biological variation of urinary N-acetyl-β-D-glucosaminidase: practical and clinical implications. Clin Chem 35:560–563
84b. Strigini F, Melis GB, Gasperini M, Ronca G, Palmieri L, Fioretti P (1989) Urinary excretion of N-acetyl-β-D-glucosaminidase and alanine aminopeptidase during pregnancy. Int J Gynaecol Obstet 28:9–12
85. Tapson JS, Hodson AW, Marshall SM, Wilkinson R (1988) Urinary enzyme excretion after donor nephrectomy. Nephron 48:126–131
86. Thompson KJ, Saunders NJ, Simpson JG, Whiting PH (1989) Renal structure and function in streptozotocin-diabetic rats treated with cyclosporin A. Br J Exp Path 70:405–414
86a. Tohyama C, Mitane Y, Kobayashi E, Sugihira N, Nakano A, Saito H (1988) The relationship of urinary metallothionein with other indicators of renal dysfunction in people living in a cadmium-polluted area in Japan. J Appl Toxicol 8:15–21
87. Walker RJ, Duggin GG (1988) Drug nephrotoxicity. Ann Rev Pharmacol 28:331–345
88. Watts GF, Vlitos MAJ, Morris RW, Price RG (1988) Urinary N-acetyl-β-D-glucosaminidase excretion in insulin dependent diabetes mellitus: relation to microalbuminuria, retinopathy and glycaemic control. Diabete et Metabolisme 14:653–658
89. Wellwood JM, Ellis BG, Price RG, Hammond K, Thompson AE, Jones NF (1975) Urinary N-acetyl-β-D-glucosaminidase activities in patients with renal disease. Brit Med J 3:408–411

90. Whiting PH, Peterson J, Power DA, Stewart RD, Catto GRD, Edwards N (1983) Diagnostic value of urinary N-acetyl-β-D-glucosaminidase, its isoenzymes and the fractional excretion of sodium following renal transplantation. Clin Chim Acta 130:369–376
91. Wrigge P, Malyusz M (1988) Enzymurie nach Kontrastmittelgabe bei hypertonen Ratten. Röntgenblätter 41:332–335
92. Yoneytama M, Sharma RP, Kleinschuster SJ (1986) Cytotoxicity of citrinin in cultured kidney epithelial cell systems. Exotoxicology and Environmental Safety 11:100–111
93. Yoshida M, Sunaga M, Hara I, Katsumata M, Minami M (1989) Elevation of urinary N-acetyl-β-D-glucosaminidase and β-galactosidase activities in workers with long-term exposure to aromatic nitro-amino compounds. Bull Environ Contam Toxicol 43:1–8
94. Yuen CT, Kind PRN, Price RG, Praill PFG, Richardson AC (1984) Colorimetric assay for N-acetyl-β-D-glucosaminidase (NAG) in pathological urine using the ω-nitrostyryl substrate: the development of a kit and the comparison of the manual procedure with the automated fluorimetric method. Ann Clin Biochem 21:295–300
95. Yuen CT, Corbett CRR, Kind PRN, Thompson AE, Price RG (1987) Isoenzymes of urinary N-acetyl-β-D-glucosaminidase (NAG) in patients with renal transplants. Clin Chim Acta 164:339–350

15 Extrarenal Diseases

U. Burchardt and J.E. Scherberich

In this section the current status of enzymuria is summarized as observed in several primary, nonrenal diseases. Reasons for an altered enzyme excretion are variable and depend at least in part on the underlying primary disease. The possible pathophysiological processes are reviewed in Chaps. 3 and 11. In the various nonrenal diseases considered below, hyperenzymuria indicates secondary renal involvement.

15.1 Heart Diseases

In 1961 the reports first appeared on elevated urinary aspartate aminotransferase (ASAT) activity in 10 [1] and 16 patients [2] with acute myocardial infarction. The augmented levels were two to five times above normal limits. The activity was elevated in the first 24 h after infarction, reached a peak at 24–48 h, and usually returned to normal within 72–92 h. In two cases [2] urinary ASAT activity remained elevated for 2–3 days longer than the serum ASAT activity.

Similar excretion courses were demonstrated for alanine aminopeptidase (AAP) [3], dipeptidylpeptidase IV (see Fig. 1), adenylate kinase [4], and lactate dehydrogenase [5].

The increased excretion rates may reflect a transitory hypoxia in the kidney after infarction with consequently increased enzyme release from tubular cells. Increase of urinary LDH, seen in patients with transmural infarction [5], suggests that cytosolic enzymes are released from tubule cells by a blebbing process (see Sect. 3.2.4). Additionally, enzymes released from heart muscle appear in urine if their molecular mass is low, as they are filtered through the glomerular barrier but not reabsorbed by the impaired tubules. A typical example is adenylate kinase [4].

Increased urinary enzyme excretion coinciding with the onset of infarction seems to offer interesting insights into the pathophysiological conditions in the acute phase of myocardial infarction. Increased oxidative damage to polyunsaturated lipids of myocardial and renal tissue cell membranes (lipid peroxidation) due to an increase of free radicals may induce cell damage in the heart and kidney.

Urinary Enzymes
K. Jung/H. Mattenheimer/U. Burchardt (Eds.)

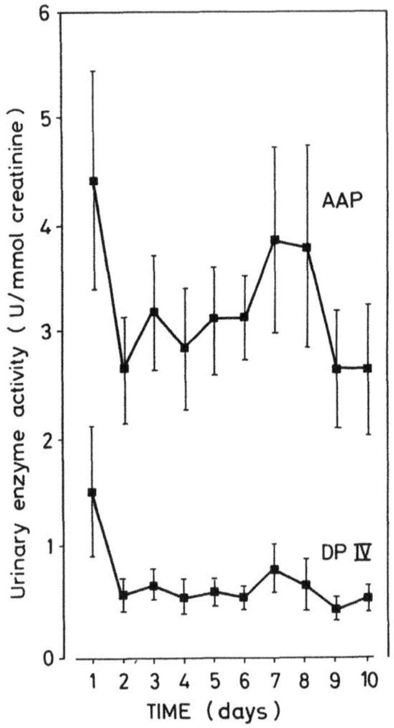

Fig. 1. Excretion of urinary alanine aminopeptidase (AAP) and dipeptidyl peptidase IV (*DP IV*) after onset of acute myocardial infarction ($n = 14$, mean $\pm$ SD)

Seven days after infarction, another elevation of enzyme excretion may occur (Fig. 1). This may reflect a circaseptan (about 7-day) rhythm of the kidney. A circaseptan bioperiodicity in rejection of human kidney transplants was demonstrated by De Vecchi et al. [6].

In 7 patients suffering from chronic congestive heart failure of various genesis, hyperenzymuria (AAP 7/7, β-glucuronidase 5/7) [3] was observed.

15.2 Hypertensive Vascular Disease

A variety of tubular markers are excreted by patients with arterial hypertension. Urinary *N*-acetyl-β-D-glucosaminidase (NAG) activity is significantly elevated in urine specimens of patients with renovascular and benign essential hypertension or with hypertension due to renal parenchymal disease [7, 8].

The incidence of elevated urinary NAG activity in arterial hypertension without a clinically detectable renal disease is approximately 20%–25% [9, 10]. Mansell et al. noted it in 49% of 137 patients with hypertension [7]. NAG is increased in more than 88% of urine specimens from patients with primary malignant or renovascular hypertension [9, 11]. Elevated activities of other renal enzymes are reported in hypertensive patients, including LDH, alkaline phosphatase (AP), β-glucuronidase, and AAP [10]. The urinary level of AAP in stage II hypertensive patients was higher than that in the normotensive or stage I hypertensive subjects.

Different results have been described when the enzyme excretions were correlated to the systolic blood pressure. Scherberich et al. [12] could not show a correlation between urinary NAG and blood pressure, probably because of the influence of antihypertensive drugs. In contrast, Kohno et al. [13] found that systolic blood pressure and renal blood flow are correlated to urinary NAG ($r = 0.43$ and $r = -0.61$, respectively) but not to AAP. In contrast, no significant correlation was found between diastolic blood pressure, age, sex, race, and creatinine clearance [14, 15]. However, in another clinical trial involving 32 patients with various forms of arterial hypertension, urinary excretion of AAP was significantly correlated with systolic blood pressure ($r = 0.55$).

According to Maruhn et al. [10, 11], the urinary enzyme pattern consisting of β-galactosidase, NAG, β-glucuronidase, gamma-glutamyltransferase (GGT), AP, and leucine arylamidase differs considerably in benign essential hypertension, malignant essential hypertension, and renal artery stenosis. The patterns allowed a correct reclassification of 86% of patients according to the clinical diagnosis [11].

Ando and Fujita [16] showed that urinary kallikrein and plasma renin activity were increased in 46 young patients with borderline hypertension. In hypertensive subjects with normal plasma renin activity, urinary aldosterone levels correlate to urinary kallikrein values ($r = 0.48$). However, no correlation was found in patients with high renin activity in plasma. Only in the early stage of hypertensive disease is kallikrein excretion increased, but it declines when hypertensive disease develops [16, 17]. Kallikrein excretion is highly correlated with the renal blood flow [17]. Thus, Levy et al. [18] have concluded that the kallikrein-kinin system contributes to the regulation of renal blood flow.

In hypertensive patients lysozymuria is uncommon. It only occurs in patients with decreased glomerular filtration rate (below 40 ml/min) [19].

Antihypertensive treatment reduces the output of various enzymes; for instance, urine NAG activity drops significantly. The reduction may occur within a few months [8, 20]. Alderman et al. [14] followed 80 patients with essential hypertension for 1 year and found that they attained their ultimate blood pressure reduction within 3 months. Whereas the urinary NAG level had not declined significantly by 3 months, it did drop significantly at 1 year (from 60-43 to 45-28 $nmol \cdot h^{-1} \cdot mg^{-1}$ of urinary creatinine).

15.3 Disorders of the Hepatobiliary System

Renal disorders are frequent complications of liver diseases [21]. In 1966 Mason et al. [22] first described increased urinary activity of AAP, leucine aminopeptidase, and β-glucuronidase in icteric surgical patients. The finding of high urinary enzyme output in jaundice was confirmed by Balcells et al. [23] The increase of enzyme activity in urine was independent of its serum concentration. In 12 out of 13 patients with infectious hepatitis, AAP excretion was increased (maximal 13× above upper normal limit) [24] (Fig 2.). Electrophoresis showed AAP from the kidney but not from serum [24]. The release of enzyme from the brush border membrane may be increased as much as 15-fold for a month in patients suffering from liver cirrhosis and/or malignant tumours associated with liver metastasis [25]. An increased de novo synthesis of enzymes in the tubule cells during icteric diseases is hypothesized.

Bile duct obstruction causes an increase in bile acid concentration in serum and urine. Segmental perfusion of rat jejunal mucosa supports the concept that the increased urinary activities of membrane enzyme in cholestasis may be due to solubilization through bile salts [1, 21].

Following experimental bile duct ligation in Wistar rates successive changes in the proximal tubules of the kidney can be visualized. The alterations suggest an adaptive process of the cell to the increased bilirubinemia. It is thought that

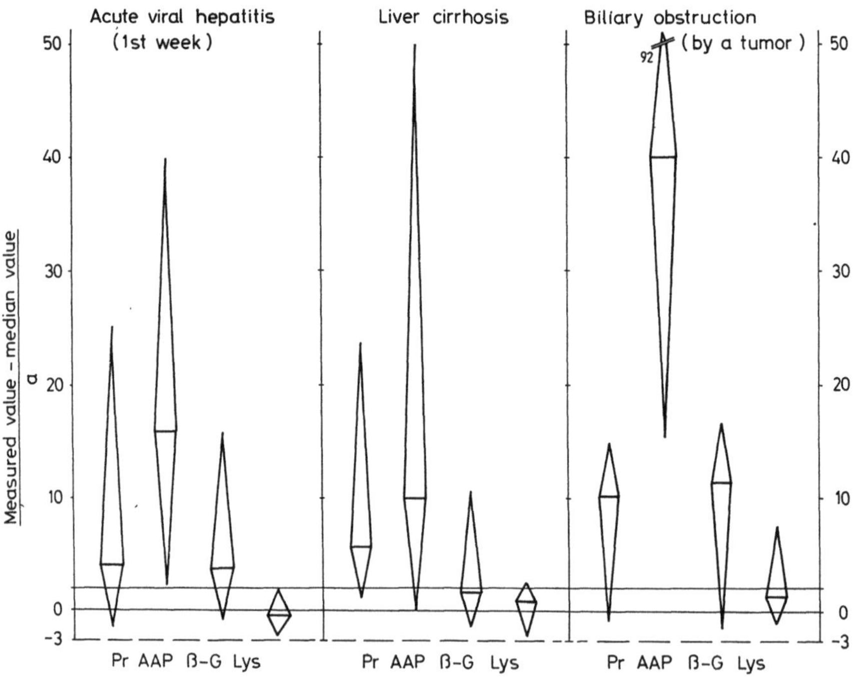

Fig. 2. Pattern of urinary protein (*Pr*) and enzyme excretion (*AAP*, alanine aminopeptidase; *β-G*,*β*-glucuronidase; *Lys*, lysozyme) in acute viral hepatitis, liver cirrhosis, and malignant biliary obstruction [24]. Enzyme activities are expressed per urinary creatinine level. Deviations from the medium value of a group of healthy persons are expressed in multiples of *a*, a distribution value defined as follows: a_1 = half the distance between the 50% percentile and the 98% percentile, and a_2 = half the distance between the 50% percentile and the 2% percentile

unconjugated bilirubin from the peritubular plexus may enter the tubular cell and that it is conjugated by glucuronyl transferase. When uptake and conjugation of bile pigment exceeds the secretion rate, intracellular accumulation occurs. This may cause segregation in autophagic vacuoles into the tubule lumen together with the enzymes [26].

Gekle and Kühner [27] demonstrated a significant increase of urinary enzyme excretion (LDH, AP) in hyperbilirubinemic newborns. The enzymuria was reversible during the recovery period. The urinary excretion of GGT and α-glucosidase were determined in patients with liver cirrhosis [28].

The enzyme activities on average were massively increased in cirrhotic compared with control patients, particularly in those with a reduced glomerular filtration rate. It appears that in patients with liver cirrhosis there is enzyme excretion which is not linked to impaired liver function, but rather to impaired renal perfusion and glomerular filtration [28].

15.4 Diseases of the Pancreas

Renal injury is one of the common and most serious complications of inflammatory diseases of the pancreas, and it leads quite often to insufficiency of the kidney.

Many factors are considered to cause renal injury:

- Pancreatic enzymes (liberated into the circulation), mainly the proteolytic or lipolytic ones, directed against tissues of other organs may lead to liberation of increased amounts of fat and protein degradation products. These substances, partially toxic, are filtered through the glomerulum, producing an overload of lysosomes (see Sect. 3.1.3) followed by release of enzymes into the urine.
- Liberated pancreatic enzymes may act directly by digesting particular components of the nephron.
- Shock itself seems to be rather seldom the primary cause of renal complications.

Acute pancreatitis is associated with increased excretion of tubular enzymes as well as filtered low molecular mass enzymes. Figure 3 demonstrates changes in urinary enzyme output in the course of a case of acute pancreatitis. Increased urinary amylase excretion is widely used to support a diagnosis of acute pancreatitis [29]. It is well-known that such increases are also seen in other disorders like burns, diabetic ketoacidosis [30], severe hepatic insufficiency, duodenal perforation [31] or in the postoperative period [32]. Other low molecular mass proteins, such as β_2-microglobulin [33, 34], lysozyme [35], ribonuclease [36, 37], lipase [36], immunoreactive trypsin [36, 38], and elastase 1 [39] are also excreted in higher amounts by patients with pancreatitis. Increased serum levels of elastase 1 have been found in patients with acute

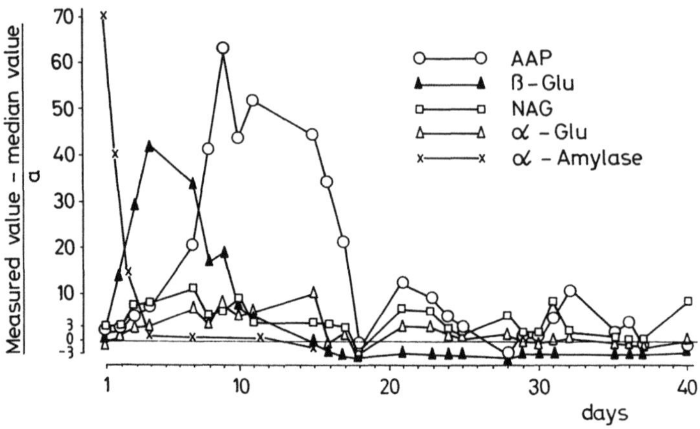

Fig. 3. Enzymuria in a patient suffering from acute pancreatitis (*AAP*, alanine aminopeptidase; *NAG*, *N*-acetyl-*β*-D-glucosaminidase; *β-Glu*, *β*-glucuronidase; *α-Glu*, *α*-glucosidase) [25]. Data were expressed as documented in Fig. 2

pancreatitis, in whom the activity of this enzyme remains altered for a longer period of time than amylase and immunoreactive trypsin [39]. It might be suggested that in connection with pancreatitis other low molecular mass substances, e.g., peptides and amino-acids, are released and that such substances interfere with reabsorptive and digestive tubular function.

It has been shown that aminoacids, in particular lysine, when infused into healthy subjects, induce tubular proteinuria by inhibition of tubular protein reabsorption [40, 41]. The increased amylase clearance/creatinine clearance (C_{am}/C_{cr}) ratio in chronic and acute pancreatitis seems to be due to a decrease in amylase reabsorption in the renal tubules rather than to an enhanced amylase filtration in the glomeruli [30, 31, 42–45]. Elevated clearance for amylase in relation to creatinine is not consistent in pancreatic diseases and was seen in only 7 of 10 patients with acute pancreatitis [34]. In patients with parotitis and increased salivary type amylase, the C_{am}/C_{cr} ratio is not increased (Fig. 4) [46]. Differences in clearances of pancreatic and salivary amylase which are identical in size but have a different charge may be a consequence of differences in charge-related glomerular filtration [47] or reabsorption.

Lipase (molecular mass 40000–46000 daltons), in contrast to amylase, is completely reabsorbed by the proximal tubules after glomerular filtration. In the initial stages of acute pancreatitis a measurable clearance of lipase was found in 9 of 10 patients. However, regression analysis failed to show a significant correlation between lipase and β_2-microglobulin clearance [34].

Fabris et al. [37, 38] investigated the role of circulating free trypsinogen (molecular mass about 24000 daltons) and renal tubular dysfunction in affecting trypsin plasma-urine transfer. Nine patients with pancreatic cancer and 15 with chronic pancreatitis had increased urinary immunoreactive trypsin excretion,

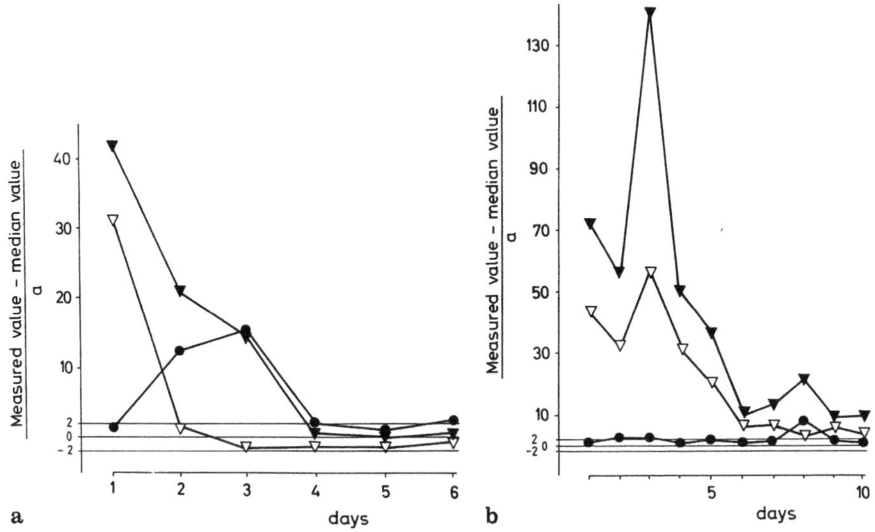

Fig. 4a. Amylase/creatinine clearance ratio (●) in a patient suffering from acute pancreatitis (▽ amylase in serum, ▼ amylase (creatinine) in urine (for calculation of data see Fig. 2) [46]. **b** Amylase/creatinine clearance ratio in a patient suffering from parotitis epidemica [46]. Enzyme activities were related to urinary creatinine level. Deviations from the median value of a group of healthy persons were expressed in multiples of *a*

regardless of the type of disease, when output of α-glucosidase or GGT was elevated.

The mechanism responsible for the selectivity and the changed selectivity of reabsorption for low molecular mass proteins in the proximal tubules of the kidney [40] is not fully understood (see Sect. 3.1.1).

The primary role of tubular damage in increasing the excretion of some low molecular mass enzymes is confirmed by the correlations found with urinary tubular cell derived enzymes and low molecular mass enzymes in pancreatic diseases:

α-glucosidase: ribonuclease	$r = 0.25$ $2p < 0.05$	[36]
GGT: ribonuclease	$r = 0.34$ $2p < 0.01$	[36]
GGT: elastase 1 clearance	$r = 0.47$ $2p < 0.01$	[39]
GGT: immunoreactive trypsin	$r = 0.64$ $2p < 0.001$	[37]
α-glucosidase: immunoreactive trypsin	$r = 0.63$ $2p < 0.001$	[37]

Investigations of urinary enzyme excretion reveal the kind of tubular alteration in pancreatitis. They are consistent with observations on renal enzymatic changes in cats during acute pancreatitis [48].

15.5 Malignant Diseases

Urinary enzyme excretion in renal and urinary tract tumours is described in Sect. 12.5. Not only patients with renal and bladder tumours but also most patients with disseminated cancer have proteinuria and/or enzymuria.

In a study of Rudman et al. [49] 24-h urine protein averaged 80 mg in normal subjects and in the range about 200 mg in patients with acute myelocytic leukemia, stage IV Hodgkin's disease, or metastatic cancer of the colon, breast, ovary, and pancreas. The proteinuria differs in composition from that in nephrotic syndrome or other types of intrinsic renal disease. In cancer patients proteinuria arises because abnormal proteins accumulate in the plasma. A saturation of renal tubular reabsorption by filtered, typical tumour low molecular mass proteins [50–52] leads secondarily to excretion of greater proportions of normal low molecular mass plasma proteins.

Morgan et al. [53] observed elevated activities of arylsulfatase B in 24-h urine from 172 out of 243 patients with colorectal cancer. Urinary AAP and NAG were increased in only 4 out of 23 patients with testicular cancer prior to treatment [54]. In a study of Goldberg and Rutenberg [55], 16 of 35 patients with cancer had elevated leucine aminopeptidase activity. During the course of treatment with cytostatics or irradiation, enzyme excretion increased transiently [56, 57]. In patients with uterine carcinoma a peak of enzyme excretion was found on the 9th day after radioactive contact therapy [58].

Markedly increased quantities of urinary lysozyme ranging up to 4.9 g per day have been found in patients with monocytic and monomyelocytic leukemia [59, 60], which are due to an increased turnover of the granulocytic series. Augmented lysozymuria occurred after X-ray irradiation of the enlarged spleen [61].

Lysozyme accumulates in enlarged lysosomes of the proximal tubule cells. Tubular cell alteration with subsequent loss of tubular cell enzymes (including lysozyme) is the consequence. Proteinuria and enzymuria are frequent findings in patients with plasmocytoma (multiple myeloma) [49, 52, 62]. While a nephrotic syndrome is very rare, enhanced excretion of microglobulins (< 50 kDa) independent of Bence Jones proteinuria is common [63, 64]. The tubular pattern of proteinuria in patients with multiple myeloma may correspond to pathohistological data of interstitial fibrosis in more than 80% of the cases studied [64]. Scherberich et al. [65] have studied the excretion of tubular marker enzymes and brush border related antigens in 26 patients with IgG and IgA myeloma, Bence Jones type myeloma (L-chain disease), and in 1 patient with nonsecretory plasmocytoma [65]. Enzymuria (AAP, GGT, and AP) was significantly increased in these patients, depending on the type of paraprotein. Patients with light chain proteinuria had a higher AAP excretion compared with those without Bence Jones proteinuria (Fig. 5).

However, this finding was only true for the total data calculated. The excretion of tubular enzymes was not correlated with the patient's individual

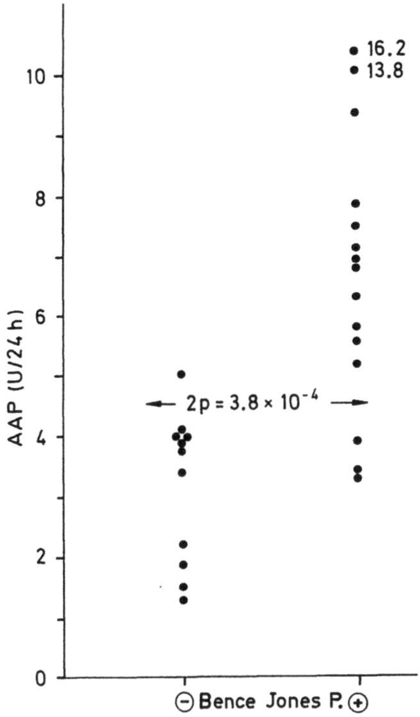

Fig. 5. Excretion of alanine aminopeptidase (AAP) in 26 patients suffering from multiple myeloma. Some 15 patients with Bence Jones proteinuria excreted the enzyme in a significantly higher amount compared with 11 patients without monoclonal light-chains in urine (Wilcoxon-Mann-Whitney test) [65]

concentration of L-chains in urine (Spearman's rank correlation test). In contrast, urinary L-chain excretion and urinary NAG activity were positively correlated ($r_S = 0.51$) [65].

15.6 Endocrine Disorders

15.6.1 Hyperthyroidism

The kidney is one of the target organs of thyroid hormones. Their primary action is modulation of renal metabolism, causing a direct [66] or indirect (nucleus-mediated) [67] enzyme induction. Triiodothyronine (T_3) augments renal Na^+–K^+ ATPase by increasing the absorptive Na^+ load. Furthermore, the glomerular filtration rate rises significantly after the addition of T_3 [68]. Hemodynamically mediated glomerular hyperfiltration leads per se to increased excretion of tubular cell enzymes (see Sect. 3.1.1).

Patients with hyperthyroidism excrete with the urine increased quantities of AAP, partly also of GGT, alkaline phosphatase, and arylsulfatase A [69, 70].

The serum thyroxine level and AAP in urine were positively correlated in 42 patients with hyperthyroidism ($r = 0.54$). In all cases treatment of the hyperthyroidism induced a drop in AAP excretion [70, 71]. Assessment of AAP in urine yields additional information for treating hyperthyroidism [69]. Ford et al. [72] found that the albumin: creatinine, NAG: creatinine, and retinol binding protein: creatinine ratios were increased in random urine specimens from hyperthyroid patients. They also observed a significant drop in the ratios in hyperthyroid patients compared with the same patients when euthyroid. These increased ratios were not caused by decreased creatinine excretion in hyperthyroid patients [72] because enzyme activities uncorrected for creatinine concentration were also significantly higher [69, 72].

High oral uptake of T_3 and thyroxine results in transient hyperenzymuria with peak values after 6 h. At 36 h later, urinary enzyme activities have returned to the values of normal controls (Fig. 6) [70].

15.6.2 Diabetes Mellitus

The diabetic metabolic derangement interferes markedly with the basement membrane metabolism and tubular reabsorption and secretion. The pathobiochemical mechanism of release of tubular cell enzymes into urine in patients with diabetes mellitus is still unclear, however.

Several studies have been consistent in demonstrating that many enzyme activities may be elevated in urine of diabetic patients. Special interest is directed

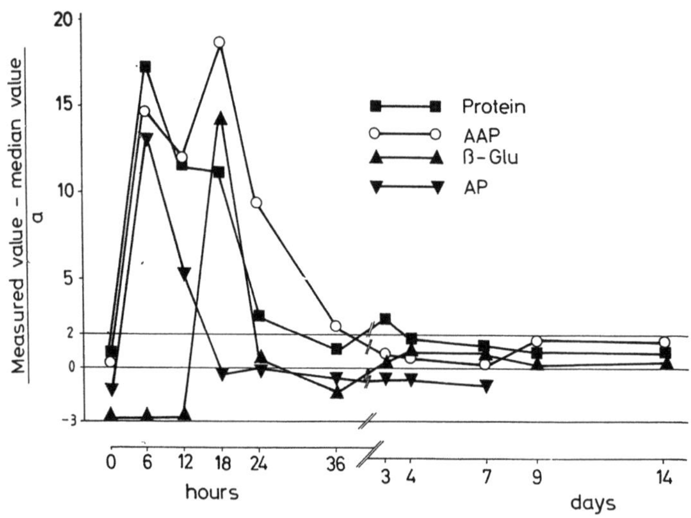

Fig. 6. Excretion of protein and urinary alanine aminopeptidase (*AAP*), *β*-glucuronidase (*β-Glu*), and alkaline phosphatase (*AP*) in a female patient after intake of 2 mg triiodothyronine and 8 mg thyroxine in an attempted suicide (according to [69])

Table 1. Summary of selected data concerning enzymuria and diabetes mellitus

Diabetes mellitus with clinical proteinuria	Elevated NAG and AAP activities, significantly higher compared with patients without proteinuria
Urinary NAG/AAP	Positive correlation ($r = 0.7$)
In diabetic patients without clinical proteinuria:	
Microalbuminuria/NAG	Moderately correlated
Microalbuminuria/AAP	Closely correlated
Diabetes mellitus without proteinuria but retinopathy	Not significant elevated NAG activity, but significant rise of urinary AAP activity
Duration of diabetes mellitus	Good correlation with microalbuminuria, Negative correlation to trehalase activity in urine
High hemoglobin A_{1C}, poor glycemic control	AAP and NAG in urine: better correlation compared with duration of disease
Mean arterial blood pressure in diabetic patients	No significant correlation to urinary NAG level

to the enzymes of glycoprotein metabolism. Elevated urine levels were reported of acid maltase, maltose riboflavin glucosyltransferase [73], NAG [74–76], α-glucosidase, trehalase, β-glucuronidase [77, 78], LDH [79], lysozyme [80], AP, and AAP [81, 82] (Table 1).

The most data are available for NAG. Newly diagnosed diabetis showed significantly higher serum and urine NAG activities than non-diabetic controls [83]. The enzyme activity of serum and urine specimens was significantly reduced after appropriate treatment. The treatment included either dietary restrictions or administration of insulin [75, 76]. Several authors [84, 85] were unable to demonstrate a significant association with blood glucose or with glycosylated serum proteins, however. NAG activity was unrelated to the degree of glucosuria or serum glucose concentration [75]. In addition, there was no correlation between NAG excretion and mean arterial blood pressure (Table 1) [74].

Significant correlations were found between NAG and hemoglobin (HbA_1) ($r = 0.50$) and albuminuria ($r = 0.45$) (Table 1). Three-way analysis of variance revealed that the relationship of raised urinary NAG level to microalbuminuria and "poor" glycemic control were statistically independent [74].

Most authors failed to demonstrate an association between urinary enzymes and diabetic retinopathy [78]. The relation of raised urinary NAG to microalbuminuria and poor long-term glycemic control raises the possibility that this enzyme might fulfil a role in studying patients at risk of renal microangiopathy [86] (Fig. 7). A prospective study suggested that urinary activity alone did not predict the development of diabetic nephropathy [87]. Urinary NAG and AAP were measured in 207 diabetic patients; 14 had type I and 193 had type II [88]. Patients with proteinuria had higher urinary NAG and AAP levels compared with normal controls. NAG and AAP activities increased more significantly in those with high HbA_1 levels than in long-term patients, NAG excretion by

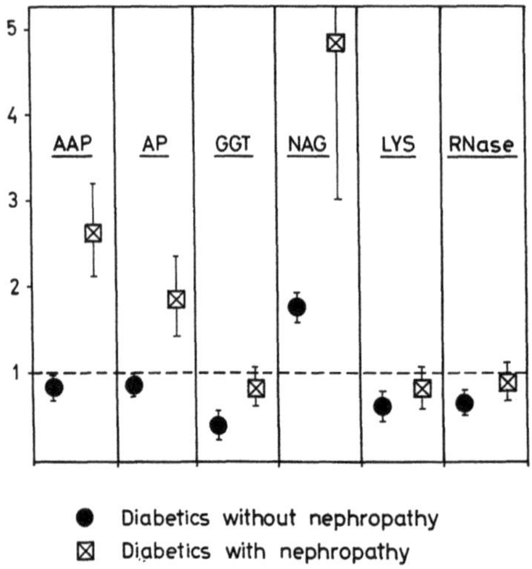

Fig. 7. Urinary enzyme activity (U/mol creatinine) in diabetic patients with and without nephropathy (multiples of upper limit). *LYS*, lysozyme; *RNase*, ribonuclease; *AAP*, alanine aminopeptidase; *AP*, alkaline phosphatase; *GGT*, gamma-glutamyltransferase; *NAG*, *N*-acetyl-β-D-glucosaminidase (according to [86])

patients with retinopathy was slightly but not significantly greater than by those without. In contrast, AAP activity was significantly raised in diabetic patients with retinopathy compared with those without (23.0–1.5 U/g creatinine vs 17.4–0.8, $p < 0.01$) (Table 1). The authors also conclude that urinary NAG and AAP levels can be used to diagnose the onset and to follow the development of diabetes-associated microvascular complications. Under these conditions AAP turned out to be a more sensitive parameter than NAG.

Increased urinary AAP and AP values were measured in patients with pathological glucose tolerance [81].

Patients suffering from diabetic nephropathy tend to show a greater increase in activity of AAP in urine after application of a contrast media (diatrizoate) compared with healthy controls or diabetic patients without nephropathy [3, 82, 89].

In children with diabetes mellitus the exercise-induced rise in urinary NAG level was a more sensitive indicator in respect of nephropathy than differences in the excretion of albumin [90]. However, in adult patients with type I diabetes, tubular enzymes (AAP, AP, GGT), albumin, and β_2-microglobulin values exhibited a highly significant increase [91].

Urinary trehalase and maltase activities were elevated in patients with type II diabetes. Furthermore, the urinary trehalase level was raised when the disease was of short duration (1–4 years), whereas over a relatively long duration (more than 13 years), the activity remains rather low (Table 1) [78].

In streptozocin-diabetic rats the most remarkable microscopic finding in the kidney is glycogen storage in the cells of the straight cortical parts of the distal tubuli.

In diabetic Wistar rats, AAP and GGT activities in urine were lowered. The urinary activities of NAG, acid phosphatase, LDH, and pyruvate kinase were higher than in fasted, nondiabetic rats [92, 93].

15.7 Disorders of the Connective Tissue

15.7.1 Rheumatoid Arthritis

Systemic manifestations of rheumatoid arthritis may also involve the kidney [94, 95]. A decreased glomerular filtration rate, independent of the administration of antirheumatoid drugs, in 244 patients with rheumatoid arthritis indicate "autochthonous" renal lesions [96]. When the urinary NAG levels were measured, a highly significant difference was detected between the mean urinary NAG excretion of patients with arthritis and that of matched controls [97]. A positive correlation was noted between urinary NAG excretion and indices of disease activity, and a negative one between NAG and the hemoglobin level [98]. Significantly higher levels of NAG activity were found in patients with rheumatoid arthritis than in those with degenerative joint disease [99]. Analysis of the results shows that salicylates, gold [100], and other antiinflammatory drugs were responsible for some of the difference but suggest that the rheumatoid disease itself causes renal damage [99, 101]. There was no significant correlation between serum, synovial fluid, and urine NAG levels [97].

The pathophysiological basis of enhanced enzymuria in patients with rheumatoid arthritis is the topic of several studies. Some 25% of patients with rheumatoid arthritis and clinical signs of renal involvement (hematuria and/or proteinuria; $n = 39$) had mild mesangioproliferative glomerulonephritis [102], which was found in half of the patients in whom both hematuria and proteinuria were present. However, all rheumatoid arthritis patients received nonsteroidal antiinflammatory drugs, and 33 had been treated with gold salts. The histological data differ greatly from those reported by Brun et al. [103] 21 years earlier. Interstitial fibrosis of the kidney was found in 46% of 76 patients with rheumatoid arthritis compared with 3% of controls [104].

Scherberich et al. [105, 106] noted increased excretion of tubular marker proteins AAP, GGT, and AP and brush border SGP 240 glycoprotein in 80 patients with rheumatoid arthritis (Fig. 8) compared with controls. Urinary AAP and the concentrations of the major brush border surface antigen (SGP 240) were significantly correlated ($r_S = 0.5$).

15.7.2 Rheumatic Fever

A correlation was observed between the activity of soluble arylsulfatases A and B in urine of children with rheumatic fever and the degree of intensity of this disease. During an episode of rheumatic fever the activity of both enzymes was higher than in urine of healthy children, and during remission this activity decreased. Remissions with persistence of chronic inflammatory foci were characterized by lack of a fall in the activity of both enzymes [107].

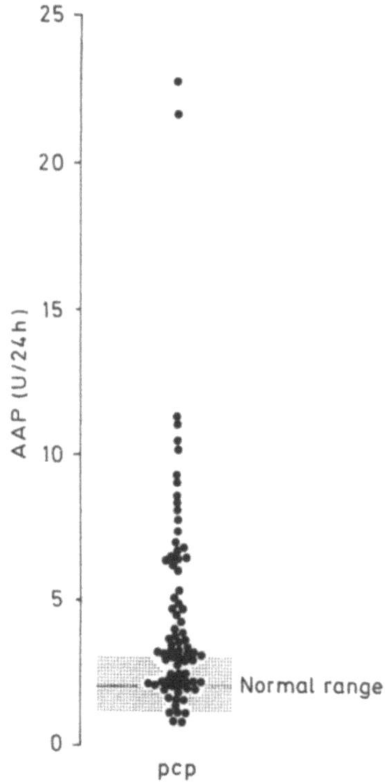

Fig. 8. Excretion of tubular membrane bound alanine aminopeptidase (*AAP*) in urine of 80 patients with rheumatoid arthritis [105]

15.7.3 Systemic Lupus Erythematosus

In patients with systemic lupus erythematosus the activities of LDH and AAP were increased in connection with lupus nephritis and/or nephrotic syndrome [108]. In a pilot study the excretion of tubular enzymes was assessed in 37 patients with systemic lupus erythematosus [106]. Of these, 23 had clinical signs of lupus nephritis, and AAP activity in urine specimens was significantly higher compared with normal controls. Serum concentrations of complement component C_{3C} were inversely correlated with urinary AAP activity (Fig. 9).

15.8 Surgery and Injuries

Acute renal failure significantly increases postoperative mortality after major surgery. The mechanism of such renal damage is complex and often cannot be satisfactorily explained. Studies on urinary enzyme excretion have led to information about underlying mechanisms. Surgical trauma causes an increase in

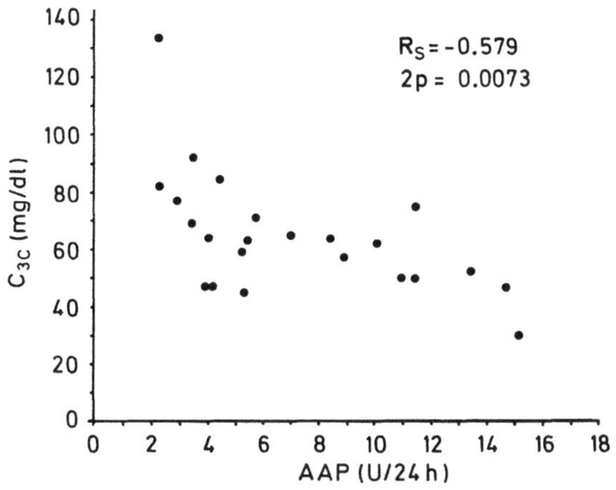

Fig. 9. Negative correlation between serum complement C_{3c} concentration and urinary excretion of tubular brush border alanine aminopeptidase (*AAP*) in 23 patients with systemic lupus erythematosus [106]

the renal excretion rate of high and low molecular mass proteins [109]. As early as 1964 Bergmann and Scheler [110] showed that cardiac surgery always results in increased urinary excretion of leucine aminopeptidase as a result of damage to the tubular integrity. Guarnieri [111] investigated 16 patients who had been operated upon using heart-lung machines and found increased enzymuria (β-glucosidase, lysozyme) and β_2-microglobinuria.

Increases in the excretion of NAG were found when the kidney was injured in the course of surgery. They showed a rough correlation with the assessment of the degree of damage [112]. Simple mobilization of the kidney without incision into its substance is associated with only a small and transient increase in enzyme excretion. The increase in AAP activity varied over a wide range from 2 to 80 times the upper limit of the normal range. Peak values were reached early after operation, often within 2–3 h [113]. Enzyme excretion rates decline on the 3rd to 4th [114] and persist until the 7th postoperative day [114, 115]. Nephrectomy or mobilization of the kidney produces no change in the excretion rate of NAG [112]. Urinary enzyme excretion is proportional to the relative clinical trauma score [109, 115]. Besides the rise in total NAG excretion, increases were shown in the percentage of the intermediate isoenzyme forms of NAG [109].

Pulmonary and cardiac surgery using heart-lung machines led to increased excretion of LDH, AAP, and NAG in 24-h pooled urine [114]. The enzymuria after cardiac surgery was generally more pronounced than after pulmonary surgery [114]. The hemoglobinuria which was observed in all of the cardiac surgery patients might be responsible for that difference. Enzyme activities were influenced by additional drug administration [114]. Hyperenzymuria was also observed after contusio renalis, kidney trauma, and extensive burn [116] but not in extracorporal shock wave lithotripsy [117–119].

15.9 Inborn Errors of Metabolism

Urinary analysis of enzyme pattern may be of value for the detection of inborn errors of metabolism. In several disorders, the lack or alteration of the appropriate enzyme (or enzymes) can be also observed in urine [120–122].

In hereditary lysosomal diseases glycosidases are absent or defective [123]. In Pompe's disease (glycogen storage disease, type II) the accumulation of glycogen is due to the deficiency of an α-glucosidase (pH optima 4/5). The use of urine for screening this disease is precluded by the preponderance of neutral α-glucosidase in urine. Therefore, only the ratio of α-glucosidase activities (pH 4.0/6.5) provides a simple and accurate screening test [124, 125].

In the tissue of patients afflicted with the Tay-Sachs disease the activity of hexosaminidase A is diminished. After incubation at 50°C it is inactivated, whereas form B retains most of its activity. The wide variation of the activity of hexosaminidase B in normal subjects prevents us from distinguishing heterozygotes from normal probands in a population. The distribution of hexosaminidases in urine and leukocytes is similar ($r = 0.96$) [124]. This indicates that the urine test is as dependable as that with leukocytes. Both tests only permit accurate differentiation between homozygotes (patients) and heterozygotes (carriers) [126].

Deficiency or abnormality of lysosomal α-*N*-acetylglucosaminidase (EC 3.2.1.50) is responsible for mucopolysaccharidosis II B (Sanfilippo B syndrome). In this syndrome an altered enzyme form was assayed in urine (previously called "Sanfilippo B corrective factor") with a calculated molecular mass of 171 000 instead of 240 000 daltons [127]. It is suggested that the disease is due to a mutation of a structural gene coding for α-*N*-acetylglucosaminidase [123, 128].

Fabry's disease, which is characterized by α-galactosidase A deficiency, causes early renal failure. Variable increases in serum and urine α-galactosidase A are measured after renal transplantation. The excretion of this hydrolase was increased by acute alkalinization and decreased by acute acidification [129].

In the group of mucopolysaccharidoses, increased excretion rates of other than affected enzymes have an intrarenal source [120] (see Sect. 3.1.3).

In metachromatic leukodystrophy Maruhn et al. [79] found a pronounced diurnal variance for arylsulfatase A, the enzyme lacking in this disease.

Ehrich et al. [130] reported on children with cystic fibrosis and high enzyme excretion values. In a child suffering from galactosemia, increased urinary NAG and AAP activity normalized during treatment with a galactose-free diet [130].

Distal ("classic" or type I) renal tubular acidosis is a syndrome characterized by a sustained metabolic hyperchloremic acidosis which is out of proportion to the impairment of glomerular filtration and by an inappropriately high pH of the urine. It was demonstrated that the increase in urinary NAG excretion, initially noted in a 3 1/2-year-old boy with this disease, disappears during citrate administration [131].

In hereditary as well as sporadic distal tubular acidosis, lysozymuria occurs [132]. Findings of Sinaniotis et al. [133] suggest that urinary NAG activity, which is elevated in renal tubular acidosis, decreases after the administration of alkali. NAG estimation may be of value in monitoring alkali therapy in these patients.

Very high values (> 340 times higher) of lysozyme/creatinine clearance were observed in five children with Fanconi's syndrome [134]. There was no overlap with any other group studied. In a 2-year-old girl with idiopathic de Toni-Debre-Fanconi syndrome (underlying disease: cystinosis) Peters et al. [135] found only the serum AAP isoenzyme, while the kidney-related isoenzyme was absent in urine. The morphological correlation is a widespread absence of the brush border in the proximal tubular epithelium [136]. The occurrence of the serum AAP in final urine also indicates a remarkable glomerular damage [135].

References

1. Sepaha GC, Bhandari CR, Vijayvargiya R (1961) Glutamic oxaloacetic transaminase in serum and urine in normal, coronary thrombosis and liver disorders. Ind J Med Res 49:68–72
2. Kalmansonn R, Kalmansonn RW (1961) Acute myocardial infarction. Urine glutamic oxalacetic transaminase activity. Calif Med 95:165–167
3. Anton D, Burchardt U, Panzner B, Neef L (1983) Harnenzymausscheidung bei kardiovaskulären Erkrankungen. Dt Gesundh Wesen 38:86–87
4. Frithz G, Ericsson P, Ronquist G (1976) Serum adenylate kinase activity in the early phase of acute myocardial infarction. Ups J Med Sci 81:155–158
5. Dietz AA, Hodges LK, Foxworthy DT (1967) Correlation of serum and urine enzyme activity in patients with acute myocardial infarction. Clin Chem 13:359–370
6. De Vecchi A, Halberg F, Sothern RB, Cantaluppi A, Ponvicelli C (1981) Circaseptan rhythmic aspects of rejection in treated patients with kidney transplant. In: Walker CA, Winget CM, Soliman KFA (eds) Chronopharmacology and chronotherapeutics. University Foundation, Tallahassee, pp 339–353
7. Mansell MA, Jones NF, Ziroyannis PN, Marson WS (1979) *N*-Acetyl-*β*-D-glucosaminidase: a new approach to the screening of hypertensive patients for renal disease. Lancet 14:803–805
8. Simon G, Altman S (1982) Increased serum *N*-acetyl-*β*-D-glucosaminidase activity in human hypertension. Clin Exper Hyper Theor Pract A4(3):355–365
9. Johnston DA, Jones N, Scoble JE, Yuen CT, Price RG (1983) The diagnostic value of urinary enzyme measurements in hypertension. Clin Chim Acta 133:317–325
10. Maruhn D (1979) Evaluation of urinary enzyme pattern in patients with kidney disease and primary benign hypertension. In: Dubach UC, Schmidt U (eds) Diagnostic significance of enzymes and proteins in urine. Huber, Bern, pp 135–149 (Curr Probl Clin Biochem 9)
11. Maruhn D, Paar D, Bock KD (1979) Lysosomal and brush border membrane enzymes in urine of patients with renal artery stenosis and with essential hypertension. Clin Biochem 12: 228–230
12. Scherberich JE, Mösbauer A, Meissner A, Bentele D, Schoeppe B, Schoeppe W (1989) Kidney and serum derivated proteins in urine of patients suffering from renal diseases or arterial hypertension. Klin Wochenschr 67 (Suppl XVII): 44–47
13. Kohno M, Kanayama Y, Yasunari K, Kawarabayashi T, Murakawa K, Takeuchi K, Inoue T, Takeda T (1985) Significance of the measurement of urinary alanine aminopeptidase and N-acetyl-*β*-D-glucosaminidase activity in evaluating patients with essential hypertension. Clin Exper Theor Pract A7:1347–1360

14. Alderman MH, Melcher L, Drayer DE, Reidenberg MM (1983) Increased excretion of urinary *N*-acetyl-*β*-glucosaminidase in essential hypertension and its decline with antihypertensive therapy. N Engl J Med 309:1213–1217
15. Simon G, Peterson S (1988) Pathophysiology of increased urinary *N*-acetyl-*β*-D-glucosaminidase activity in human hypertension: effect of cilazapril therapy. Clin Exp Hyper Theory Pract A10:767–777
16. Ando K, Fujita T (1983) Increased urinary kallikrein excretion in young borderline hypertensive patients. Clin Exper Hyper Theor Pract A5:903–918
17. Maier M, Jerabek J, Reissert G, Höltzl E, Binder BR (1988) Correlation of two different assays for urinary kallikrein in normotensive and hypertensive subjects. Clin Chim Acta 178:127–140
18. Levy SB, Lilly JJ, Frigon RP, Stone RA (1977) Urinary kallikrein and plasma renin activity as determinants of renal blood flow. J Clin Invest 60:129–138
19. Jensen H (1973) Renal handling of muramidase in patients with kidney diseases or hypertension. Scand J Clin Lab Invest 32:111–115
20. Metz-Kurschel U, Kurschel E, Anlauf M, Bock KD (1987) Harnenzymexkretion bei Patienten mit primärer arterieller und sekundärer renovaskulärer Hypertonie vor und während antihypertensiver Behandlung. Klin Chem 18:192
21. Epstein M (ed) (1983) The kidney in liver disease. Elsevier, New York
22. Mason EE, Charnigoy FA, Gulessarion HP, Tector AJ (1966) Enzymuria in icteric surgical patients. Surg Gyn Obst 122:333–339
23. Balcells A, Sandoval DP, Vázques JJ (1967) Correlations entre la enzymuria y otras constantes biologicas. Med Clin 49:369–373
24. Burchardt U, Peters JE, Thulin H (1972) Harnenzymbestimmungen bei hepatobiliären Erkrankungen. Ber Ges Inn Med 8:271–273
25. Cuartero RA, Carril JN (1974) Behavior of the enzymes gammaglutamyl-transpeptidase, leucine aminopeptidase, lactate dehydrogenase and alpha-hydroxybutyrate dehydrogenase in the urine of hepatobiliary patients. Rev Esp Enferm Apar Dig 44:365–374
26. De Vos R, de Wolf-Peeters C, Desmet V (1972) Histochemical changes in rat kidney after bile duct obstruction. Beitr Pathol 145:315–324
27. Gekle D, Kühner U (1972) Untersuchungen über den nephrotoxischen Einfluß des Bilirubins auf die Urinenzymausscheidung Neugeborener. Monatsschr Kinderheilkd 120:304–307
28. Gatta A, Amodio P, Frigo A, Merkel C, Milani L, Zuin R, Ruol A (1981) Evaluation of renal tubular damage in liver cirrhosis by urinary enzymes and beta-2-microglobulin excretion. Eur J Clin Invest 11:239–243
29. Heer M, Pei P, Streuli R, Bühler H, Amann R (1983) Pankreatitis—Diagnostik am Krankenbett mittels Urinamylase-Test-Tape. Schweiz Med Wochenschr 113:1950–1952
30. Andriulli A, Bergia R, Masoero G (1979) Amylase to creatinine clearance ratio in renal disease. Gastroenterology 77:86–90
31. Berger GMB, Cowlin J, Turner TJ (1976) Amylase/creatinine clearance ratio and urinary excretion of lysozyme in acute pancreatitis and acute duodenal perforation. S Afr Med J 50:1559–1561
32. Gross JB, Levitt MD (1979) Postoperative elevation of amylase/creatinine clearance ratio in patients without pancreatitis. Gastroenterology 77:497–499
33. Karlsson FA, Jacobson G (1979) Renal handling of *β*-2-microglobulin, amylase and albumin in acute pancreatitis. Acta Chir Scand 145:59–63
34. Mogensen CE, Solling K (1977) Studies on renal tubular protein reabsorption:partial and near complete inhibition by certain amino acids. Scand J Clin Lab Invest 37:477–482
35. Klopocki T, Wysokinska-Borowicz A, Sawicki Z, Zenkteler J, Krawczynski J (1974) Lysozymuria as an index of renal injury occurring in the course of pancreatitis. J Clin Chem Clin Biochem 12:575–577
36. Fabris C, Meani A, Farini R, Vianello D, Del Favero G, Picolli A, Bonvicini P, Brosolo P, Naccarato R (1983) Urinary ribonuclease excretion in pancreatic disease. J Clin Chem Clin Biochem 21:573–575
37. Fabris C, Benini L, Del Favero G, Cavallini G, Basso D (1987) Molecular size distribution of immunoreactive trypsin and renal tubular dysfunction:role in trypsin plasma-urine transfer. Enzyme 37:174–181
38. Fabris, C, Piccoli A, Farini R, Del Favero G, De Besi T, Ceriotti G (1981) Trypsin plasma-urine transfer—preliminary study. Clin Chim Acta 114:101–105
39. Fabris C, Basso D, Benini L, Meggiato T, Del Favero G (1989) Urinary elastase 1 in chronic pancreatic disease. Enzyme 42:80–86

40. Maack T, Johnson V, Kan ST, Figneiredo J, Sigulem D (1979) Renal filtration, transport, and metabolism of low molecular weight proteins: a review. Kidney Int 16:251–270
41. Murraty WR, Mackay C (1977) The amylase creatinine clearance ratio in acute pancreatitis. Br J Surg 64:184–193
42. Fukumoto K, Wakabashi A, Takeda Y, Saeki M, Saitoh S, Amtsu T, Tada H, Tsukiyama J, Itahashi T, Ohshiba S (1981) The amylase clearance/creatinine clearance ratios in acute pancreatitis. Bull Osaka Med School 27:55–63
43. Marten A Beales D, Elias E (1977) Mechanisms and specificity of increased amylase ratio in pancreatitis. Gut 18:703–708
44. Munch R, Buhler H, Kehl O, Ammann R (1987) Lipasuria in acute pancreatitis:result of tubular dysfunction? Pancreas 2:333–338
45. Noda A (1972) Renal handling of amylase: evidence for reabsorption by stop-flow analysis. Metabolism 21:351–356
46. Burchardt, U (1983) Harnenzymausscheidung bei Erkrankungen der inneren Organe. Z Ges Inn Med 38:565–570
47. Wetzels IFM, Hafkenscheid JCM, Hessels M, Holtsma AJ (1988) Renal clearance of pancreatic and salivary amylase relative to creatinine clearance in patients with renal disease and proteinuria. Clin Chem 34:589–591
48. Snop S, Olowska L, Kusmicki Z, Barcew B (1980) Renal enzymatic changes during acute pancreatitis. Pol Tyg Lek 25:1821–1824
49. Rudman D, Chawla RK, Hendrickson LJ (1976) Isolation of a novel glycoprotein (EDC1) from the urine of a patient with acute myelocytic leukemia. Cancer Res 36:1837–1846
50. Batuman, V, Sastrasin M, Sastrasin S (1986) Light chain effects on alanine and glucose uptake by renal brush border membrane. Kidney Int 30:662–665
51. Bohle A (1983) Pathologie der Myelomniere. Verh Dtsch Ges Pathol 67:440–446
52. Clyne DH, Pesce AJ, Thompson RE (1979) Nephrotoxicity of Bence Jones proteins in the rat:importance of protein isoelectric point. Kidney Int 16:345–352
53. Morgen LR, Samuels MS, Thomas W, Krementz T, Meeker W (1975) Arylsulfatase B in colorectal cancer. Cancer 36:2337–2345
54. Diener U, Knoll E, Ratge D, Langer B, Wisser H (1982) Urinary excretion of alanine aminopeptidase and *N*-acetyl-*β*-D-glucosaminidase during sequential combination chemotherapy. J Clin Chem Clin Biochem 20:615–619
55. Goldbarg JA, Rutenberg AM (1958) The colorimetric determination of leucine aminopeptidase in urine and serum of normal subjects and patients with cancer and other diseases. Cancer 11:283–287
56. Burchardt U, Franke M, Krauss J, Barth A (1986) Renale Dipeptidylpeptidase-IV-Ausscheidung bei medikamenteninduzierten Nierenalterationen. Z Urol Nephrol 79:587–593
57. Kurschel E, Metz-Kurschel U, Hofmann W, Niederle N (1987) Untersuchungen zur Erfassung der Nephrotoxizität von humanen Alpha-2b-Interferon unter besonderer Berücksichtigung der Analyse von Harnenzymen bei Patienten mit chronischer myeloischer Leukämie. Klin Wochenschr 65:667–672
58. Dreller E, Burchardt U (1983) Renale Enzymausscheidung bei Tumoren und Tumortherapie. Z Ges Inn Med 38:593–599
59. Ohta H, Yasuma A (1974) Lysozyme activity in haematologic and non-hematologic disorders with special reference to reactive monocytosis associated with chronic infections and inflammatory reactions. Tohoku J Exp Med 114:15–26
60. Osserman EF, Lawlor DP (1966) Serum and urinary lysozyme (muramidase) in monocytic and monomyelocytic leukemia. J Exp Med 124:921–951
61. Tischendorf FW, Ledderose G, Müller D, Wilmanns W (1972) Heavy lysozymuria after X-irradiation of the spleen in human chronic myelocytic leukaemia. Nature 235:274–275
62. Coward RA, De Lamore JW, Mallick NP, Robinson LE (1984) Urinary *N*-acetyl-*β*-D-glucosaminidase as an indicator of tubular damage in multiple myeloma. Clin Chim Acta 138:293–298
63. Ronco, Mougenot B, De Meyer-Brasseur M (1987) Pathophysiologische Aspekte der myelominduzierten Tubulopathie. Nieren Hochdruckkrankh 16:120–130
64. Thönes W, Baum HP, Schneider HM (1987) Die Niere bei Gammopathie und Amyloidose. Nieren Hochdruckkrankh 16:107–115
65. Scherberich JE, Basting K, Walter F, Mondorf W, Mitrou P, Schoeppe W (1981) Excretion of specific tubular antigens as an indicator of kidney damage in patients with malignant monoclonal gammopathy. Kidney Int 20:162–163

66. Davis PJ, Blas SD (1981) In vitro stimulation of human red blood cell Ca ATPase by thyroid hormone. Biochem Biophys Res Com 99:1073–1080
67. Oppenheimer JH, Koerner D, Schwartz HL (1972) Specific nuclear triiodothyronine binding sites in rat liver and kidney. J Clin Endocrinol Metab 35:330–333
68. Burchardt U, Mertens E, Klaua M (1980) Effektiver renaler Plasmafluβ und Glomerularfiltratmenge vor und während medikamentöser Therapie der Hyperthyreose. Dtsch Gesundh Wesen 35:105–106
69. Burchardt U, Peters JE, Thulin H, Hempel RD, Gründig CA (1974) Enzymausscheidung im Harn bei Hyperthyreose. Z Ges Inn Med 29:719–722
70. Burchardt U, Hempel RD, Neef L (1978) Zur diagnostischen Wertigkeit von Enzymveränderungen im Serum und Harn bei Hyperthyreose. Z Med Labor Diagn 19:38–42
71. Hempel RD, Burchardt U, Höpfner JB, Neef L, Peters JE (1976) Aminopeptidasen im Serum und Harn bei Schilddrüsenüberfunktion. Z Ges Inn Med 31:261–265
72. Ford HC, Lim WC, Chisnall WN, Pearce JM (1989) Renal function and electrolyte levels in hyperthyroidism: urinary protein excretion and the plasma concentrations of urea, creatinine, uric acid, hydrogen ion and electrolytes. Clin Endocrin 30:293–301
73. Kai M, Kamada T, Baba Y, Shitomoto M, Setojama S, Otsuji S (1980) Elevated levels of urinary acid maltase and maltose: riboflavin glucosyltransferase activities in diabetic patients. Clin Chim Acta 108:259–266
74. Watts GF, Vlitos MAJ, Morris RW, Price RG (1988) Urinary *N*-acetyl-β-D-glucosaminidase excretion in insulin-dependent diabetes mellitus: relation to microalbuminuria, retinopathy and glycaemic control. Diabete Metab 14:653–658
75. Whiting PH, Ross IS, Borthwick L (1979) Serum and urine *N*-acetyl-β-D-glucosaminidase in diabetics on diagnosis and subsequent treatment, and stable insulin dependent diabetics. Clin Chim Acta 92:459–463
76. Whiting PH, Ross IS, Borthwick LJ (1983) *N*-acetyl-β-D-glucosaminidase levels and the onset of diabetic microangiopathy. Ann Clin Biochem 20:15–19
77. Maruhn D, Rosenstiel K, Hennewig K, Paar D (1974) Urinary excretion of β-glucuronidase, *N*-acetyl-β-D-glucosaminidase, and trehalase in diabetes mellitus. J Clin Chem Clin Biochem 12:263–265
78. Nakano M, Iguchi A, Kurimoto H, Sakamoto N (1987) Elevation of urinary trehalase and maltase activities with maturity-onset diabetes mellitus. J Clin Biochem Nutr 3:125–133
79. Hemmingsen L, Hiby N, Kragh-Sorensen P (1970) Protein and LDH-isoenzyme pattern of the urine from patients with diabetes mellitus determined by disk-electrophoresis. Diabetologia 6:512–518
80. Farr M, Wardle EN, Wilkinson K (1976) Lysozymuria in diabetes. Br Med J 1:624–625
81. Jungmann E, Mondorf AW, Scherberich JE, Sperzel RK, Schöffling K (1982) Schädigung des tubulären Apparates der Niere bei pathologischer Glukose-Toleranz. Münch Med Wochenschr 124:231–233
82. Stammberger K, Thulin H, Schanzenbach H (1977) Alaninaminopeptidaseausscheidung im Urin nach Amidotrizoat-Applikation bei Diabetikern. Dtsch Gesundh Wesen 32:542–543
83. Wiesmann U, Peheim E, Colombo JP (1989) Ausscheidung von *N*-acetyl-β-D-Glukosaminidase im Urin bei Kindern mit Typ I juvenilen Diabetes mellitus und von Patienten während der Behandlung mit dem Aminoglykosid Amikacin. Klin Wochenschr 67 (Suppl XVII): 40–43
84. Perchichizz G, Cucinotta D, Fera R, Caesare ED, Campos Sonadrito G (1987) Correlation between urinary activity of *N*-acetyl-β-D-glucosaminidase (NAG) and albumin excretion rate in type II (non-insulin dependent) diabetic subjects. Acta Diabetol Lat 24:149–155
85. Shrka J, Perusicora J, Stolba P, Stirbor V, Par J (1987) Comparison of *N*-acetyl-β-glucosaminidase and albuminuria with clinical findings of microangiopathy in type I diabetes mellitus. Clin Chim Acta 166:135–141
86. Jung K, Pergande M, Schimke E, Ratzmann KP (1989) Harnenzyme und niedermolekulare Proteine als Indikatoren der diabetischen Nephropathie. Klin Wochenschr 67 (Suppl XVII): 27–30
87. Agardh CD, Tallroth A, Hultberg B (1987) Urinary *N*-acetyl-β-D-glucosaminidase activity does not predict development of diabetic nephropathy. Diabetes Care 10:604–606
88. Shimojo N, Kitahashi S, Naka K, Fujii A, Okuda K, Tanaka S, Fujii S (1987) Comparison of *N*-acetyl-β-D-glucosaminidase and alanine aminopeptidase activities for evaluation of microangiopathy in diabetes mellitus. Metabolism 36:277–280
89. Burchardt U, Mampel E (1974) Das Verhalten der Alanin-aminopeptidase-Ausscheidung nach Visotrastapplikation als diagnostischer Test. Z Ges Inn Med 29:853–855

90. Brouhard BH, Allen K, Sapire D, Travis LB (1985) Effect of exercise on urinary *N*-acetyl-*β*-D-glucosaminidase activity and albumin excretion in children with type I diabetes mellitus. Diabetes Care 8:466–472
91. Schoeppe W, Scherberich JE, Hoffmann G, Rosak C (1988) Glomeruläre und tubuläre Markerproteine im Harn bei Gesunden und Typ-I-Diabetikern unter standardisierten Belastungsbedingungen. Akt Endokrin Stoffw 9:89–90
92. Ellis NE, Brouhard BH, LaGrone L (1984) Urinary *N*-acetyl-*β*-D-glucosaminidase in streptozotocin—induced diabetic rats. Biochem Med 31:303–310
93. Grötsch H, Hropot M, Kief H, Klaus E (1986) Enzymuria in streptozotocin—diabetic rats. J Clin Chem Clin Biochem 24:533–539
94. Kuznetsky KA, Schwartz MM, Lohmann LA, Lewis EJ (1986) Necrotizing glomerulonephritis in rheumatoid arthritis. Clin Nephrol 26:257–264
95. Pollak VE, Pirani CL, Steck IE, Kark RM (1962) The kidney in rheumatoid arthritis. Studies by renal biopsy. Arthritis Rheumatism 5:1–16
96. Sörensen AWS (1963) The kidney function in relation to drug consumption in patients with rheumatoid arthritis. Acta Rheum Scand 9:122–140
97. Dieppe PA, Doyle DV, Burry HC, Tucker SM (1976) Renal disease in rheumatoid arthritis. Br Med J 1:611–612
98. Rautenstrauch H, Knoll E, Wisser H (1979) Tubuläre Nephropathie bei chronischer Polyarthritis. Z Rheumatol 38:252–256
99. Dieppe PA, Tucker SM, Burry HC, Bresnihan FB (1975) The use of *β*-*N*-acetylglucosaminidase excretion to investigate renal disease in rheumatoid arthritis. Rheum Rehabil 14:226–230
100. Merle LJ, Merlef J, Reidenberg MM, Camacho MT, Jones BR (1980) Renal injury in patients with rheumatoid arthritis treated with gold. Clin Pharmacol Therap 28:216–222
101. Knoll E, Wisser H, Rautenstrauch H (1980) Vergleichende Untersuchung zur diagnostischen Wertigkeit von Diskelektrophorese der Urinproteine und *N*-Acetylglucosaminidaseausscheidung zur Erkennung von tubulären Nierenschädigungen bei chronischer Polyarthritis. J Clin Chem Clin Biochem 18: 53–58
102. Helin H, Korpela M, Mustonen J, Pasternak A (1986) Mild mesangial glomerulopathy—a frequent finding in rheumatoid arthritis with hematuria or proteinuria. Nephron 42:224–230
103. Brun C, Olsen S, Raaschou F, Sorensen AWS (1965) Renal biopsy in rheumatoid arthritis. Nephron 2:65–81
104. Ramirez G, Lambert R, Bloomer AH (1981) Renal pathology in patients with rheumatoid arthritis. Nephron 29:124–126
105. Scherberich JE, Sniehotte, Miehlke K, Schoeppe W (1987) Nierenbeteiligung bei rheumatoider Arthritis. Nieren Hochdruckkrankh 16:69–77
106. Scherberich JE, Krippner H, Kammer R, Miehlke K, Sniehotta KP, Schoeppe W (1982) Beurteilung einer Nierenalteration bei Patienten mit Lupus erythematodes und primär chronischer Polyarthritis mit Hilfe organspezifischer Gewebsparameter (abstr). Schweiz-Dtsch Rheumakongress, Basel, 5–9 Okt 1982
107. Dzialoszynki L, Golenia A, Materna A, Milewicz T (1975) The activity of soluble arylsulfatases A and B in the urine of children with acute rheumatic fever. Ped Pol 50:1273–1278
108. Rovensky J, Cebecauer L, Tovarek J, Zitnan D (1975) Urinary activity of lactate dehydrogenase and aminoacylarylamidases in systemic lupus erythematosus. Cas Lek Cesk 114:826–831
109. Wallenkamp GHJM, Vree BT, Guelen PJM, Jongmann-Nix B (1983) The effect of surgery on the renal excretion of *β*-microglobulin. Clin Chim Acta 129:27–37
110. Bergmann H, Scheler F (1964) Der Nachweis tubulärer Funktionsstörungen der Niere durch Bestimmung der Aminopeptidase—Aktivität im Harn. Klin Wochenschr 42:275–279
111. Guarnieri GF (1978) Urinary enzymes (*β*-glucosidase and lysozyme), protein pattern and *β*-microglobulin as indices of tubular damage. In: Enzymes in Health and Disease. Scientific and Medical International Society of Clinical Enzymology, London, pp 186–192
112. Kind PRN (1982) *N*-Acetyl-*β*-D-glucosaminidase in urine of patients with renal disease, and after renal transplants and surgery. Clin Chim Acta 19:89–97
113. Corbett CRR (1980) Urinary enzyme after transplantation and other surgical operations. Thesis, Cambridge University
114. Hartmann HG, Jutzler GA, Isringhaus H, Volkmer J (1983) Renal tolerance of cefotaxime after cardiothoracic surgical intervention. Evaluation of quantitative enzyme analysis. Clin Trials J 20:327–339
115. Price RG, Dance N, Richards B, Cattell WR (1970) The excretion of *N*-acetyl-*β*-D-glucosaminidase and *β*-galactosidase following surgery to the kidney. Clin Chim Acta 27:65–72

116. Liebert H (1988) Relevance of urinary enzyme determinations for the diagnosis and course—control of kidney injuries and secondary kidney damages—results of prospective study. Langenbecks Arch Chir 373:270–286
117. Ackaert KSJW, Schröder FH (1989) Effects of extracorporeal shock wave lithotripsy (ESWL) on renal tissue. Urol Res 17:3–7
118. Jung K, Brien B, Kirschner P, Schröder K, Schmiedel T (1989) Excretion of urinary enzymes after extracorporeal shock wave lithotripsy. Clin Chem 35:2000–2001
119. Kitada S, Kuramoto H, Kumazawa H, Yamaguchi A, Nakasu H, Hara S (1989) Effects of extracorporeal shock wave lithotripsy on urinary excretion of *N*-acetyl-beta-D-glucosaminidase. Urol Int 44:35–37
120. Erickson RP, Sandman R, Epstein CJ (1975) Lack of relationship between blood and urine levels of glycosaminoglycans and lysosomal enzymes. Biochem Med 12:331–339
121. Kress, BC, Miller AL (1979) Urinary lysosomal hydrolases in mucolipidosis II and mucolipidosis III. Biochem J 177:409–415
122. Strecker G, Michalski JC (1979) Sialidosis, a new type of inborn disease. In: Dubach UC, Schmidt U (eds) Diagnostic significance of enzymes and proteins in urine. Huber, Bern, pp 370–382 (Curr Probl Clin Biochem 9)
123. Hickman S, Neufeld EF (1972) A hypothesis for I-cell disease: defective hydrolases that do not enter lysosomes. Biochem Biophys Res Commun 49: 992–999
124. Navon R, Padeh B (1972) Urinary test for identification of Tay-Sachs genotypes. J Ped 80:1026–1030
125. Soyama K, Ono E, Shimada N, Tanaka K, Kusunoki T (1977) Urinary alpha-glucosidase analysis for the detection of the adult form of Pompe's disease. Clin Chim Acta 77:61–67
126. Saifer A, Amorosa J, Perle G (1976) Urinary hexosaminidase analysis for the identification of Tay-Sachs genotypes. Clin Chim Acta 67:315–319
127. Salvatore D, Bonatti S, Di Natale P (1984) Lysosomal alpha-*N*-acetylglucosaminidase: purification and characterization of the human urinary enzyme. Bull Mol Biol Med 9:11–121
128. Di Natale P, Murino P, Pontarelli G, Salvatore D, Andria G (1982) Sanfilippo B syndrome (MPS III B): altered residual alpha-*N*-acetylglucosaminidase activity in an unusual sibship. Clin Chim Acta 122: 135–143
129. Berty RM, Adler S, Basu A, Glew RM (1990) Effect of acid base changes on urinary hydrolases in Fabry's disease after renal transplantation. J Lab Clin Med 115:696–703
130. Ehrich JHH, Wurster U, Lütge M, Mertz J, Doehring E, Oemar B (1986) Urinary enzymes and renal diseases in children. Monatschr Kinderheilkd 134:832–839
131. Purvis ML, Buckalew VM (1974) Lysozymuria in distal renal tubular acidosis. Nephron 13:472–478
132. Morris RC, Ueki I, Sebastian A, Morris E (1967) Lysozymuria in acidification defects of the proximal nephron. Clin Res 15:142
133. Sinaniotis CA, Kaukautsakis P, Spyridis P (1989) Estimation of urinary *N*-acetyl-β-D-glucosaminidase activity for monitoring therapy of distal renal tubular acidosis. Acta Paediatr Scand 78:453–454
134. Barratt TM, Crawford R (1970) Lysozyme excretion as a measure of renal dysfunction in children. Clin Sci 39: 457–465
135. Peters JE, Schneider J, Unger G, Böhme J, Fukala E, Haschen RJ (1970) Alaninaminopeptidase (Aminosäurearylamidase) im Harn bei De Toni-Debré-Fanconi-Syndrom. Clin Chim Acta 29:576–578
136. Manz F, Waldherr R, Fritz HP, Lutz P (1984) Idiopathic de Toni-Debré-Fanconi syndrome with absence of proximal tubular brush border. Clin Nephrol 22:149–157

Part 4
Urinary Enzymes in Animal Experiments

16 Animal Models

E. Fuchs and H. Grötsch

16.1 Introduction

Measurement of urinary enzyme excretion as a test for the integrity of the renal parenchyma has become a valuable procedure in animal experiments, particularly in nephrotoxicity studies. Since the number of drugs with known nephrotoxic side effects increased from 60 in 1980 [3] to about 150 in 1988 [7], the analysis of urinary enzymes is included in many protocols for drug-testing experiments. This non-invasive method permits monitoring of very early stages of renal injury and allows observation of the same animal over longer periods of time.

A priority task of laboratories involved in drug testing is the development and improvement of appropriate methods, including the selection of adequate animal models. Reviewing the literature of the past 20 years reveals that most of the nephrotoxicity studies in combination with enzymuria were done with rats. There are only a few publications reporting on experiments with cats, dogs, rabbits or primates [1, 10, 17, 20, 27]. There are several reasons for preferring rats as experimental animals in nephrotoxicity studies: they are easy to handle and tolerate after adaption, single housing in metabolic cages, which makes urine collection easy. Rat strains are highly standardized, and several are available with metabolical/physiological impairments. The renal physiology of rats is better understood than that of other experimental animal. Rats—though providing a different renal handling of proteins as compared with humans—react with significant pathological enzymuria when substances are administered which also induce increased enzyme excretion in humans.

In order to rule out toxicological effects of newly developed drugs, these compounds have to be tested according to governmental regulations with at least two different mammalian species. In most cases, industrial laboratories do not publish the results of their tests in dogs, cats, rabbits or primates. This policy may offer an additional explanation for the lack of data on enzymuria in other mammals than rats. For this book, members of the working group "Clinical Chemistry of Laboratory Animals" of the German Society of Clinical Chemistry have collected—in addition to values in rats—data on enzymuria in dogs and primates which were used in the control groups of pharmacological and toxicological studies. These data are summarized in Chap 17. Renal target sites

Urinary Enzymes
K. Jung/H. Mattenheimer/U. Burchardt (Eds.)

of drugs may be recognized by enzyme excretion patterns that are characteristic for subcellular structures and, in some instances, for particular nephron segments (see Chap. 2).

In the following, major factors are addressed which affect enzymuria in mammals and which should be weighed in the planning and execution of nephrotoxicity studies. As a result of the above-mentioned preference for rats as experimental animals, most of the examples derive from studies with this rodent species. However, one has to consider that factors affecting enzymuria in rats are also relevant in other mammals. General guidelines concerning analytical methods, duration and number of sampling methods, statistical analysis and other issues were the topics of two symposia [12, 22].

16.2 Factors Affecting Enzymuria

Physiological factors affecting enzymuria are discussed in detail in Chap. 5; they are grouped into invariable factors (age, sex, race, genetic) and variable factors (body weight, muscle mass, nutrition, physical activity, biorhythms, climate). Experiments with animals allow us to control most of these: animals can be selected by age, sex, strain and genetic determinants; nutrition can be standardized and environmental factors regulated; known biorhythms should be considered.

16.2.1 Animal Handling and Environment

Special attention must be paid to the effect that the handling of animals may have on urine production and enzymuria. These include catherization to collect urine directly from the bladder, a procedure often recommended but performed only with complications, e.g. in dogs (pers. comm. Dr. Rebel), or transfer of animals from their home cage to a metabolic cage, which may cause isolation or restraint stress. The excretion, for example, of alanine aminopeptidase (AAP), alkaline phosphatase (AP) and *N*-acetyl-β-D-glucosaminidase (NAG) was increased in rats kept in metabolic cages during the first 5 days of long-term experiments, after which the animals adapted to the new situation, and the activities of the enzymes returned to control levels. In contrast, excretion of gamma-glutamyltransferase (GGT), glutamate dehydrogenase (GLDH) and aspartate aminotransferase (ASAT) was not affected [19]. Consequently, an adaptation period of several days must be allowed. Furthermore, dietary factors must be considered as they may enhance or diminish the effects of test substances (for references see [2]).

Despite the efficient standardization of the environmental conditions in modern animal houses throughout the year, significant changes were reported of several blood and organ variables in two sub-strains of rats investigated in different seasons. The levels and circadian patterns of the parameters changed during the seasons with respect to their absolute concentrations, amplitudes and phasing [15]. Accordingly, it was observed that male and female rats excrete more urinary NAG, GGT and acid phosphatase in winter than in other seasons [9]. These manifest variations have to be taken into account in the planning and evaluation of studies. The response to drugs may also alter with the seasons.

16.2.2 Age and Sex

Aging reduces the effectiveness and the capacity of the body to compensate for external influences and to stabilize homeostatic processes. These age-induced events include changes in the sensitivity to drugs [5].

A biochemical study on the age-related changes in the activity of certain enzymes suggests that the kidney in aged rats of both sexes derives its energy from metabolic pathways other than oxidative metabolism [25]. The activity of renal GGT increases with development and growth; it is absent or low in the fetal kidney and reaches adult levels after about 40 days [26]. The urinary excretion of GGT and AAP increases in older rats [8, 23].

The effect of uranyl nitrate on the excretion of lactate dehydrogenase (LDH), leucine aminopeptidase (LAP) and AP in newborn, adult and old rats (Wistar Jena strain) revealed an age-dependence of the time course of the nephrotoxicity [4]. The fact that the kidneys of old rats are more sensitive to the toxic effects of gentamicin than those of younger animals [16] may have implications in geriatric medicine if the human kidney reacts similarly. The relevance of this result from animal studies is supported by the observation that the incidence of renal toxicity is higher in older than in younger patients undergoing gentamicin therapy [6].

Sex-dependence of enzymuria and of sensitivity to toxins was found with Wistar rats. The excretion of AAP and GGT was higher in males than in females, while the excretion of NAG was equal in both sexes [8]. The tubulus toxicity of aminoglycosides was higher in male than in female animals [21].

16.2.3 Circadian Rhythms

The nephrotoxicity of a drug may depend on circadian rhythms. This was shown, for example, by the response of female Fischer rats to cisplatin. Urinary NAG activity was used to monitor the circadian rhythm of tubular sensitivity to the drug, which changed significantly with the time of administration. A statistically-significant effect of time of injection upon tolerance was found. Differences of three- to eight fold in the survival rate at 50% mortality and a nearly three-fold difference in long-term survival dependent on the circadian timing of the cisplatin administration were observed [11, 13]. Today, it is well established that human and rodent tolerance for cisplatin, as well as for 20 other anti-neoplastic drugs, varies with the time of application over a 24-h period

[14]. Drug effects depend partially on the metabolic state of their target tissue(s), and circadian rhythmic responses to drugs may well be coupled with respective changes in metabolism. In addition to circadian rhythms of enzymuria, infradian changes have been observed [8].

16.3 Pre-analytical Guidelines

Pre-analytical guidelines, as they are set forth in this section, reflect the numerous influences on enzymuria described in the preceding paragraphs. The protocol for nephrotoxicity studies and the handling and treatment of animals must conform with governmental regulations on experimentation with animals. Animals are selected by species, strain, age, body weight and sex, as each of these can bias enzymuria and susceptibility to nephrotoxins; biorhythms (infradian, circadian and seasonal) must be considered in regard to both enzymuria and response to treatment. The selection of enzymes to be included in a study is affected by the target site of the drug (nephron segment and/or subcellular structure).

Complementary to measurements of urinary enzymes, other diagnostic parameters of urine should be evaluated: volume, osmolality, pH, creatinine, electrolytes, glucose, total protein and individual proteins (albumin, α_1-microglobulin). The protein selectivity of the nephron may be assessed by sodium dodecyl sulphate polyacrylamide gel electrophoresis. Ideally, a histopathological study of the kidney should be included at the end of an experiment [18, 24, 28, 29; see Chap. 4].

Animals, when transferred into metabolic cages, need several days for recovery from the isolation and restraint stress, during which excretion of some enzymes increases. Precautions are necessary to avoid contamination of urine in metabolic cages. However, the activity normally found in feces and debris is low for most enzymes. A known exception is glutamate dehydrogenase, which is excreted in feces at a rate a thousand times higher than in urine [4]; contamination with feces could therefore be a problem.

Cooling of the urine and/or the addition of a preservative (thymol or sodium azide) is recommended during the collection periods in order to avoid bacterial growth, which can easily occur in the environment of an animal house. The preparation of the urine for enzyme analysis is covered in Chap. 6.

16.4 Conclusions and Outlook

Enzymuria is a sensitive indicator for the nephrotoxic effects of drugs. In the planning, execution and evaluation of animal experiments, factors must be considered which affect enzymuria and the response to a nephrotoxic insult. The potential of this approach may not yet have been fully exploited because studies are generally conducted with healthy animals. It is suggested for the future to include in drug studies animal strains with pathophysiological defects or with impaired renal function. Results from such studies may help to bridge the gap between experimentation in animals and application in humans, since patients receiving these drugs are not healthy but have impaired body functions.

References

1. Adelman RD, Sprangler WL, Beasom F, Ishizaki G, Conzelman GM (1981) Fursemide enhancement of netilimicin nephrotoxicity in dogs. J Antimicrob Chemother 7:431–440
2. Alt J (1988) Proteinurie des Versuchstieres. Parey, Berlin Hamburg (Schriftenreihe Versuchstierkunde H. 13)
3. Berscheid G, Grötsch H, Hropot M, Klaus E (1983) Enzymurie der Ratte unter Gabe von Azneimitteln. J Clin Chem Clin Biochem 21:239–240
4. Bräunlich H, Fleck C (1981) Urinary enzyme excretion as an indicator of nephrotoxicity in dependence on age. Exp Path 20:182–187
5. Coper H, Schulze G (1986) Altersbedingte Änderungen in der Empfindlichkeit für Arzneimittel. Internist 27:53–59
6. Cronin RE (1979) Aminoglycoside nephrotoxicity: pathogenesis and prevention. Clin Nephrol 11:251–256
7. Dukes MNG (ed) (1988) Meyler's side effect of drugs. 11th edn. Elsevier, Amsterdam
8. Grötsch H, Hropot M, Klaus E, Malerczyk V, Mattenheimer, H (1985) Enzymuria of the rat: biorhythms and sex differences. J Clin Chem Clin Biochem 23:343–347
9. Grötsch H, Hropot M, Klaus E, Wesely J (1990) Influences of seasonal biorhythms on urinary excretion of enzymes and other parameters. J Clin Chem Clin Biochem 28:285–286
10. Hardy ML, Hsu RC, Short CR (1985) The nephrotoxic potential of gentamicin in the cat enzymuria and alterations in urine concentrating capability. J Vet Pharm Ther 8:382–392
11. Hrushesky WJM, Levi FA, Halberg F, Kennedy BJ (1982) Circadian stage dependence of cis-diamminedichlorplatinium lethal toxicity in rats. Cancer Res 42:945–949
12. Jung K, Burchardt U (1987) Harnenzyme im Experiment und in der Klinik. J Clin Chem Clin Biochem 25:823–828
13. Levi F, Hrushesky WJM, Blomquist CH, Lakuta DJ, Haus E, Halberg F, Kennedy BJ (1982) Reduction of cis-diammiedichloroplatinium nephrotoxocity in rats by optimal circadian drug timing. Cancer Res 42:950–955
14. Levi F, Boughattas NA, Blazsek I (1987) Comparative murine chronotoxicity of anticancer agents and related mechanism. Ann Rev Chronopharmacol 4:283–331
15. Mayersbach v H (1977) Seasonal influences on laboratory rats living under highly standardized environmental conditions. J Interdiscipl Cycle Res 8:385
16. McMartin DN, Engel SG (1982) Effect of aging on gentamicin nephrotoxicity and pharmacokinetics in rat. Res Commun Chem Pathol Pharmacol 38:193–207
17. Pierce RJ, Price RG, Marsden AM, Fowler JSL (1979) Urinary enzyme assays in toxicological studies in the rat and marmoset. Curr Probl Clin Biochem 9:201–214
18. Plummer DT, Noorazar S, Obatomi DK, Haslam JD (1986) Assessment of renal injury by urinary enzymes. Uremia Invest 9:97–102
19. Schirrow R, Wesslau C, Jung K (1987) Zur Ausscheidungsrhythmik verschiedener Harnenyzme bei Ratten. Nieren- und Hochdruckrankh 16:407
20. Schultz SG, Lavelle KJ, Swain R (1982) Nephrotoxicity of radiocontrastmedia in ischemic renal failure in rabbits. Nephron 32:113–117
21. Schulz E, Brodersen HP, Sack K (1983) Verhalten der Harnausscheidung von Tubuluszellen und Malatdehydrogenase in Abhängigkeit von Körpergewicht, Geschlecht und Stamm. J Clin Chem Clin Biochem 21:245–246

22. Simane ZJ, Breuer J (1983) Harnenzyme und andere Urinparameter im Tierversuch. J Clin Chem Clin Biochem 21:237–247
23. Simane ZJ, Helm W (1984) The output of some urinary enzymes in rat as related to sex and age. Rev Brasil Anal Clin 16:97
24. Stonard MD, Gore CW, Oliver GJA, Smith IK (1987) Urinary enzymes and protein patterns as indicators of injury to different regions of the kidney. Fundam Appl Toxicol 9:339–351
25. Talesara CL,Mohini L (1978) Age-related changes in the activity of certain enzymes in the kidney of rat. Indian J Exp Biol 16:183–186
26. Tate SS, Meister A (1975) Identity of malate-stimulated "glutaminase" with γ-glutamyltranspeptidase in rat kidney. J Biol Chem 250:4619–4624
27. Wachsmuth ED (1981) Nephrotoxicity of cefotiam (CGP 14221/E) in rats and rabbits. Arch Toxicol 48:135–138
28. Weber MH (1988) Urinary protein analysis. J Chrom 429:315–344
29. Zbinden G, Fent K, Thouin MH (1988) Nephrotoxicity screening in rats; general approach and establishment of test criteria. Arch Toxicol 61:344–348

17 Reference Values

H. Grötsch, E. Fuchs, and Z.J. Simane

Members of the working group "Clinical Chemistry of Laboratory Animals" of the German Society of Clinical Chemistry have collected values from laboratory animals which were used in the control groups of pharmacological and toxicological studies. The authors emphasise that the compiled values should be understood as preliminary values of laboratory animals. Data from other working groups are left out of consideration; through this a better comparability should be achieved. For the collection, preparation and analytical procedure of the urine samples, the published recommendations were used (see Table 1). Phenomena such as seasonal biorhythms [3] or any connection between the protein turnover were disregarded. A listing of different enzymes was made for dogs, rabbits, rats and rhesus monkeys.

In general, the animals had free access to food and water. Laboratory IV described the additional application of 2 × 15 ml water by pharynx tube. It has to be noted that this additional fluid did not result in an elevated diuresis. Multiple data from laboratories I, III and IV indicate the effect of changing urinary excretion rates as observed in rats with changing age indicated by the body weight.

Concerning the problem of incomplete collection of urine in dogs, the calculation of enzymatic activities was performed from spontaneous samples to U/mol creatinine (Table 3).

Table 1. Methods used by members (laboratories I-VII) of the working group "Clinical Chemistry of Laboratory Animals" of the German Society of Clinical Chemistry. Numbers in brackets refer to methods listed in References

Laboratory	Preparation of urine	Enzymes				
		AAP	AP	GGT	LDH	NAG
I	GF	[2]		[6], [7]		[4]
II	GF			[6]; [7]	[1]	[4]
III	GF	[5]		[6], [7]	[1]	[4]
IV	GF	[2]		[6], [7]		[4]
V	D		[1]	[6], [7]	[1]	[4]
VI	GF	[5]		[6], [7]		[8]
VII	GF	[5]		[6], [7]		[8]

The enzymes AAP, AP, GGT and LDH were determined at 25°C; NAG was measured at 37°C. GF, Gel filtration; D, dialysis 2 h against distilled water; AAP, alanine aminopeptidase; AP, alkaline phosphatase; GGT, gamma-glutamyltransferase; LDH, lactate dehydrogenase; NAG, *N*-acetyl-β-D-glucosaminidase

Urinary Enzymes
K. Jung/H. Mattenheimer/U. Burchardt (Eds.)

Table 2. Excretion of urinary enzymes in males and females of two rat strains

Enzymes	Laboratory	Male			Female			Strain
		Enzyme (U/24 h)	BW(g)	*n*	Enzyme (U/24 h)	BW(g)	*n*	
Alanine aminopeptidase	Ia	0.142 ± 0.035	165 ± 13	5	0.048 ± 0.020	143 ± 4	5	WS
	Ib	0.245 ± 0.060	250 ± 19	5	0.049 ± 0.007	178 ± 6	5	WS
	IIIa	0.227 ± 0.0082		20				WS
	IIIb	0.353 ± 0.127		20				WS
	IVa	0.135 ± 0.080	145 ± 3	10				WS
	IVb	0.230 ± 0.090	170 ± 6	10				WS
Gamma-glutamyl transferase	Ia	3.160 ± 1.100	165 ± 13	5	1.264 ± 0.383	143 ± 4	5	WS
	Ib	7.164 ± 1.528	250 ± 19	5	1.714 ± 0.234	178 ± 6	5	WS
	II				4.313 ± 1.846		105	SD
	IIIa	3.305 ± 1.310		20				WS
	IIIb	6.891 ± 2.078		20				WS
	IVa	1.220 ± 0.557	145 ± 3	10				WS
	IVb	2.060 ± 0.469	170 ± 6	10				WS
N-acetyl-β-D-glucosaminidase	Ia	0.165 ± 0.029	165 ± 13	5	0.157 ± 0.029	143 ± 4	5	WS
	Ib	0.192 ± 0.064	250 ± 19	5	0.173 ± 0.009	178 ± 6	5	WS
	II				0.255 ± 0.108		105	SD
	IIIa	0.243 ± 0.119		20				WS
	IVa	0.162 ± 0.028	145 ± 3	10				WS
	IVb	0.092 ± 0.029	170 ± 6	10				WS
Lactate dehydrogenase	II				0.092 ± 0.059		105	SD
	IIIa	0.063 ± 0.032		20				WS
	IIIb	0.099 ± 0.059		20				WS
	IVa	0.078 ± 0.037	145 ± 3	10				WS
	IVb	0.046 ± 0.021	170 ± 6	10				WS

WS, Wistar; SD, Sprague-Dawley. Data are given as mean ± SD. For reasons of comparison, body weight (BW) in grams (g) is included.

Table 3. Excretion of urinary enzymes in male rats (Wistar; $n = 12$) and male rabbits (White Newsealand; $n = 18$)

Species	Urinary enzymes				Body weight	Urine collection
	GGT	LDH	AP	NAG	(g)	period (h)
RAT	0.776 (0.671/1.347) U/18 h	0.055 (0.017/0.075) U/18 h	0.159 (0.080/0.139) U/18 h	0.073 (0.059/0.090) U/18 h	166–225	18
Rabbit	1.4 (1.2/1.8) U/24 h	1.6 (1.1/2.7) U/24 h			2200–3600	24

Results are from laboratory V. Activity expressed as U per urine collection period. Data are presented as median values and (in brackets) the lower and upper quartiles.
AP, alkaline phosphatase; GGT, gamma-glutamyltransferase; LDH, lactate dehydrogenase; NAG, *N*-acetyl-β-D-glucosaminidase

Table 4. Excretion of urinary enzymes in male (n = 12) and female (n = 14) beagle dogs

Enzyme	Male	Female
Alanine aminopeptidase	0.59 ± 0.26	0.40 ± 0.17
Gamma-glutamyltransferase	7.82 ± 9.49	1.16 ± 0.35
N-acetyl-β-D-glucosaminidase	1.14 ± 0.67	0.25 ± 0.19
Lactate dehydrogenase	3.11 ± 2.49	2.12 ± 1.58

Activity expressed in U/mol creatinine. Values are given as means ± SD. Data are from laboratories VI and VII

Table 5. Excretion of urinary enzymes in one male and one female rhesus monkey

Enzyme	Male	Female
Alanine aminopeptidase	0.44 ± 0.14	0.42 ± 0.31
Gamma-glutamyltransferase	3.07 ± 0.96	5.80 ± 1.55
N-acetyl-β-D-glucosaminidase	0.71 ± 0.27	1.67 ± 0.92
Urine (ml)	303 ± 76	965 ± 99

Data are from laboratory I. Activity expressed in U/24 h. Results represent means ± SD from an 11-day collection period. For reasons of comparison, urine volume/24 h (ml) is also given

References

1. Bergmeyer HU, Büttner H, Hillmann G, Kreutz FH, Lang H, Laue D, Rick W, Schmidt E, Schmidt FW, Stamm D, Szasz G (1970) Empfehlungen der Deutschen Gesellschaft für Klinische Chemie. Standardisierung von Methoden zur Bestimmung von Enzymaktivitäten in biologischen Flüssigkeiten. Z Klin Chem Klin Biochem 8:658–660
2. Berscheid G, Grötsch H, Hropot M, Klaus E, Mattenheimer H (1983) Enzymuria of the rat: the preparation of urine for enzyme analysis. J Clin Chem Clin Biochem 21:799–804
3. Grötsch H, Hropot M, Klaus E, Wesely J (1990) Influence of seasonal biorhythms on urinary excretion of enzymes and other parameters. J Clin Chem Clin Biochem, 28:285–286
4. Maruhn D (1976) Rapid colorimetric assay of β-glactosidase and N-acetyl-β-glucosaminidase in human urine. Clin Chim Acta 73:453–461
5. Mattenheimer H, Frölke W, Grötsch H, Maruhn D, Simane Z (1988) Recommendation for the measurement of alanine aminopeptidase in urine. J Clin Chem Clin Biochem 26:635–644
6. Szasz G (1970) γ-Glutamyltranspeptidase-Aktivität im Urin. Z Klin Chem Klin Biochem 8:1–8
7. Szasz G, Weimann G, Stähler F, Wahlefeld AW, Persijn JP (1974) New substrates for measuring γ-glutamyl transpeptidase activity. Z Klin Chem Klin Biochem 5:228
8. Yakata M, Sugita O, Sakai T, Uchiyama K, Wada K, (1983) Basic studies on assay of NAG activity in urine and its variability in course of renal diseases. Jap J Clin Path Suppl 56:90–101

18 Influence of Different Substances on Urinary Enzyme Excretion

18.1 X-Ray Contrast Media

A.S. Bhargava, R. Hofmeister, and P. Günzel

18.1.1 Introduction

Ever since the development of the first ionic monomeric contrast medium in 1952 [31], tri-iodinated benzoic acid derivatives have been of great importance as diagnostic X-ray contrast media. One has to differentiate between two main classes of these agents, the ionic monomer (e.g. amidotrizoate) and dimer (e.g. ioxaglate) class and non-ionic monomer (e.g. iohexol, iopamidol, iopromide, iosimide and metrizamide) and dimer (e.g. SH H 340 AB: 5,5'-oxayldiimino-bis-[2,4,6,-triiodo-*N*,*N'*-bis-(2,3-dihydroxypropyl)-*N*, *N'*-dimethyl-isophthalamide]; 300 mg iodine/ml) class. In pre-clinical studies (with experimental animals) as well as in clinical studies (with humans) the aspect of nephrotoxicity is of special importance because of the almost total renal elimination of these water soluble contrast media [34]. However, in these studies, mostly serum urea nitrogen, serum creatinine, proteinuria and cellular components in urine have been used as parameters for detecting nephrotoxicity, in addition to some cases of histomorphological examination of the kidneys in animal experiments. Since determinations of urinary enzymes have been shown to be sensitive in predicting renal lesions in animals and humans [4, 18–21, 32], many X-ray contrast media have now been studied in animal experiments using rats, rabbits and dogs as the animal subjects and including enzyme determinations. This section deals with the effect of some contrast media on the excretion of various enzymes in urine with regard to their localization in different segments of the nephron and their correlation with other parameters such as serum urea nitrogen and creatinine level as well as histomorphological changes in animal species. Many enzymes from different subcellular locations which have been used to assess the early kidney changes in animal species after contrast media application are presented in Table 1.

18.1.2 Enzymuria in Rats Following Contrast Media Administration

18.1.2.1 *Effect on Normal Rats*

The effects of a single i.v. application of some ionic and non-ionic contrast media on normal intact rats on the urinary enzyme excretion, serum urea nitrogen and

Urinary Enzymes
K. Jung/H. Mattenheimer/U. Burchardt (Eds.)

Table 1. Enzymes used for the detection of early renal changes in animal species after treatment with contrast media

Cellular fraction	Enzyme	Abbreviation	Enzyme code no.
Cytosol	Lactate dehydrogenase	LDH	1.1.1.27
	Aspartate aminotransferase[a]	ASAT	2.6.1.1
	Alanine aminotransferase	ALAT	2.6.1.2
Brush border	γ-Glutamyltransferase	GGT	2.3.2.2
	Alanine aminopeptidase	AAP	3.4.11.2
	Alkaline phosphatase	AP	3.1.3.1
Lysosome	N-Acetyl-β-D-glucosaminidase	NAG	3.2.1.30

[a] Also present in mitochondria

creatinine levels and histological alterations are summarized in Table 2.

As is evident from this table, an increase in LDH and GGT in urine samples collected in the first 18 or 24 h following very high i.v. doses of contrast media as in the case of SH H 340 AB and iopromide was observed [3, 5, 15]. This increase in enzymuria correlated well with a slight increase in serum urea nitrogen and creatinine and histomorphological changes consisting of vacuolisation and necrosis of the proximal tubules.

At a low dosage of contrast media (1.75 g I/kg), Thomsen et al. [27, 28] have shown an increase in urinary enzyme activities (GGT, LDH and NAG) when the urine samples were collected over the first 2 h after administration of an ionic monomer (diatrizoate). The other two non-ionic monomers (iohexol and ioxilan) which were included in this study showed a less pronounced increase in urinary LDH and GGT activities in samples collected over the first 2 h when compared with diatrizoate. Of the two non-ionic monomers, iohexol induced more enzymuria than ioxilan, indicating that ioxilan was the better tolerated contrast medium for tubular cells on the basis of enzyme excretion in urine. This increase in enzymuria was not detectable in urine samples collected from 2 to 22 h after application. However, neither increase in serum urea nitrogen and creatinine levels nor histological abnormalities were observed on day 8 of the experiment. This acute enzymuria, as discussed by the authors, may be due to the composite effect of osmolality, hydrophilicity and the charge of the contrast media.

A typical example of enzymuria associated with kidney damage after a single i.v. application of a non-ionic dimer (SH H 340 AB) in rats is presented in Table 3. A drastic increase in urinary LDH and GGT followed by ASAT was observed in samples collected over the first 24 h at the highest dose of 9 gI/kg applied at a volume of 30 ml/kg [3]. The histological examination of the kidneys 24 h after application of the contrast medium revealed vacuolisation of the proximal tubular epithelium, a phenomenon called "osmotic nephrosis" associated with

Table 2. Effect of single intravenous application of contrast media in normal rats

Contrast media	Dose (gI/kg)	Urine collection period per day	Enzyme activity expressed in relation to:	Enzymes measured in urine (↑ ≙ increase)	Correlation with blood chemistry (serum urea nitrogen, serum creatinine) and histology	References
Dimer						
SH H 340 AB (non-ionic)	9	24 h	urine volume over collection period	↑ LDH > ↑ GGT > ↑ ASAT	No increase in urea nitrogen and creatinine at 24 h but vacuolisation and necrosis of proximal tubules	4
SH H 340 AB (non-ionic)	7.5	18 h on days 2, 3, 4 and 9	urine volume over collection period	↑ LDH > ↑ GGT; ↑ AP ↑ AAP > ↑ ALAT > NAG	Slight increase in urea nitrogen on day 2 and vacuolisation of proximal tubules and cell regeneration on day 9	15
Monomer						
Diatrizoate (ionic)	1.75	2 h and 2-22 h up to 8 days	creatinine excretion in urine	↑ GGT > ↑ LDH > ↑ NAG		
Iohexol (non-ionic)	1.75			↑ GGT > ↑ LDH; NAG	No increase in serum urea and creatinine and no tubular vacuolisation on day 8	27, 28
Ioxilan (non-ionic)	1.75			↑ GGT > ↑ LDH; NAG in first 2 h; diatrizoate > iohexol > ioxilan; no increase in urine samples collected from 2-22 h		
Iohexol (non-ionic)	5.0, 7.5, 10.0, 12.5	18 h on days 2, 3, 4 and 8	urine volume over collection period	↑ NAG (from 5.0 gI/kg onwards); ↑ LDH (12.5 gI/kg) AP; GGT in first 18 h	No increase in urea nitrogen and creatinine up to day 8 but dose-dependent vacuolisation of proximal tubules (maximum being 48 h after application)	16
Iosimide (non-ionic)	10.0, 12.5	18 h before treatment and 18, 42, 66 h and on day 7 after treatment	urine volume over collection period	LDH; GGT	No change in blood chemistry and histology after iosimide and iopamidol and very slight increase in urea nitrogen and creatinine at 18 h and 42 h at 12.5 gI/kg and increase in frequency of necrosis of proximal tubules from 10 gI/kg after iopromide at 48 h	5
Iopamidol (non-ionic)	10.0, 12.5			LDH; GGT		
Iopromide (non-ionic)	7.5, 10.0, 12.5			↑ LDH > ↑ GGT (from 10 gI/kg) in first 18 h		

Table 3. LDH, GGT and ASAT in urine after a single intravenous application of SH H 340 AB (mean ± standard deviation; $n = 5$)

Treatment			Enzyme excretion in the first 24 h urine (mU/24 h)		
Compound	Dose (gI/kg)	Applied volume (ml/kg)	LDH	GGT	ASAT
Normal saline	Control	30	197 ± 89	1309 ± 263	82±27
SH H 340 AB	1	3.3	136±22	1488±872	34±16
SH H 340 AB	3	10	282±175	1645±884	105±95
SH 340 AB	9	30	1940±942**	10271 ± 3712**	311±197*

Significantly different from control (Dunnett test): $^{**}P < 0.01$; $^{*}P < 0.05$

necrosis of tubular epithelium at the highest dose (Fig. 1). At this time (24 h), serum urea nitrogen and creatinine levels were still normal.

In a separate study [15], the effect of a single i.v. application of SH H 340 AB (7.5 gI/kg) and mercuric chloride (1 mg/kg) on six urinary enzymes was studied in rats over 9 days. LDH and GGT were found to be more sensitive indicators

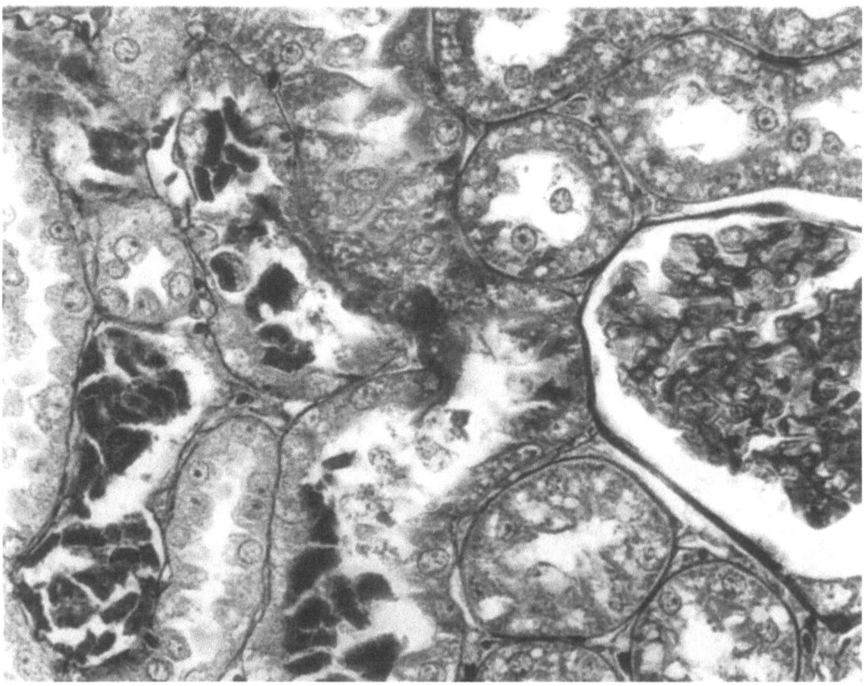

Fig 1. Vacuolisation of proximal tubular cells and necrosis of the tubular epithelium 24 h after a single i.v. application of a non-ionic dimeric contrast medium (SH H 340 AB) in rats at a dose of 9 gI/kg. body wt H&E; ×400

than other enzymes (AP, AAP, ALAT and NAG) in predicting the proximal tubule damage in rats.

Experiments with a non-ionic monomer (iohexol) in rats suggest that the determination of urinary NAG may be useful in detecting contrast medium-induced vacuolisation of proximal tubules. Hofmeister et al. [16] demonstrated a dose-dependent increase in urinary NAG (in first 18-h samples) and vacuolisation of the proximal tubules (maximum at 48 h) at doses of 5 gI/kg and above, whereas an increase in LDH was only observed at the highest dose of 12.5 gI/kg (see Table 2).

Histological examination of the kidneys in this study showed that the vacuolisation of the proximal tubular cells is reversible even after extremely high doses of iohexol. The tiny increase in urinary LDH at the highest dose of 12.5 gI/kg may be due to very slight tubular damage which could not be observed in animals killed 48 h after iohexol application. The correlation between urinary NAG excretion and the appearance of the vacuoles in the proximal tubular cells (Fig. 2) seems to imply that lysosomes may play some role in the development of the latter.

It is known from the literature that rat renal proximal tubular cells, when compared with distal nephron structures, contain more lysosomes and that NAG activity is highest in the proximal tubules [12]. The vacuolisation obviously induces an increased expression of lysosomal NAG followed by a rise in urinary excretion.

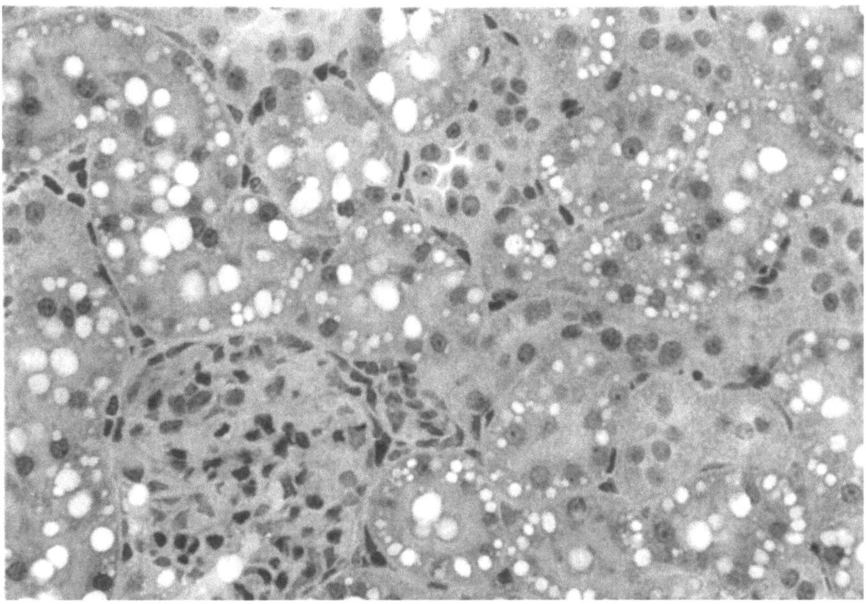

Fig 2. Vacuolisation of proximal tubular cells 48 h after single i.v. administration of iohexol (12.5 gI/kg body wt) in rats. H&E; ×300

It has been shown by electron microscopy studies using an ionic dimeric contrast medium that the appearance of vacuoles in the proximal tubules in rats is not associated with any degenerative cellular ultrastructure [2].

The experiment with iohexol in rats demonstrates that the differential diagnosis between vacuolisation of the proximal tubules and actual damage may be achieved in toxicological studies with contrast media by the determinations of LDH and NAG activities in rat urine.

An increase in urinary NAG without any subsequent increase in LDH activity may indicate just reversible vacuolisation of the proximal tubules, while a subsequent increase in LDH or GGT activity indicates, in addition, the presence of cellular damage in the proximal tubules.

18.1.2.2 *Effect on Dehydrated and Pre-injured Rats*

Some of the risk factors for contrast media-induced renal changes are pre-existing renal insufficiency due to diabetes mellitus or myeloma, advanced age, dehydration or volume depletion and large doses of contrast media [29]. Recently, some investigations have been carried out to explore the effect of a few contrast media by means of a urinary profile of enzymes under a few of these risk conditions in rats.

The effect of contrast medium application in water-deprived animals and in animals with pre-injury of the kidney brought about by sodium maleate and in a renal hypertonic model (in rats) is summarized in Table 4.

A single i.v. dose of diatrizoate (1.75 gI/kg) caused a greater increase in tubular brush border enzyme GGT in dehydrated rats than in hydrated rats in a double cross-over study conducted on days 1 and 28. On the other hand, the effect of diatrizoate on the cytosolic LDH was independent of the state of hydration as no difference in increase of urinary LDH between dehydrated and hydrated animals could be shown in either series (days 1 and 28).

Obviously, the dehydration state increases the concentration of the contrast medium within the tubules, which may result in temporary damage to the tubular brush border. These data suggest that attention should be paid to the state of hydration of the patient, especially when the high osmolar ionic contrast medium diatrizoate is used.

In an effort to examine the effect of contrast media on the diseased kidney, Thomsen et al. [30] used an i.v. application of sodium maleate 2 h prior to giving contrast medium, which is known to induce instantaneous dehydration with tubular and glomerular dysfunction closely resembling Fanconi's syndrome. The contrast medium and the doses used were the same as in a previous study in normal rats (see also Table 2). The excretion of urinary NAG was higher following diatrizoate injection in sodium maleate-pretreated animals when compared with saline, iohexol or ioxilan treatment. However, no significant difference could be observed between iohexol and ioxilan in this model while in normal rats iohexol produced more enzymuria than ioxilan. Thus, the expectation that such a model would be useful in differentiating better tolerance of

Table 4. Effect of single, i.v. application of contrast media in preinjured rats (in all experiments performed, enzyme activity is expressed in relation to creatinine in urine)

Effect of contrast media on the kidneys in various models	Contrast media or reference compound	Dose (gI/kg)	Urine collection period per day	Enzymes measured in urine ($\uparrow \hat{=}$ increase)	Correlation with blood chemistry (serum urea nitrogen, serum creatinine) and histology	References
Dehydration:						
–hydrated rats (free access to tap water)	Diatrizoate (ionic monomer)	1.75	Up to 3 h after application on	$\uparrow$ GGT > $\uparrow$ LDH	(no blood chemistry and histology were performed)	29
–dehydrated rats (water was withdrawn for 24 h before application of contrast medium)			days 1 and 29 (double cross-over study at an interval of 28 days)	$\uparrow$ GGT > $\uparrow$ LDH (GGT more pronounced in dehydrated than in hydrated rats in the first 3 h in both series)		
Pre-injured kidney:						
–with i.v. application of sodium maleate (400 mg/kg) 2 h prior to contrast medium application	Diatrizoate (ionic monomer)	1.75	Cumulated excretion over 4 days except 2 h/day (feeding interval)	$\uparrow$ LDH > $\uparrow$ GGT > $\uparrow$ NAG (more pronounced than iohexol and ioxilan)	(no blood chemistry was performed) –histology on day 8 showed no findings	30
	Iohexol (non-ionic monomer)	1.75		$\uparrow$ LDH > $\uparrow$ GGT; NAG		
	Ioxilan (non-ionic monomer)	1.75		$\uparrow$ LDH; GGT; NAG (diatrizoate > iohexol = ioxilan		
Renal hypertonic model:						
–left kidney (normal)	Ioxaglate (ionic dimer)	1.12	For 1 h after application from left and right urethra via catheter	$\uparrow$ NAG; $\uparrow$ GGT in normal and sclerotic kidneys after all three contrast media, but slightly more pronounced in ioxaglate-treated animals. No difference between iohexol and iopamidol. Mannitol also showed increase in NAG and GGT	(no blood chemistry and histology were performed)	33
–right kidney (stenosis of arteria renalis, 24 weeks before contrast media application inducing nephrosclerosis)	Iohexol (non-ionic monomer)	1.05				
	Iopamidol (non-ionic monomer)	1.05				
	Mannitol	Osmotic equivalent dose				

contrast media in pre-existing kidney damage could not be fulfilled. Probably, sodium maleate alone had such a pronounced effect on the enzyme excretion that it prevented the demonstration of the small quantitative differences between the contrast media.

A similar kind of approach, using i.v. application of mercuric chloride to induce pre-injury of the kidney in rats and then comparison of the effects of different contrast media on the excretion of various enzymes in the urine, failed to demonstrate differences among the contrast media. In this study, the pre-injury caused by mercuric chloride alone was so pronounced that no further potentiation of damage after contrast media application could be achieved as demonstrated by biochemical parameters including enzymuria and histomorphological examination [14].

Thus, in experimental models, beside the mechanism of pre-injury many aspects such as the extent and reproducibility of damage, animal-to-animal variations and the time interval between the pre-injury and contrast medium application as well as the urine collection period for enzyme profile and the optimal time for necropsy (histological examination of the kidneys) produce limitations on the study of this aspect of contrast media-induced nephropathy.

Wrigge and Malyusz [33] studied the effect of the ionic low osmolar contrast medium ioxaglate and the non-ionic iohexol and iopamidol as well as mannitol at equiosmolar doses in a renal hypertonic model in rats (Table 4). All three contrast media produced enzymuria (NAG and GGT), the ionic ioxaglate being more active than the two non-ionic contrast media. However, mannitol at an equiosmolar dose was almost as effective as ioxaglate in inducing enzymuria.

18.1.3 Enzymuria in Rabbits Following Contrast Media Administration

18.1.3.1 *Effect on Normal Rabbits*

The effects of various ionic and non-ionic monomer contrast media in normal rabbits and the ionic dimer iodipamide in rabbits with pre-injured kidneys are presented in Table 5. A transient increase in urinary LAP and AP was observed after injection of both contrast media (metrizamide and iothalamate) at clinical doses as well as mannitol at a molar equivalent dose. However, no significant difference could be observed between the two contrast media [11].

Rygaard et al. [22] examined the effect of three contrast media in rabbits, which were also tested in rats at the same doses (see also Table 2). The only effect on urinary enzyme excretion was the elevation of GGT after diatrizoate in rabbits in contrast with the findings in rats in which all three contrast media produced enzymuria (diatrizoate > iohexol > ioxilan). In spite of differences in the urine collection period (2 h in rat and 24 h in rabbit studies), the magnitude of the increase of enzymuria, if similar in rabbits and rats, should have been detected in the 24 h samples. One explanation may be the difference in the overall sensitivity to contrast media between the species, rats being more sensitive than rabbits, though no disturbance of kidney function or kidney damage could be demonstrated in this study on the basis of serum urea nitrogen and creatinine levels as well as histological examination.

On the other hand, rabbits were found to be more sensitive than rats when three non-ionic monomer contrast media (iosimide, iopamidol and iopromide) were tested after a single i.v. application [5]. The effect of these contrast media on GGT excretion in the urine and serum urea nitrogen level in normal rabbits is shown in Fig. 3.

A good correlation between enzymuria (GGT in first 24 h urine samples) and a rise in serum urea nitrogen level as well as in serum creatinine concentration (42-h and 72-h samples) was observed at high doses of contrast medium for iopamidol and iopromide. Histological examination of the kidneys also revealed a good correlation between these parameters and the appearance of necrosis of the proximal tubules.

The urinary LDH activity as a parameter for detecting nephrotoxicity in rabbits was less important because of large inter-individual variations after treatment with contrast medium in this study. It is known from the literature [24] that GGT in rabbits is exclusively located in the proximal tubules of isolated single nephrons. Thus, urinary GGT activity can be used for predicting damage to this segment, as also shown in an experiment with contrast media application in rabbits [5].

As already mentioned, in normal rats iopromide produced an enzymuria followed by necrosis of the proximal tubules in a few animals (see Table 2) only at almost lethal doses (10 gI/kg), and the other two compounds, iosimide and iopamidol, produced no changes. These results suggest that the sensitivity of animal species may differ from contrast medium to contrast medium.

18.1.3.2 *Effect on Pre-injured Rabbits*

Ischaemic injury was produced in rabbits by occluding the renal arteries for 90 min, and the effect of iodipamide on the excretion of AAP and NAG in the urine was investigated (see Table 5) [23]. Animals receiving contrast medium just after the ischaemic injury exhibited a distinct rise in enzymuria, and it was not possible to distinguish between the effect of ischaemia alone and that of contrast medium alone on the basis of the serum creatinine level and enzymuria. Furthermore, animals which received iodipamide just after induction of the ischaemic injury showed a more pronounced effect than animals which received the contrast media 1 and 3 days after the ischaemic injury.

18.1.4 Enzymuria in Dogs Following Contrast Media Administration

Excretion of LDH, ASAT, creatinine kinase and catalase was studied after injection of 0.3 ml/kg of several contrast media (diatrizoate, iothalamate and acetrizoate) and hypertonic solutions of mannitol and sodium chloride into the

Table 5. Effect of single i.v. application of contrast media in normal intact and pre-injured rabbits

Contrast media and reference substances	Dose (gI/kg)	Urine collection period per day	Enzyme activity expressed in relation to:	Enzymes measured in urine ($\uparrow$ $\triangleq$ increase)	Correlation with blood chemistry (serum urea nitrogen, serum creatinine) and histology	References
In normal rabbits:						
Monomer						
Metrizamide (non-ionic)	0.7	24 h over 3 days (LAP only for first 24 h)	Creatinine excretion in urine	$\uparrow$ LAP, $\uparrow$ AP	(blood chemistry and histology were not performed)	11
Iothalamate (ionic)	0.7			$\uparrow$ LAP, $\uparrow$ AP		
Mannitol	Molar equivalent dose			$\uparrow$ LAP > $\uparrow$ AP (maximum 24 h after application)		
Diatrizoate (ionic)	1.75	24 h on days 1, 3, 5	Creatinine excretion in urine	$\uparrow$ GGT (in first 24 h); LDH, NAG	Blood chemistry (24, 72, 120 h after application) and histology (120 h after appl.) showed no findings	22
Iohexol (non-ionic)	1.75			GGT, LDH, NAG		
Ioxilan (non-ionic)	1.75			GGT, LDH, NAG		
Iosimide (non-ionic)	2.5, 5.0, 7.5, 10.0	24 h before treatment and 24, 48, 72 h and on day 7 after treatment	Urine volume over collection period	$\uparrow$ GGT (from 5.0 gI/kg onwards)[a]		
Iopamidol (non-ionic)	1.0, 2.5, 5.0, 7.5			$\uparrow$ GGT (from 5.0 gI/kg onwards)	Good correlation with serum urea, nitrogen and creatinine (48 h and 72 h) and necrosis of proximal tubules in iopamidol- and iopromide-treated animals from 5.0 gI/kg onwards	5
Iopromide (non-ionic)	1.0, 2.5, 5.0, 7.5			$\uparrow$ GGT (from 5.0 gI/kg onwards) (over first 2 days) (LDH large variation and no correlation to serum urea nitrogen)		

in pre-injured rabbits:						
Dimer						
Iodipamide (ionic) on ischaemic injury by occluding both renal arteries for 90 min with clamps (groups of animals with occlusion alone and normal rabbits with contrast media injection were included in this study)	3.36 ml/kg (52% Meg. Iodipamide ≙ 0.87 gI/kg) either just prior to removal of clamps or 24 h or 3 days after ischaemic injury	24 h over 4 days	Creatinine excretion in urine	↑ AAP; ↑ NAG (max. 24 h–72 h) in all groups but more pronounced in animals of the group treated just prior to removal of clamps	Good correlation with serum creatinine (max. 48 h/72 h) and tubular vacuolisation and necrosis in all groups but more pronounced in animals of group treated just 90 min after of ischaemic injury	23

LAP, Leucine aminopeptidase; in this case synonymous with alanine aminopeptidase
[a] The increase in urinary GGT after iosimide from 5 gI/kg was not dose-dependent

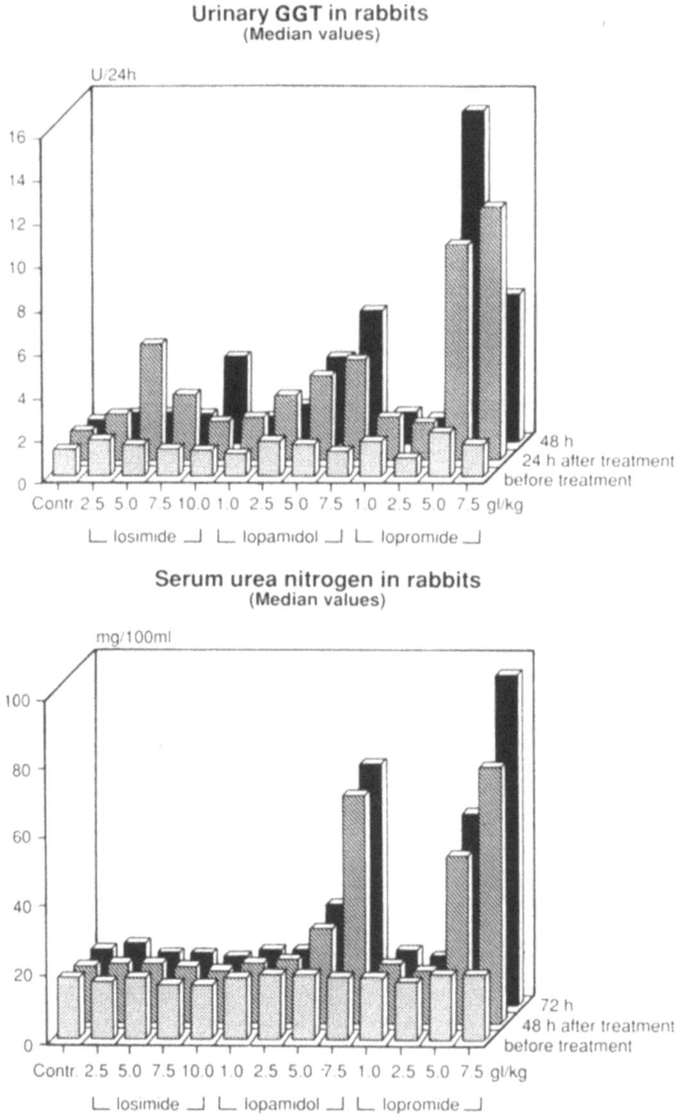

Fig. 3. Effect of single i.v. application of contrast media on gamma-glutamyltransferase (*GGT*) excretion in urine and serum urea nitrogen in rabbits.

renal artery of anesthetized dogs. In addition, diatrizoate was also tested at a dose of 0.75 ml/kg [26]. The urine samples were collected via cannulation of the ureters during the injection of contrast media and 0–3, 3–5, 5–10, 10–20 and 30–60 min after application. All contrast media as well as the hypertonic solutions mannitol and sodium chloride caused enzymuria consisting of mainly catalase, LDH and ASAT in this model. The determinations of LDH isoenzymes in the urine samples collected over the first 3 min after administration of two contrast media showed an increase in output of LDH_1 and LDH_2 without haematuria; thus in some cases measurement of the LDH isoenzymes may be useful in studying the effect of different compounds on kidney changes.

The same group [25] has also studied the effect of a contrast medium consisting of diatrizoate to which gelatin was added to make a 6% solution in renal arteriography in the above-mentioned model in dogs using enzymuria as the indicator. The low dose of 0.08 ml/kg which was sufficient for radiography showed less enzymuria than a higher dose of 0.3 ml/kg.

18.1.5 Discussion (with Regards to the Experience in Humans)

Most of the contrast media, when administered i.v. at high doses, induce vacuolisation of the proximal tubular epithelium in normal animals. This vacuolisation is a reversible process, and it has been recently shown by electron microscopy that the appearance of vacuoles in the proximal tubules in rats is not associated with any degenerative cellular ultrastructure [2].

The determination of urinary enzymes in animal studies is useful in predicting early renal changes in the proximal tubules after administration of contrast media. The urinary enzymes LDH and GGT in rats, GGT in rabbits and LDH and catalase in dogs were sensitive indicators of such changes. In addition, determinations of lysosomal NAG may be helpful with regard to analysing the vacuolisation of the proximal tubules induced by contrast media in rats. Furthermore, a differential diagnosis between vacuolisation and actual cellular damage of proximal tubules may be achieved by the measurement of LDH or GGT and NAG in rats. Actual cellular damage is associated with an increase in urinary LDH and GGT.

At low doses of contrast media, an increase in enzymuria similar to that following a molar equivalent dose of mannitol can be observed. This transient increase is not associated with any change in serum urea and creatinine levels or histological damage in the kidney in rats, rabbits and dogs. At very high doses (lethal) of contrast media, enzymuria is followed by an increase in serum urea nitrogen and creatinine levels and by vacuolisation and necrosis of the proximal tubules in normal intact animals.

The comparison of effects observed in rats and rabbits suggests that for certain contrast media rabbits are more sensitive than rats, and for others, rats may be more sensitive than rabbits with regard to enzymuria and renal changes. Thus, there is no particular species which can be considered as ideal for reproducing the effects to be expected in human subjects. Therefore, the findings observed in the most sensitive species should be taken into account for risk assessment for humans.

Many of the contrast media such as iopamidol, iothalamate, iohexol, ioxaglate when tested in diagnostic doses up to 0.8 gI/kg in humans (normal subjects as well as patients with different nephropathies), demonstrated an increase in enzymuria, mostly in the first 24 h after application [1, 6, 7, 9, 10, 13, 17]. In all these studies diatrizoate was included to compare the effect with the above-mentioned contrast media.

In almost all these studies, only diatrizoate, an ionic monomer contrast mediun, showed a slightly more pronounced increase in excretion of some of these enzymes such as AAP, GGT, AP, LDH and NAG when compared with other ionic monomer (iothalamate), non-ionic monomer (iopamidol, iohexol) or ionic dimer (ioxaglate) contrast media. The other contrast media showed no clear difference in their pattern of enzyme excretion. However, these changes were not accompanied by any disturbance of renal function. Burchardt [6] has demonstrated an increase in the excretion of AAP in the urine in humans after i.v. application of mannitol, dextran and diatrizoate and suggests "osmotic nephropathy" as the cause of the enzymuria. Thus, in humans a slight increase in enzymuria, as also observed after mannitol application, may occur after injection of certain contrast media at diagnostic doses without any disturbance of renal function.

On the other hand, contrast media-induced renal dysfunction has been reported to occur in 15%–42% of patients with azotemia when the limit of the amount of the recommended contrast medium is exceeded. Therefore, it has been proposed to adjust the amount of contrast material to the severity of the azotemia. By utilizing an empiric formula based on the serum creatinine concentration, the maximal amount of contrast medium (limit) to be administered in patients with renal disease can be calculated [8].

References

1. Bani E, Federighi F, Ghio R, Marchitiello M, Galigani P, Palla R (1985) The use of iohexol in pediatric urography: a comparative study with meglumine diatrizoate. Int J Pediatr Nephrol 6:271–274
2. Battenfeld R (1988) Licht- und elektronenmikroskopische Untersuchungen der osmotischen Nephrose nach Applikation eines Röntgenkontrastmittels. Dissertation, Tierärztliche Hochschule, Hannover
3. Bhargava AS, Siegmund F (1983) Ergebnisse diagnostischer Untersuchungen zur Beurteilung der Nierenverträglichkeit eines Röntgenkontrastmittels bei der Ratte. J Clin Chem Clin Biochem 21:240
4. Bhargava AS, Khater AR, Günzel P (1978) The correlation between lactate dehydrogenase activity in urine and serum and experimental renal damage in the rat. Toxicol Lett 1:319–323
5. Bhargava AS, Hofmeister R, Siegmund F, Schöbel C, Günzel P (1990) Effect of three non-ionic contrast media on rats and rabbits with regard to renal changes (interspecies comparison). Arzneimittel-forsch Vol. 40:822–829
6. Burchardt U (1975) Alaninaminopeptidaseausscheidung mit dem Harn und osmotische Nephropathie. Z Inn Med 30:65–69
7. Cavaliere G, Arrigo G, D'Amico G, Bernasconi P, Schiavina G, Dellafiore L, Vergnaghi D (1987) Tubular nephrotoxicity after intravenous urography with ionic high-osmolal and non-ionic low-osmolal contrast media in patients with chronic renal insufficiency. Nephron 46:128–133
8. Cigarroa RG, Lange RA, Williams RH, Hillis LD (1989) Dosing of contrast material to prevent contrast nephropathy in patients with renal disease.Am J Med 86:649–652
9. Donadio C, Tramonti G, Giordani R, Lucchetti A, Calderazzi A, Sbragia P, Bianchi C (1988)

Glomerular and tubular effects of ionic and nonionic contrast media (diatrizoate and iopamidol). Contrib Nephrol 68:212–219
10. Gale ME, Robbins AH, Hamburger RJ, Widrich WC (1984) Renal toxicity of contrast agents:iopamidol, iothalamate and diatrizoate. AJR 142:333–335
11. Golman K, Holtas S (1980) Proteinuria produced by urographic contrast media. Invest Radiol 15:61–66
12. Guder WG, Ross BD (1984) Enzyme distribution along the nephron. Kidney Int 26:101–111
13. Hartmann HG, Jutzler GA, Bambauer R, Keller HE, Maruhn D (1984) Enzymbestimmungen im Harn zur Beurteilung der Nierenverträglichkeit des wasserlöslichen Röntgen-Kontrastmittels Iopamidol. Radiologe 24:442–445
14. Hofmeister R (1988) Untersuchungen zur Nephrotoxizität von ionischen und nichtionischen Röntgenkontrastmitteln bei Ratten mit vorgeschädigten und nicht vorgeschädigten Nieren unter besonderer Berücksichtigung der Harnenzymdiagnostik. Dissertation, Free University of Berlin, Department of Veterinary Medicine
15. Hofmeister R, Bhargava AS, Günzel P (1986) Value of enzyme determinations in urine for the diagnosis of nephrotoxicity in rats. Clin Chim Acta 160:163–167
16. Hofmeister R, Bhargava AS, Günzel P (1990) The use of urinary N-acetyl-β-D-glucosaminidase (NAG) for the detection of contrast-media-induced "osmotic nephrosis" in rats. Toxicol Lett 50:9–15
17. Jevnikar AM, Finnie KJC, Dennis B, Plummer DT, Avila A, Linton AL (1988) Nephrotoxicity of high- and low-osmolality contrast media. Nephron 48:300–305
18. Maruhn D, Paar D, Hartmann HG, Bock KD, Bomhard E, Lorke D (1981) Enzyme patterns of rat urine in folate-induced acute renal failure. In: Brown SS, Davies DS (eds) Organ-directed toxicity chemical indices and mechanisms (IUPAC). Pergamon, Oxford, pp 69–73
19. Mondorf AW, Breier J, Hendus J, Scherberich JE, Mackenrodt G, Shah PM, Stille W, Schoeppe W (1978) Effect of aminoglycosides on proximal tubular membranes of the human kidney. Eur J Clin Pharmacol 13:133–142
20. Price RG (1982) Urinary enzymes, nephrotoxicity and renal disease. Toxicology 23:99–134
21. Raab WP (1972) Diagnostic value of urinary enzyme determinations. Clin Chem 18:5–25
22. Rygaard H, Dorph S, Thomsen HS, Mygind T, Nielsen H, Larsen S, Skaarup P, Hemmingsen L, Holm J (1988) Effects of intravenous injection of diatrizoate, iohexol or ioxilan on renal size, urine profiles and blood profiles in the rabbit. Acta Radiol 29:491–494
23. Schultz SG, Lavelle KJ, Swain R (1982) Nephrotoxicity of radiocontrast media in ischemic renal failure in rabbits. Nephron 32:113–117
24. Shimada H, Endou H, Sakai F (1982) Distribution of gamma-glutamyl-transpeptidase and glutaminase isoenzymes in the rabbit single nephron. Jpn J Pharmacol 32:121–129
25. Sovak M, Talner LB, Rushmer HN, Dunn CR (1973) Adhesive contrast material in the canine upper urinary tract. Invest Radiol 8:61–67
26. Talner LB, Rushmer HN, Coel MN (1972) The effect of renal artery injection of contrast material on urinary enzyme excretion. Invest Radiol 7:311–322
27. Thomsen HS, Dorph S, Mygind T, Sovak M, Nielsen H, Rygaard H, Larsen S, Skaarup P, Hemmingsen L, Holm J (1988) Intravenous injection of ioxilan, iohexol and diatrizoate, Effects on urine profiles in the rat. Acta Radiol 29:131–136
28. Thomsen HS, Dorph S, Mygind T, Hemmingsen L, Holm J, Larsen S, Nielsen H, Rygaard H, Skaarup P (1988) Urine profiles following intravenous diatrizoate, iohexol, or ioxilan in rats. Invest Radiol 23:168–170
29. Thomsen HS, Hemmingsen L, Dorph S, Skaarup P (1988) Effects on urine profiles of diatrizoate in hydrated and dehydrated rats, A double cross-over study. Acta Radiol 29:731–735
30. Thomsen HS, Dorph S, Mygind T, Nielsen H, Rygaard H, Larsen S, Skaarup P, Hemmingsen L, Holm J (1988) Do contrast media aggravate Fanconi's Syndrome in rats? A comparison of diatrizoate, iohexol and ioxilan. Invest Radiol 23:164–167
31. Wallingford VH (1952) 3-carboxylic-acylamino-2,4,6-triiodobenzoic acids and the ethyl ester and nontoxic salts. U.S. patent 2, 611, 786:23. September
32. Watanabe M, Nomura G, Hirata M, Imai K, Koizumi H (1980) Studies on the validity of urinary enzyme assay in the diagnosis of drug-induced renal lesions in rats. Toxicol Pathol 8:22–33
33. Wrigge P, Malyusz M (1988) Enzymurie nach Kontrastmittelgabe bei hypertonen Ratten. Röntgenblätter 41:332–335
34. Zurth C (1984) Mechanism of renal excretion of various x-ray contrast materials in rabbits. Invest Radiol 2:110–115

18.2 Aminoglycosides and Cephalosporins

H. Grötsch and Z.J. Simane

Among antibiotics, nephrotoxicity has been recognized as a major complication, mainly in aminoglycosides and cephalosporins. In animal models, the damage to the tubular cells of kidney (pars convoluta and pars recta) caused by these drugs can certainly be recognized by histological examination. The early stages of such damage, however, cannot be indicated by routine techniques such as serum chemistry (urea nitrogen, creatinine) or urinalysis (total protein, osmolality) because of their poor sensitivity which can be derived from the enormous functional reserve of the kidney. Therefore, the attention of a number of investigators has been attracted to more sensitive noninvasive tests for nephrotoxicity. In this connection, the measurement of urinary excretion of renal enzymes has been found helpful both in monitoring toxicity during clinical application of nephrotoxic antibiotics and in experimental animal models. Among all laboratory animals the rat is best suited for such investigations, since the enzyme distribution along the tubule is similar to that of man: damage to different parts of the kidney will result in similar changes of urinary enzyme pattern in the rat and in man [30, 41, 43, 48].

18.2.1 Aminoglycosides

As shown by morphological examination in rats, the nephrotoxic effects of high doses of aminoglycosides are characterized by development of tubular necrosis, particularly in the proximal part. In addition, damage to Malpighi bodies and nonspecific glomerular nephrosis can occur. More extended injuries such as proliferative alterations or necrosis of the loop are rare [47].

The uptake of aminoglycosides into the cell seems to be due to adsorptive pinocytosis and specific reversible binding in the brush border region [40]. A selective accumulation of gentamicin in the lysosomes of the proximal tubular cells was detected by autoradiography [51]. Electron-microscopic studies [29] demonstrated the formation of phagolysosomes inside lysosomes. The first histological alteration induced by gentamicin on kidney ultrastructure is the appearance of lamellar structure deposition in the lysosomes of tubular cells. The interference of gentamicin with the catabolism of phospholipids leads to further cellular disturbance and necrosis.

The tubular damage in rats treated with various doses of gentamicin was proportional to the amount of administered antibiotic. The correlation of morphological changes in tubuli and increased output of the urinary enzymes confirms their tubular origin in gentamicin nephrotoxicity [22, 61]. However, although enzymuria is an early sign of kidney injury by aminoglycosides, the urinary enzyme level has not been shown to be a reliable predictor of the severity of such effects [31].

Aminoglycoside antibiotics promote increased urinary excretion of lysosomal enzymes as well as membrane-bound enzymes from proximal tubular cells. Lysosomal enzymuria may be derived from exocytosis of phagolysosomes with release of lysosomal enzymes into the tubular lumen or from frank proximal tubular cell necrosis (see Sect 3.1.3). The increased urinary excretion of membrane-bound enzymes may reflect accelerated metabolism of brush border epithelium.

Studies concerning effects of aminoglycoside antibiotics on the urinary enzyme output are summarized in Table 1. Nearly all referred papers describe similar tubulotoxic and glomerulotoxic behaviour but the effects depend on chemical structure and dosage design which is related to renal accumulation of the drug [8, 44]. Thus, in a comparative study of gentamicin and netilmicin nephrotoxicity in dogs, unequal rates of enzymuria increased at a very early stage after administration of both drugs had indicated a lower nephrotoxicity of netilmicin; only later could this observation be confirmed by electron microscopy [2]. Sack et al. [44, 48] have designed a model to quantify and compare nephrotoxic potentials of aminoglycosides in humans, rats and guinea pigs. The relative toxicity was estimated by counting the excreted epithelial cells and by

Table 1. Urinary enzymes used for detection of aminoglycoside nephrotoxicity in experimental animals

Drug	Species	Enzyme	Ref.
Amikacin	Rat	ASAT, HEX, LDH, MDH NAG	[31, 33, 38, 64]
Bekanamycin	Rat	ASAT, LDH, MDH	[9]
Butirosin	Rat	ASAT, LDH, MDH	[9]
G-418	Rat	ASAT, LDH, MDH	[7]
Gentamicin	Rat	AAP, AP, FUC, α-GAL, β-GAL, GGT, GLU, GLUC, HEX, LAP, LDH LYS, MAN, MDH, NAG	[1, 6, 13–15, 18, 27 28, 30, 32, 33, 35, 36, 39, 40, 42, 44, 45 54, 62]
	Dog	β-GAL, GLUC, LYS	[1–3]
Gentamicin B	Rat	ASAT, LDH, MDH	[7]
Kanamycin	Rat	AcP, AP, ASAT, GGT, HEX, LAP, LDH, LYS, MDH, NAG	[15, 33, 60, 64]
Neomycin	Rat	GGT, GLUC, MDH	[16, 29]
Netilmicin	Rat, dog	HEX, LYS, MDH, NAG	[2, 15, 33, 34]
Puromycin	Rat	AcP, AP, ASAT, GGT, LAP, LDH	[60]
Rifampicin	Rat	GGT, MDH	[19]
Sisomicin	Rat	ASAT, HEX, LDH, MDH	[33, 44]
Streptomycin	Rat	HEX	[33]
Tobramycin	Rat	AP, ASAT, HEX, LAP, LDH, MDH	[33, 34, 44, 45, 46, 62]

AAP, alanine aminopeptidase; AP, alkaline phosphatase; ASAT, aspartate aminotransferase; AcP, acid phosphatase; FUC, α-fucosidase; α-GAL, α-galactosidase; β-GAL, β-galactosidase; GGT, γ-glutamyltransferase; GLU, β-glucosidase; GLUC, β-glucuronidase; HEX, β-hexosaminidase; LAP, leucine aminopeptidase; LD H, lactate dehydrogenase; LYS, lysozyme; MAN, α-mannosidase; MDH, malate dehydrogenase; NAG *N*-acetyl-β-D-glucosaminidase;

measuring the urinary output of malate dehydrogenase, lactate dehydrogenase and aspartate aminotransferase. The parallelism between the excretion rate of epithelial cells and enzymuria in rats was demonstrated in a study with various dosages of the aminoglycoside butirosin [9]. A relationship between the dosage of gentamicin and the increase of urinary output of *N*-acetyl-*β*-D-glucosaminidase (NAG) in rats was reported by Carver et al. [14]. In contrast, Mozer et al. [38] described in vitro and in vivo inhibition of NAG by amikacin in rat and human urine. A gentamicin-induced, dose-related increase in urinary excretion of further lysosomic glycosidases (α-mannosidase, α-fucosidase, *β*-galactosidase, *β*-glcosidase, and *β*-glucuronidase), and in brush border enzyme alkaline phosphatase in rats, was described by Patel et al. [39]. In this study the enzymuria was elevated significantly by day 3, whereas proteinuria, urine osmolality and tests reflecting glomerular filtration were later indicators of nephron damage. Similar observations were reported by Wellwood et al. [61].

The effects of aminoglycosides seem to depend on age. Gentamicin nephrotoxicity in female rats, as measured by urinary excretion rate of tubulus cells and malate dehydrogenase, was more pronounced in old than in younger individuals. Very young rats (90 g) responded only with slight and nonsignificant increases in the cellular and enzyme output. This observation was consistent with lower renal concentration of gentamicin in younger animals [35]. The age-related toxicity of gentamicin in rats was also confirmed by measurement of urinary *β*-galactosidase [36].

Generally, there exists a correlation between the renal concentration of the aminoglycoside and the excretion rate of enzymic nephrotoxicity markers. Gilbert et al. [22] described a phase-like accumulation of gentamicin in the kidney of rats; he explained it by biorhythm in protein turnover. A diurnal fluctuation of the renal excretion of aminoglycosides (gentamicin, sisomicin, tobramycin, netilmicin, and amikacin) has been found to be synchronous with the output rate of both lysosomal and brush border enzymes in humans [11, 12, 37, 50]. The enzyme elimination occurred at intervals, which might possibly have been caused by impairments of cells cycles.

Circadian variations in the renal toxicity of gentamicin in rats, as reflected by urinary γ-glutamyltransferase, alanine aminotransferase and NAG, were reported by Pariat et al. [40].

The aminoglycoside susceptibility was reduced under pathological conditions such as metabolic acidosis [25] but particularly in individuals suffering from kidney damage. As compared with control or treated animals, or those infected with Eschesichia coli, pyelonephritic rats showed, after treatment with low gentamicin doses, significant increase in urinary *β*-galactosidase and protein associated with severe changes in diuresis and osmolality. The increase in the gentamicin toxicity correlated with the antibiotic levels in cortex and medulla which were more pronounced in infected animals than in the normals [6]. The diagnostic value of urinary enzyme for detecting drug-induced tubular lesions in individuals with preexisting renal disease was pointed out in studies on gentamicin in humans [12] and on tobramycin in rats [46].

The nephrotoxic potential of aminoglycosides might be influenced by concurrent administration of some substances such as dextran [11], diuretics [1, 26], mercuric chloride [18], polyvinyl alcohol [12], X-ray contrast media [11], cephalothine [20], carbenicillin [50] or metilcillin [62]. Clinical observations on enhanced nephrotoxicity of aminoglycosides, if they are given in combination with furosemide, were supported by measurement of urinary enzymes in rats and dogs [1, 3, 26]. On the other hand, various cephalosporin derivatives protected kidneys from gentamicin nephrotoxicity [17, 32, 53]. Vice versa, gentamicin and tobramycin reduced the renal tolerance against cephalotine [62]. The protective effect of cephalosporins may involve drug interaction within the renal cortex [31] but unfortunately, animal studies have not resolved the problem of aminoglycoside–cephalosporin interaction completely and its mechanism is still unclear [4].

18.2.2 Cephalosporins

Cephaloridine is the only representative of this substance class possessing relatively high nephrotoxicity. The newer congeners belong to the group of well-tolerated antibiotics and there are no relevant differences between various derivatives in this respect [49]. Cephaloridine has been reported to cause renal tubular necrosis in man and experimental animal [5]. Light microscopic examinations showed early alterations in the proximal tubular brush border, followed later by frank necrosis. In rabbit, the substance produces a dose-dependent lesion of the second segment of the proximal tubule that results in early disruption of structure and function of the cell membrane [52]. The intrinsic toxicity of cephaloridine for the proximal tubular cells depends upon its peculiar transport characteristics [49]. In a rat model using probenecid as transport inhibitor, the tubular proximal toxicity was directly related to the local intracellular concentration resulting from its active transport. This unique relationship between active transport and cytotoxicity constitutes evidence for cephaloridine's selectivity as a pure tubular toxin [56].

The nephrotoxic effects of cephalosporins are generally less strong than those of aminoglycosides, and high doses are necessary to induce renal damage in laboratory animals, particularly in rats. A comparison of renal tolerance towards various cephalosporin derivatives is therefore only possible at a very high dosage level [49].

Animal studies using urinary enzymes for assessing the nephrotoxicity of cephalosporins are summarized in Table 2. Generally, a more or less evident increase in urinary enzyme activity was observed. Its extent depends on the tubulotoxic potential of the drug and on its dose. The increase in urinary enzymes, measured in parallel, was mostly synchronous. However, Grötsch [24] reported a distinctly different behavior of urinary γ-glutamyltransferase and NAG in rats after cefadizime. In the first 24 h after application cefadizime

Table 2. Urinary enzymes used for detection of cephalosporin nephrotoxicity in experimental animals

Drug	Species	Enzymes	Ref
Cefamandole	Rat	ASAT, LDH, MDH	[10]
Cefazedone	Rat	AP, ASAT, LAP, LDH, MDH	[10, 42, 58]
Cefodizime	Rat, dog	GGT, NAG	[24]
Cefotiam	Rat, rabbit	AAP, AP, LDH	[58, 59]
Cefoxitin	Rat	ASAT, LDH, MDH	[49]
Cephacetril	Rat	MDH	[47]
Cefazolidine	Rat	FUC, NAG	[32]
Cephaloridine	Rat	AAP, ALD, AP, FUC, LAP, LDH, MDH, NAG	[32, 54, 59]
Cephalotin	Rat	ASAT, FUC, LDH, MDH, NAG	[32, 62]
EMD 29 946	Rat	ASAT, LDH, MDH	[10]
HR 756	Rat	ASAT, LDH, MDH	[49]

AAP, alanine aminopeptidase; ALD, aldolase; AP, alkaline phosphatase; ASAT, aspartate aminotransferase; FUC, α-Fucosidase; GGT, γ-glutamyltransferase LAP, Leucine aminopeptidase; LDH, lactate dehydrogenase; MDH, malate dehydrogenase; NAG, *N*-acetyl-β-D-glucosaminidase

(1600 mg/kg i.v.) led to an approximately 9-fold increase in γ-glutamyltransferase excretion with a following decrease below the control values. The histological and morphological examinations give evidence that the reduced output of γ-glutamyltransferase has to be seen in connection with the formation of new cells which are partially resistant to the nephrotoxin. In the same animals a continuous increase in excretion rate of NAG was observed, after day 4, which could be explained by lysosomal storage of cefadizime and following cell destruction.

The changes in urinary output of renal enzymes due to nephrotoxin administration can be influenced by circadian rhythms. Wachsmuth and Wirz [59] reported that administration of cephaloridine to rats at night caused more severe renal damage, i.e., a stronger response of urinary lactate dehydrogenase, than the same treatment in the morning.

The interaction between cephalosporin derivatives and aminoglycosides has been mentioned above. Unlike gentamicin, the renal toxicity of cephaloridine (100 mg/kg) in rats was not enhanced by concurrent furosemide (50 mg/kg) [57].

References

1. Adelman RD, Conzelman G, Spangler W, Ishizaki G (1977) Furosemide potentiated gentamicin nephrotoxicity: early detection by monitoring of urinary enzymes. Kidney Int 12:538
2. Adelman RD, Conzelman G, Spangler W, Ishazaki G (1979) Comparative nephrotoxicity of gentamicin and netilmycin: functional and morphological correlation with urinary enzyme activities. Curr Probl Clin Biochem 9:166–182

3. Adelman RD, Spangler WL, Beasom F, Ishizaki G, Conzelman GM (1979) Furosemide enhancement of experimental gentamicin nephrotoxicity: comparison of functional and morphological changes with activities of urinary enzymes. J Infect Dis 140:342–352
4. Appel GB, Neu HC (1977) The nephrotoxicity of antimicrobial agents. II N Engl J Med 296:722–728
5. Atkinson RM, Currie JP, Davis B, Pratt DAH, Sharpe HM, Tomich EG (1966) Acute toxicity of cephaloridine, an antibiotic derived from cephalosporin C. Toxicol Appl Pharmacol 8:398–406
6. Beauchamp D, Poirier A, Bergeron MG (1985) Increased nephrotoxicity of gentamicin in pyelonephritic rats. Kidney Int 28:106–113
7. Beck H, Sack K (1979) Experimentelle Untersuchungen zur Nierenverträglichkeit der Aminoglykoside Gentamicin B und O-2-Amino-2-deoxy-6-C-methyl-α-D-glucopyranosyl-(1 → 4)-garamin (G-418). Arzneimittelforschung 29:1695–1700
8. Beck H, Züllich B, Marre R, Sack K (1979) Tierexperimentelle Untersuchungen zur Nephrotoxizität von 7 Aminoglycosiden bei Langzeitbehandlung. Arzneimittelforschung 29:935–939
9. Beck H, Eikenberg P, Sack K (1980) Experimentelle Untersuchungen zur Nierenverträglichkeit der Aminoglykoside Butirosin und Bekanamycin. Arzneimittelforschung 30:288–294
10. Beck H, Nierhoff N, Sack K (1980) Tierexperimentelle Untersuchungen zur Nephrotoxizität neuer Cephalosporin-Antibiotika: Cefamandol, EMD 29 645 und 29 946. Arzneimittelforschung 30:660–666
11. Burchardt U, Peters JE, Neef L, Thulin H, Gründig A, Haschen RJ (1977) Der diagnostische Wert von Enzymbestimmungen im Harn. Z Med Lab Diag 18:190–212
12. Burchardt U, Schinköthe G, Müller G, Neef L, Krosch H (1978) Ausscheidungskinetik von Enzymen und Protein mit dem Harn bei Applikation therapeutischer Gentamycindosen. Schweiz Med Wochenschr 108:1541–1545
13. Carver MP, Monteiro-Riviere NA, Brown TT, Riviere JE (1985) Dose-response studies of gentamicin nephrotoxicity in rats with experimental renal dysfunction. II. Polyvinyl alcohol glomerulopathy. Toxicol Appl Pharmacol 80:264–273
14. Carver MP, Shy-Modjeska JS, Brown TT, Rogers RA, Riviere JE (1985) Dose-response study study of gentamicin nephrotoxicity in rats with experimental renal dysfunction. I. Subtotal surgical nephrectomy. Toxicol Appl Pharmacol 80:251–263
15. Cojocel C, Dociu N, Ceacmacudis E, Baumann K (1984) Nephrotoxic effects of aminoglycoside treatment on renal protein reabsorption and accumulation. Nephron 37:113–119
16. Conzelman GM, Gribbi DH (1973) Urinary excretion of β-glucuronidase after administration of neomycin to rats. Toxicol Appl Pharmacol 26:158–160
17. Dellinger P, Murphy T, Pinn V, Barza M, Weinstein L (1976) The protective effect of cephalotin against gentamicin-induced nephrotoxicity in rats. Antimicrob Agents Chemother 9:172–178
18. Dzuriková V, Krechňáková A (1981) Nefrotoxicita gentamicínu a ortuti v experimente na potkanoch. Bratisl Lek Listy 76:21–28
19. Fent K, Mayer E, Zbinden G (1988) Nephrotoxicity screening in rats: a validation study. Arch Toxicol 61:349–358
20. Fillastre JP, Laumonier R, Humbert G, Dubois D, Metayer J, Delpech A, Leroy J, Robert M (1973) Acute renal failure associated with combined gentamicin and cephalotin therapy. Br Med J 1:396–397
21. Gibey R, Dupond JL, Alber D, Leconte des Floris R, Henry JC (1981) Predictive value of urinary N-acetyl-β-D-glucosaminidase (NAG), alanine-aminopeptidase (AAP) and β-D-microglobulin (β_2M) in evaluating nephrotoxicity of gentamicin. Clin Chem Acta 116:25–34
22. Gilbert DN, Houghton DC, Bennett WM, Plamp CE, Reger K, Porter GA (1979) Reversibility of gentamicin nephrotoxicity in rats: recovery during continuous drug administration. Proc Soc Exp Biol Med 160:99–103
23. Gray JE, Purmalis A, Purmalis B, Mathews J (1971) Ultrastructural studies of the hepatic changes brought about by clindamycin and erythromycin in animals. Toxicol Appl Pharmacol 19:217
24. Grötsch H (1984) Enzymurie und Nierenschädigung. In: Experimentelle Pathologie. Symposium, 23 Nov. Hoechst, Frankfurt/M pp 41–49
25. Hsu CH, Kurtz TW, Easterling RE, Weller IM (1974) Potentiation of gentamicin nephrotoxicity by metabolic acidosis. Proc Soc Exp Ther Med 146:894–897
26. Kahn T (1977) Effect of furosemide on gentamicin and netilmycin nephrotoxicity. Kidney Int 12:527
27. Kluve WM, Hook JB (1978) Functional nephrotoxicity of gentamicin in the rat. Toxicol Appl Pharmacol 45:163–175

28. Kolb R, Mautner Markhof C, Zekert F (1976) Verhalten von Harnenzymen nach Vitamin A, Gentamycin und Klemmung der Nierenarterien. Wien Klin Wochenschr 88:463–467
29. Kosek JC, Mazze RI, Cousins MJ (1974) Nephrotoxicity of gentamicin. Lab Invest 30:48–57
30. Koseki C, Endou H, Sudo J, Shimada H, Sakai F (1980) Evaluation of nephrotoxic site in rat proximal tubule: Intrarenal distribution of three enzymes and effects of mercuric chloride and gentamicin on their excretion into urine. Nippon yakungaku zasshi 76:59–69
31. Luft FC, Patel V (1978) Lysosomal acid hydrolases as urinary markers of aminoglycoside nephrotoxicity in the rat. In: Fillastre JP (ed) Nephrotoxicity. Masson, New York, pp 127–141
32. Luft FC, Patel V, Yum MN, Kleit SA (1976) Nephrotoxicity of cephalosporin-gentamicin combinations in rats. Microb Agents Chemother 9:831–839
33. Luft F, Bloch R, Yum M, Sloan RM (1977) Comparative aminoglycoside nephrotoxicity in rats. Kidney Int 12:528
34. Marre R, Abraham M, Freisleben H, Sack K (1979) Netilmicin und Tobramycin: vergleichende tierexperimentelle Untersuchungen zur Pharmakokinetik, Nierenverträglichkeit und therapeutischen Effektivität. Arzneimittelforschung 29:940–945
35. Marre R, Tarara N, Louton T, Sack K (1980) Age-dependent nephrotoxicity and the pharmacokinetics of gentamicin in rats. Eur J Pediatr 133:25–29
36. McMartin DN, Engel SG (1982) Effect of aging on gentamicin nephrotoxicity and pharmacokinetics in rats. Res Commun Chem Pathol Pharmacol 32:193–207
37. Mondorf AW, Breier J, Hendus J, Scherberich JE, Mackenrodt G, Shah PM, Stille W, Schoeppe W (1978) Effect of aminoglycosides on proximal tubular membranes of the human kidney. Eur J Clin Pharmacol 13:133–142
38. Mozer JL, Gibey R, Henry JC (1989) Inhibition de l'activité de la N-acetyl-β-D-glucosaminidase (NAG) urinaire et de ses isoenzymes par l'amikacine. C R Soc Biol (Paris) 183:166–173
39. Patel V, Luft FC, Moo Nahm Yum, Patel B, Zeman W, Kleit SA (1975) Enzymuria in gentamicin-induced kidney damage. Antimicrob Agents Chemother 7:364–369
40. Pariat C, Courtois P, Cambar J, Piriou A, Bouquet S (1988) Circadian variations in the renal toxicity of gentamicin in rats. Toxicol Lett 40:175–1802
41. Porter GA, Bennett WM (1989) Drug induced renal effects of cyclosporine, aminoglycoside antibiotics and lithium: extrapolation of animal data to man. In: Bach PH, Lock EA (eds) Nephrotoxicity: in vitro to in vivo: animals to man. Plenum, New York, pp 147–170
42. Raab W (1979) Investigations on the action of cefazedone and gentamicin, alone and combined, on renal enzyme excretion in rats. Arzneimittelfoschrung 29:433–435
43. Raab W (1980) Nephrotoxicity of drugs, evaluated by renal enzyme excretion studies. Arch Toxicol [suppl] 4:194–200
44. Sack K, Freisleben H (1975) Experimentelle Untersuchungen zur potentiellen Nephrotoxizität von Tobramycin, Gentamicin und Sisomicin. Infection 3:40S–49S
45. Sack K, Wilhelm J (1973) Vergleichende Untersuchungen zur Nephrotoxizität von Aminoglykosiden: Gentamycin und Tobramycin. Verh Dtsch Ges Inn Med 79:676–678
46. Sack K, Wilhelm J (1975) The diagnostic value of enzymuria, cell excretion, and proteinuria in experimental renal diseases. Urol Res 3:31–40
47. Sack K, Züllich B (1978) Differenzierung der nephrotoxichen Potenzen von Aminoglykosiden und Cephalosporinen im Tierexperiment. Therapiewoche 28:2107–2114
48. Sack K, Freisleben H, Züllich B, Beck H, Schulz E (1976) Nebenwirkungen von Aminoglycosiden: Nephrotoxizität. Infection 4:231–238
49. Sack K, Lepere A, Schwieder G (1978) Nierenverträglichkeit von Cephalosporinantibiotika: Cefoxitin und HR 756. Med Welt 29:1233–1236
50. Schinköthe G, Burchardt U, Anton D, Müller G, Patsch R (1982) Medikamenteninduzierte Nephropathie und Enzyme renalen Ursprungs im Harn. Z Urol Nephrol 75:571–577
51. Silverblatt FJ, Kuehn C (1979) Autoradiography of gentamicin uptake by the rat proximal tubule cell. Kidney Int 15:335–345
52. Silverblatt F, Turck M, Bulger R (1970) Nephrotoxicity due to cephaloridine: a light and electrone microscopic study in rabbits. J Infect Dis 122:33–44
53. Sugerman A, Brown RS, Silva P, Rosen S (1983) Features of gentamicin nephrotoxicity and effect of concurrent cephalotin in the rat. Nephron 34:239–247
54. Terhorst B, Lymberopoulos S, Buss H (1971) Tierexperimentelle und klinische Untersuchungen zur Frage der Nephrotoxizität von Cephaloridin. Med Welt 22:1565–1569
55. Tune BM (1975) Relationship between the transport and toxicity of cephalosporins in the kidney. J Infect Dis 132:189–194
56. Tune BM, Wu KY, Kempson RL (1977) Inhibition of transport and prevention of toxicity of

cephaloridine in the kidney. Dose-responsiveness of the rabbit and the guinea pig to probenecid. J Pharmacol Exp Ther 202:433–471
57. Tune BM, Wu KY, Longerbeam DF, Kempson RL (1977) Transport and toxicity of cephalosporidine in the kidney. Effect of flurosemide, p-aminohippurate and saline diuresis. J Pharmacol Exp Ther 202:472–478
58. Wachsmuth ED (1981) Nephrotoxicity of cefotiam (CGP 14221/E) in rats and rabbits. Arch Toxicol 48:135–156
59. Wachsmuth E, Wirz H (1979) Relevance of enzyme evaluation in 24th urine to rat kidney injury caused by i.v. cephaloridine injection. Curr Probl Clin Biochem 9:88–104
60. Watanabe M, Nomura G, Hirata M, Imai K, Koizumi H (1980) Studies on the validity of urinary enzyme assay in the diagnosis of drug-induced renal lesions in rats. Toxicol Pathol 8:2233
61. Wellwood JM, Lowell D, Thompson AE, Tighe JR (1976) Renal damage caused by gentamicin: a study of the effects on renal morphology and urinary enzyme excretion. J Pathol 118:171–182
62. Wilhelm J, Sack K (1975) Tierexperimentelle Studie zur Frage der Nephrotoxizität von Cephalotin und Cephalotin-Aminoglykosid-Kombinationen. Infection 3:89S–95S
63. Yver L, Becq-Giraudon B, Pourrat O, Sudre Y (1976) La néphrotoxicite de l'association methilcillin-gentamicine. Sem Hop Paris 52:1903–1907
64. Züllich B, Sack K (1978) Amikacin und Kanamycin. Vergleichende Untersuchungen zur Nephrotoxizität. Arzneimittelforschung Res 28:271–278
65. Züllich B, Sack K (1979) Experimental investigations of the renal tolerance of cefazedone. Arzneimittelforschung 29:430–433

18.3 Diuretics

H. Grötsch and M. Hropot

Enzymuria was reported in laboratory animals after the administration of diuretics [1, 4, 5]. There are only a few references in human medicine concerning that topic [3, 6]. The influence of diuretics in combination with certain drugs has been described more often. Examinations performed in animals as well in humans were carried out with furosemide, at that time a very new diuretic with effects never seen before with other diuretic drugs. Investigators attempted to find answers to questions concerning the transient increased proteinuria, enzymuria and glomerular filtration rate. It has been shown in human trials that the intravenous administration of 120 mg furosemide [6] increases the permeability of the glomerular capillary walls for macromolecules. β-Glucuronidase rose by about 600%, N-acetyl-β-D-glucosaminidase (NAG) by 91% and α-amylase by only 37%. However, the tubular origin of the increased enzyme excretion after furosemide-induced protein excretion correlates with renal prostaglandin (PGE_2) production. Similar results were reported by Guarnieri et al. [3], who examined the influence of furosemide on the excretion of NAG, β-glucosidase, lysozyme and β_2-microglobulin, as well as by Brennan et al. [1], who determined the excretion of ribonuclease and lysozyme in addition to measuring the proteinuria. He concluded that there was a transient impairment of protein reabsorption in the tubules.

While Hohenegger [5] only reported on results concerning the influence of furosemide after a single dose administration in rats, Helmchen [4] first described the reaction of furosemide in rats after a single i.p. administration of 50–200 mg/kg body weight. Furosemide causes acute epithelial lesions in the pars recta of the proximal tubules in the rat kidney. These lesions are considered to be due to an increased intracellular furosemide accumulation. Such cellular lesions may be more pronounced in a status of dehydration due to furosemide-induced fluid loss which is also dose-dependent.

Grötsch et al. [2] described the effect of daily treatment with high- and low-ceiling diuretics such as furosemide, piretanide, hydrochlorothiazide and indapamide on enzymuria in rats during a 4-week experiment. The drugs were administered orally in equi-effective doses. Analysis of the total enzyme activities as summarized in Table 1 showed in particular elevated amounts of alanine aminopeptidase (AAP) when furosemide, piretanide and indapamide were applied. In addition, indapamide causes a very uncommon excretion pattern of LDH. During the period of urine collection and the determination of the enzyme activities, this total excreted activity of LDH resulted from the first 10 days administration of indapamide. It is different from all other diuretics described in this experiment.

Histological and morphological examinations confirmed damage in the distal tubular cells of the kidney. The elevated excretion of AAP after furosemide and piretamide did not produce any histological findings. Experimental experience concerning the enzymuria due to effects of diuretics indicates that the elevated excretion of certain enzymes must not necessarily be accompanied by nephrotoxicity. The usage of equi-effective doses of diuretics allows us to recognize a possible nephrotoxic potential of a diuretic.

In conclusion, increased urinary excretion of enzymes has been reported during treatment with different diuretics. However, the diuretic-induced enzymuria need not lead to cellular damage but may indicate early changes in the integrity of renal cells.

Table 1. Total excreted enzyme activities over 25 days—treatment with diuretics in Wistar rats ($n = 7$).

	AAP	GGT	NAG	LDH
Control	9.5 ± 4.3	339 ± 106	11.0 ± 4.5	10.6 ± 7.1
Furosemide	21.2** ± 6.2	521 ± 157	19.3 ± 3.0	12.0 ± 2.7
Piretanide	18.8** ± 3.5	518* ± 126	17.8* ± 3.3	10.2 ± 2.3
Hydrochlorothiazide	14.6 ± 4.7	447 ± 86	11.6 ± 2.3	9.3 ± 3.1
Indapamide	14.6* ± 2.3	460 ± 102	14.5 ± 6.5	20.6* ± 5.1

Enzymes expressed in U/kg body weight. Values are mean ± SD, *$P < 0.05$ and **$P < 0.01$ (*U test*) vs controls

AAP, alanine aminopeptidase; GGT, gamma-glutamyltransferase; LDH, lactate dehydrogenase; NAG, *N*-acetyl-β-D-glucosaminidase

References

1. Brennan MF, Taverner PL, McKenzie RG, Macbeth WAAG (1970) Effects of furosemide upon renal excretion of protein, ribonuclease and lysozyme. Proc Univ Otago Med Sch 48:60–61
2. Grötsch H, Hropot M, Kief H, Klaus E (1984) Enzymuria in rats under treatment with diuretics. Diuretics J Puschett by Elsevier Science Publ Co Inc 514–518
3. Guarnieri G, Ianche M, Lin S (1979) Renal enzyme and protein excretion after induction of a diureses. Brit Med J:50–51
4. Helmchen U, Fischbach H, Schmidt U (1971) Akute Schaden des proximalen Tubulusepithels der Rattenniere nach einer einmaligen hochdosierten Furosemid-Gabe. Klin Wochenschr 49:1298–1300
5. Hohenegger M, Raab W (1970) Das Verhalten der Enzymaktivitäten im Harn bei Veränderung der Wasser- und Ionenausscheidung. Clin Chim Acta 28:61–65
6. Ylitalo P, Pasternack A, Kallio S, Vaenttinen T, Metsae-Ketelae T (1980) Increased urinary protein excretion after intravenous injection of furosemide in man. Acta Med Scand 208:279–283

18.4 Heavy Metals

E. Bomhard

Among the class of heavy metals those studied in more detail are cadmium, hexavalent chromium, inorganic lead, inorganic and to some extent organic mercury. All of them have some clinical and/or occupational relevance.

18.4.1 Cadmium

18.4.1.1 *Effects of Single Administration*

Although the kidney is not the major target under acute exposure conditions, relatively high parenteral doses have been shown to produce hydropic changes and acidophilic necrosis of the proximal convoluted tubules [27]. Single oral administration of 50 mg Cd/kg to rats had no effect on the excretion of 10 different urinary enzymes over the subsequent 4 days (E. Bomhard and D. Maruhn, unpublished results). After a single i.p. injection of 2.5 mg Cd/kg or s.c. injection of 2.0–9.0 mg Cd/kg, slight increases in lactate dehydrogenase (LDH), gamma-glutamyltransferase (GGT), leucine arylamidase (LAS), alanine aminotransferase (ALT), glycylproline-diaminopeptidase (GP-DAP) and decreases in glutathione S-transferase (GST), leucine aminopeptidase (LAP) and β-galactosidase (GAL) excretion were measured [5, 64, 68]. Much lower single i.p. doses (0.15 or 0.30 mg Cd/kg) bound to metallothionein caused renal tubular injury and dose-dependent increases in urinary alkaline phosphatase (AP), GGT, LDH and *N*-acetyl-β-D-glucosaminidase (NAG) excretion, with LDH being the most affected [62, 63].

18.4.1.2 *Effects of Multiple or Chronic Exposure*

The earliest sign of kidney damage after chronic cadmium exposure is increased excretion of plasma proteins (e.g. β-microglobulin, lysozyme, retinol binding protein) because of dysfunction in tubular reabsorption in humans and in experimental animals. Elevated urinary NAG, GAL and trehalase excretion has also been reported as early signs in highly exposed people [2, 24, 26, 44, 48].

In most of the animal experiments studying urinary enzyme excretion cadmium was injected repeatedly by the parenteral route. This seems to have different consequences with respect to the time course and even kind of kidney lesions, probably due to different kinetics of cadmium and/or cadmium-metallothionein complexes as compared with the oral and inhalational route relevant for the human exposure situation [11]. Thus, repeated s.c. or i.p. injections in rats (study duration ca. 3–24 weeks) have been reported to increase urinary AP, LDH, NAG and β-glucosidase (GLU) and decrease GGT and kallikrein excretion [6, 20, 30, 54, 66]. Rabbits treated s.c. with doses ranging from 1.0 to 15.0 mg Cd/kg for about 1 week or up to 3 months excreted more AP, acid phosphatase (ACP), LAP and trehalase than controls [47, 49, 69]. Chronic oral administration to rats of doses ranging from 10 to 20 ppm in drinking water decreased urinary kallikrein and increased α-glucosidase and lysozyme excretion, with GGT being unaffected [8, 19]. Chronic feeding of cadmium to rats did not affect GST and LDH excretion up to 50 ppm but increased output of both enzymes beginning from the 13th week and lasting to the final determination at 68 weeks [4]. Morphologically, the chronic lesions are characterized by tubular degeneration and interstitial oedema [27].

18.4.2 Hexavalent Chromium

The salts of hexavalent chromium, e.g. Na_2CrO_4 and $K_2Cr_2O_7$, selectively affect the cells of the convoluted portion of the proximal tubule [21]. The reported animal experiments have been mainly done in rats after parenteral administration of Na_2CrO_4 or $K_2Cr_2O_7$ in single doses ranging from 1 to 30 mg/kg. Doses below 10 mg/kg of both salts do not seem to have any consistent effect on urinary enzyme excretion [1, 15, 21, 60]. At higher doses increased output of brush border enzymes (AP, GGT, LAS, maltase), lysosomal enzymes [ACP, GAL, β-glucuronidase (GRS), NAG], cytosolic enzymes [aspartate aminotransferase (AST), GST, LDH] as well as filtered lysozyme with maximum excretion on the first 2 days after injection has been reported (Fig. 1) [5, 18, 21, 45, 60]. This time course is in good correlation with that of the

Fig. 1. Excretion (mU/24 h) of 10 urinary enzymes in percentage of controls after single i.p. administration of 10, 20 or 30 mg/kg Na_2CrO_4 to female Sprague-Dawley rats (D. Maruhn and E. Bomhard, unpublished data). *ALP*, alkaline phosphatase; *ASA*, arylsulphatase A; *GAL*, β-galactosidase; *GGT*, gamma-glutamyltransferase; *GLDH*, glutamate dehydrogenase; *GST*, glutathione S-transferase; *LAS*, leucine arylamidase; *LDH*, lactate dehydrogenase; *NAG*, *N*-acetyl-β-D-glucosaminidase; *PHI*, phosphohexoseisomerase

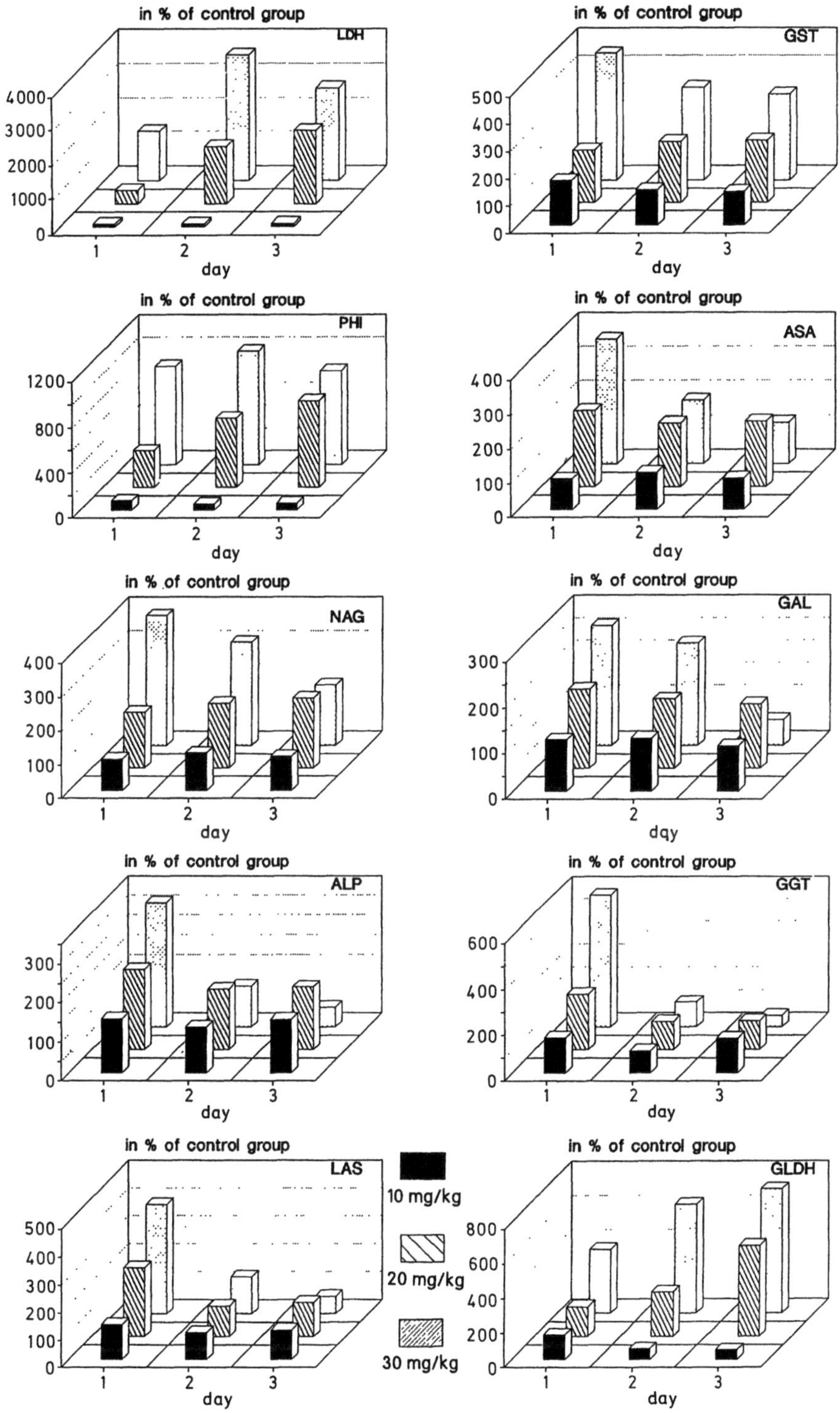
in % of control group
LDH
4000
3000
2000
1000
0
1
2
3
day
in % of control group
GST
500
400
300
200
100
0
1
2
3
day
in % of control group
PHI
1200
800
400
0
1
2
3
day
in % of control group
ASA
400
300
200
100
0
1
2
3
day
in % of control group
NAG
400
300
200
100
0
1
2
3
day
in % of control group
GAL
300
200
100
0
1
2
3
day
in % of control group
ALP
300
200
100
0
1
2
3
day
in % of control group
GGT
600
400
200
0
1
2
3
day
in % of control group
LAS
500
400
300
200
100
0
1
2
3
day
10 mg/kg
20 mg/kg
30 mg/kg
in % of control group
GLDH
800
600
400
200
0
1
2
3
day

morphological changes. LAP and urokinase excretion was comparable with controls [21, 53].

18.4.3 Inorganic Lead

The kidney is one of the most sensitive target organs to chronic Pb toxicity, with a higher susceptibility in the developing organism [17, 23]. The cytological changes include mitochondrial damage and the formation of dense intranuclear inclusion bodies in cells lining the proximal tubules [35]. Lead seems to act by an inhibition of the enzymes of haeme biosynthesis. Increased urinary excretion of NAG and decreased kallikrein excretion have been measured in people with a relatively high occupational exposure [7, 12, 41, 52]. Daily gavage treatment for 8 weeks with 15 mg Pb/kg or administration of 0.5%–2.0% lead acetate in drinking water for up to 3 months increased AST, AP, LDH, lysozyme, NAG and GST excretions in rats [16, 29, 67]. Single or repeated i.p. injections of 5 or 20 mg Pb/kg on the other hand had no significant influence on GGT, ALT and AP excretion, while LDH excretion was slightly increased [22, 46].

18.4.4 Mercury

Both organic and inorganic mercury compounds interact with sulphhydryl groups of a wide variety of enzymes, especially membrane-bound ones. Kidney damage is most prominent with inorganic compounds. The central nervous system is the primary target after organic mercury intoxication, but alterations in kidney function and morphology have also been described [42].

18.4.4.1 *Acute Experiments with Mercuric Chloride*

$HgCl_2$ has been extensively used as a model compound for acute renal failure. Cell necrosis in the proximal tubular epithelial cells, preferentially in the S2 and S3 segments, is the prominent feature [23]. Urinary enzyme excretion data after acute administration mostly via the parenteral route to rats have been described in more than 20 publications. The intoxication is characterized by an extremely enhanced output of most of the enzymes measured, especially within the first day after injection of doses as low as 0.75 mg/kg. By the oral route does at least 10 times higher are necessary to produce the same severity of effects [32]. The enzymuria is rapidly reversible despite persisting morphological changes, and even subnormal values especially of lysosomal enzymes have been reported in the regeneration phase [59]. Excretion of a large number of enzymes localized in

→

Fig. 2. Excretion (mU/24 h) of 10 urinary enzymes in percentage of controls after single i.p. administration of 0.5, 0.75 or 1.0 mg/kg $HgCl_2$ to female Sprague-Dawley rats (D. Maruhn and E. Bomhard, unpublished data). For abbreviations, see Fig. 1

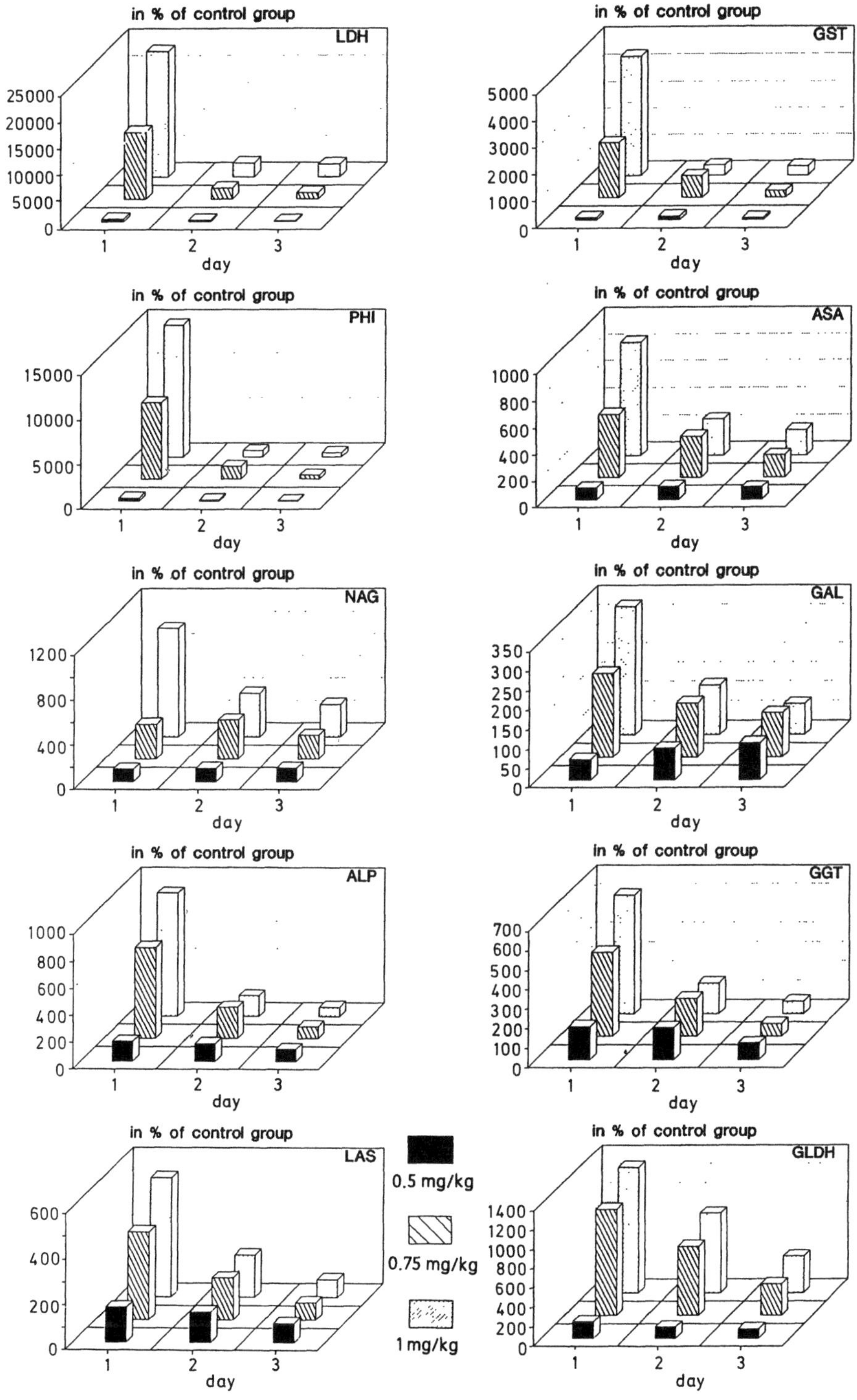

in % of control group
LDH
25000
20000
15000
10000
5000
0
1
2
3
day
in % of control group
GST
5000
4000
3000
2000
1000
0
1
2
3
day
in % of control group
PHI
15000
10000
5000
0
1
2
3
day
in % of control group
ASA
1000
800
600
400
200
0
1
2
3
day
in % of control group
NAG
1200
800
400
0
1
2
3
day
in % of control group
GAL
350
300
250
200
150
100
50
0
1
2
3
day
in % of control group
ALP
1000
800
600
400
200
0
1
2
3
day
in % of control group
GGT
700
600
500
400
300
200
100
0
1
2
3
day
in % of control group
LAS
600
400
200
0
1
2
3
day
0.5 mg/kg
0.75 mg/kg
1 mg/kg
in % of control group
GLDH
1400
1200
1000
800
600
400
200
0
1
2
3
day

the brush border (AAP, AP, GGT, LAP, LAS, maltase), cytosol [ALT, AST, fructose-1, 6-biphosphatase (FBP), GST, LDH, pyruvate kinase (PK)], lysosomes (ACP, GAL, GRS, NAG) and mitochondria (glutamate dehydrogenase; GLDH) as well as lysozyme of serum origin has been investigated. The enzymes with the highest increased output were cytosolic LDH and GST followed by the brush border enzymes AP, AAP, GGT and maltase (Fig. 2) [3, 5, 9, 13, 14, 28, 31–34, 37–40, 43, 51, 55, 56, 59, 60, 65].

18.4.4.2 *Repeated Mercuric Chloride Administration*

After repeated oral or parenteral $HgCl_2$ administration to rats for up to 15 days, excretion of the enzymes investigated (AAP, acid protease, AP, AST, GGT, LAP, LDH, lysozyme, NAG) was enhanced essentially only in the early phase of tubular damage and returned to normal levels despite continued treatment, increased numbers of tubular cells in the urine and morphological signs of a decreased tissue response [25, 32, 50, 55, 58].

18.4.4.3 *Organic Mercury Compounds*

Very few studies are available which report on enzyme excretion after intoxication with organic mercury compounds. After acute i.p. treatment of rats with doses ranging from 0.5 to 2.0 mg Hg/kg (as phenylmercuric acetate or methoxymethylmercuric chloride) AP excretion was elevated (three- to four fold) and that of GGT about eight fold at 1.0 and 2.0 mg/kg in one study, but lower at 0.5 and 1.0 mg/kg in another study despite tubular cell necrosis [10, 37]. Methylmercuric chloride was administered on days 8, 10 and 12 of gestation at 3 or 6 mg/kg, and excretion of GGT, AP and NAG was measured in urinary specimens from neonatal rats at 3, 6, 9 and 12 days after birth. Dose-related increases of the three enzyme activities (130%–200% of controls) occurred at dose levels that produced neither maternal or postnatal toxicity nor overt morphological malformation of the kidney [57]. After treatment of rats with daily i.p. injection of 1 mg methylmercuric chloride/kg for 20 days, urinary excretion of GAL and ACP appeared to be decreased, but AP was not affected [61]. After daily treatment (6 days/week) with methoxyethylmercuric chloride in doses of 0.2 mg/kg for 12 weeks or 2.0 mg/kg for 6 weeks, AST and ALT excretion was significantly increased from weeks 2 to 4 in parallel with increased cell counts in the urinary sediment. ALT and AST activity in the serum remained unchanged [36].

18.4.5 Conclusions

The kidney is the major target especially of the inorganic heavy metal salts. A large number of investigations using mainly hexavalent chromium and inorganic mercury compounds indicate a close correlation between early morphological damage and increased excretion of urinary enzymes mainly originating from the cytosol and the brush border of the proximal tubular parts. Under prolonged treatment with $HgCl_2$ a return to normal enzyme excretion has been reported, probably as a result of the regeneration of resistant cells and/or exhaustion of the enzyme pools. On the other hand, chronic nephrotoxicity by $CdCl_2$ was accompanied by an elevated output of enzymes, thus indicating the usefulness of measuring urinary enzyme activity also in chronic renal damage.

References

1. Bass NM, Kirsch RE, Tuff SA, Campbell JA, Saunders JS (1979) Radioimmunoassay measurement of urinary ligandin excretion in nephrotoxin-treated rats. Clin Sci 56:419–426
2. Bernard A, Buchet JP, Roels H, Masson P, Lauwerys R (1979) Renal excretion of proteins and enzymes in workers exposed to cadmium. Eur J Clin Invest 9:11–22
3. Bhargava AS, Khater AR, Günzel P (1978) The correlation between lactate dehydrogenase activity in urine and serum and experimental renal damage in the rat. Toxicol Lett 1:319–323
4. Bomhard E, Maruhn D, Paar D, Wehling K (1984) Urinary enzyme measurements as sensitive indicators of chronic cadmium nephrotoxicity. Contrib Nephrol 42:142–147
5. Bomhard E, Maruhn D, Vogel O (1986) Comparative investigations on the effects of acute intraperitoneal cadmium, chromium, and mercury exposure on the kidney. Uremia Invest 9:131–136
6. Bonner FW, Laurence JK, Parke DV (1980) The urinary excretion of alkaline phosphatase after the repeated parenteral administration of cadmium to rats given a high dietary supplement of zinc. Toxicol Lett 6:369–372
7. Boscolo P, Porcelli G, Cechetti G, Salimei E, Iannaccone A (1978) Urinary kallikrein activity of workers exposed to lead. Br J Ind Med 35:226–229
8. Boscolo P, Porcelli G, Carmignani M, Finelli VN (1981) Urinary kallikrein and hypertension in cadmium-exposed rats. Toxicol Lett 7:189–194
9. Braun JP, Rico AG, Bernard P, Burgat-Sacaze V, Eghbali B, Godfrain JC (1978) La gamma-glutamyl transferase urinaire en toxicologie renale chez le rat. Bases de son utilisation-interet lors de nephrite aigue mercurielle. Toxicology 11:73–82
10. Burgat-Sacaze V, Braun JP, Rico A, Bernard P, Eghbali B (1980) Methoxyethylmercury nephrotoxicity: effects on enzymuria and kidney function. Arch Toxicol 43:227–231
11. Cain K (1987) Metallothionein and its involvement in the heavy metal induced nephropathy. In: Bach PH, Lock EA (eds) Nephrotoxicity in the experimental and clinical situation. Nijhoff, Dordrecht, pp 473–531
12. Chmielnicka J, Komsta-Szumska E, Szymanska JA (1981) Arginase and kallikrein activities as biochemical indices of occupational exposure to lead. Br J Ind Med 38:175–178
13. Dierickx PJ (1989) Urinary gamma-glutamyl transferase as a specific marker for mercury after heavy metal treatment of rats. Toxicol Lett 6:235–238
14. Emanuelli G, Cestonaro G, Anfossi G, Calcamuggi G, Gatti G, Marcarino C (1982) Urinary enzyme excretion and renal lactate dehydrogenase isoenzyme pattern in acute $HgCl_2$ nephropathy of rat. Enzyme 27:89–98
15. Feinfeld DA, Bourgoignie JJ, Fleischner G, Goldstein EJ, Biempica L, Arias IM (1977) Ligandinuria in nephrotoxic acute tubular necrosis. Kidney Int 12:387–392
16. Flora SJS, Singh S, Tandon SK (1983) Role of selenium in protection against lead intoxication. Acta Pharmacol Toxicol 53:28–32
17. Fowler BA, Kimmel CA, Woods JS, McConnell EE, Grant DL (1980) Chronic low-level lead toxicity in the rat. Toxicol Appl Pharmacol 56:59–77
18. Franchini I, Mutti A, Cavatorta A, Corradi A, Cosi A, Olivetti G, Borghetti A (1978) Nephrotoxicity of chromium. Contrib Nephrol 10:98–100
19. Gatta A, Bazzerla G, Amodio P, Menon F, Angeli P, Schiaffino E, Schmid C (1989) Detection of the early steps of cadmium nephropathy—comparison of light and electron-microscopical patterns with the urinary enzymes excretion. Nephron 51:20–24

20. Girolami JP, Bascands JL, Pécher C, Cabos G, Moatti JP, Mercier JF, Haguenoer JM, Manuel Y (1989) Renal kallikrein excretion as a distal nephrotoxicity marker during cadmium exposure in rats. Toxicology 55:117–129
21. Gumbleton M, Nicholls PJ (1988) Dose-response and time-response biochemical and histological study of potassium dichromate-induced nephrotoxicity in the rat. Food Chem Toxicol 26:37–44
22. Hofmann U, Segewitz G (1975) Influence of chelation therapy on acute lead intoxication in rats. Arch Toxicol 34:213–225
23. Hubermont G, Buchet JP, Roels H, Lauwerys R (1976) Effect of short-term administration of lead to pregnant rats. Toxicology 5:379–384
23. Hultman P, Eneström S (1986) Localization of mercury in the kidney during experimental acute tubular necrosis studied by the cytochemical silver amplification method. Br J Exp Pathol 67:493–503
24. Iwata K, Katoh T, Morikawa Y, Aoshima K, Nishijo M, Teranishi H, Kasuya M (1988) Urinary trehalase activity as an indicator of kidney injury due to environmental cadmium exposure. Arch Toxicol 62:435–439
25. Josepovitz C, Levine R, Lane B, Kaloyanides GJ (1985) Contrasting effects of gentamicin and mercuric chloride on urinary excretion of enzymes and phospholipids in the rat. Lab Invest 52:375–386
26. Kawada T, Koyama H, Suzuki S (1989) Cadmium, NAG activity, and β_2-microglobulin in the urine of cadmium pigment workers. Br J Ind Med 46:52–55
27. Kawai K, Fukuda K, Kimura M (1976) Morphological alterations in experimental cadmium exposure with special reference to the onset of renal lesions. In: Nordberg GF (ed) Effects and dose-response relationships of toxic metals. Elsevier, Amsterdam, pp 343–370
28. Kempson SA, Ellis BG, Price RG (1977) Changes in rat renal cortex, isolated plasma membranes and urinary enzymes following the injection of mercuric chloride. Chem Biol Interact 18:217–234
29. Khalil-Manesh F, Gonick HC, Weiler E, Rosen V, Roche L, Mutti A, Bergamaschi E, Alinovi R, Franchi I (1989) Early indicators of lead nephropathy. In: Bach PH, Lock EA (eds) Nephrotoxicity: in vitro to in vivo: animals to man. Plenum, New York, pp 119–126
30. King LC, Clark V, Faeder EJ (1976) Effects of cadmium exposure on rat kidneys. Bull Environ Contam Toxicol 16:572–577
31. Kirschbaum BB (1979) Alanine aminopeptidase excretion after mercuric chloride renal failure. Biochem Med 21:220–225
32. Kluwe WM (1982) Development of resistance to nephrotoxic insult: changes in urine composition and kidney morphology on repeated exposures to mercuric chloride or biphenyl. J Toxicol Environ Health 9:619–635
33. Kotanko P, Gstraunthaler G, Pfaller W (1984) Harnenzyme zur nicht invasiven Diagnostik von Nierenepithelschäden im akuten Nierenversagen. Wien Klin Wochenschr 96:625–629
34. Kyle GM, Luthra R, Bruckner JV, MacKenzie WF, Acosta D (1983) Assessment of functional, morphological and enzymatic tests for acute nephrotoxicity induced by mercuric chloride. J Toxicol Environ Health 12:99–117
35. Landrigan PJ (1989) Toxicity of lead at low dose. Br J Ind Med 46:593–596
36. Lehotzky K, Bordás S (1971) Urinary and serum transaminase levels in rats with organic mercury poisoning. Acta Med Acad Sci Hung 28:139–143
37. Magos L, Sparrow S, Snowden R (1982) The comparative renotoxicology of phenylmercury and mercuric chloride. Arch Toxicol 50:133–139
38. Magos L, Sparrow S, Snowden RT (1984) Effect of prolonged saline loading on $HgCl_2$-induced renal tubular damage. Br J Exp Pathol 65:567–575
39. Maruhn D, Paar D, Bomhard E (1983) Diagnostic sensitivity of urinary enzymes in experimental kidney damage in the rat. In: Galteau MM, Siest G, Heung J (eds) Biologie prospective—5e Colloque de Pont-à-Mousson. Masson, Paris, pp 943–945
40. Maruhn D, Bomhard E, Paar D (1984) Urine enzyme patterns in the rat after damage to different parts of the nephron. In: Goldberg DM, Werner M (eds) Selected topics in clinical enzymology, vol 2. De Gruyter, Berlin, pp 429–434
41. Mason HJ, Somervaille LJ, Tennant DR, Chettle DR, Scott MC (1989) In-vivo bone lead measurements and renal effects. In: Bach PH, Lock EA (eds) Nephrotoxicity: in vitro to in vivo: animals to man. Plenum, New York, pp 113–118
42. Mottet NK (1989) A pathologist's perspective on the toxic effects of mercury compounds. Comments Toxicol 3:179–190
43. Nakano M, Itoh G (1983) Elevation of urinary trehalase in mercuric chloride-induced nephro-

toxic rabbits: urinary trehalase as a specific indicator of renal brush border damage. Chem Biol Interact 45:179–189
44. Nakano M, Aoshima K, Katoh T, Teranishi H, Kasuya M (1987) Elevation of urinary trehalase activity in patients of itai-itai disease. Arch Toxicol 60:300–303
45. Ngaha EO (1981) Renal effects of potassium dichromate in the rat: comparison of urinary enzyme excretion with corresponding tissue patterns. Gen Pharmacol 12:497–500
46. Nicholls DM, Teichert-Kuliszewska K, Kuliszewski MJ (1983) The activity of membrane enzymes in homogenate fractions of rat kidney after administration of lead. Toxicol Appl Pharmacol 67:193–199
47. Nishimura N, Oshima H, Nakano M (1986) Urinary trehalase as an early indicator of cadmium-induced renal tubular damage in rabbit. Arch Toxicol 59:255–260
48. Nogawa K, Yamada Y, Honda R, Tsuritani I, Ishizaki M, Sakamoto M (1983) Urinary N-acetyl-β-D-glucosaminidase and β_2-microglobulin in itai-itai disease. Toxicol Lett 16:317–322
49. Nomiyama K, Sato C, Yamamoto A (1973) Early signs of cadmium intoxication in rabbits. Toxicol Appl Pharmacol 24:625–635
50. Ohata H, Momose K, Takahashi A, Omori Y (1987) Urinalysis for detection of chemically induced renal damage—changes in urinary excretions of enzymes and various components caused by mercuric chloride and gentamicin. J Toxicol Sci 12:341–355
51. Ohata H, Hashimoto T, Monose K, Takashi A, Terao T (1988) Urinalysis for detection of chemically induced renal damage—establishment and application of radioimmunoassay for lysozyme of rat urine. Arch Toxicol 62:60–65
52. Ong CN, Endo G, Chia KS, Phoon WO, Ong HY (1987) Evaluation of renal function in workers with low blood lead levels. In: Foa V, Emmett EA, Maroni M, Colombi A (eds) Occupational and environmental chemical hazards. Ellis Horwood, Chichester
53. Paar D, Maruhn D, Bock KD, Bomhard E, Lorke D (1981) Untersuchungen zur Urokinaseausscheidung im Harn der Ratte nach experimenteller Nierenschädigung. In: Bluemel G, Haas S (eds) Mikrozirkulation und Prostaglandinstoffwechsel. Schattauer, Stuttgart, pp 371–374
54. Pierce RJ, Price RG, Fowler JSL (1977) The effect of cadmium administration on activities of enzymes in the urine of the rat and marmoset. Biochem Soc Trans 5:238–241
55. Planas-Bohne F (1977) The effect of mercuric chloride on the excretion of two urinary enzymes in the rat. Arch Toxicol 37:219–225
56. Raab W (1966) Experimentelle Untersuchungen über Veränderungen der "Leucinamino-peptidase"-Aktivität des Harnes. Wien Klin Wochenschr 78:364–369
57. Saillenfait AM, Brondeau MT, Zissu D, Ceaurriz de J (1989) Effects of prenatal methylmercury exposure on urinary proximal tubular enzyme excretion in neonatal rats. Toxicology 55:153–160
58. Sener S, Braun JP, Rico AG, Bernard P, Burgat-Sacaze V (1979) Urine gamma-glutamyl transferase in rat kidney toxicology: nephropathy by repeated injections of mercuric chloride. Effects of sodium selenite. Toxicology 12:299–305
59. Sternberg M, Szlamka I, Moisy M, Rebeyrotte P, Lagrue G (1974) Comparison of the urinary excretion of aspartate aminotransferase, lactic dehydrogenase, alkaline and acid phosphatase and β-galactosidase during nephrotoxic serum glomerulonephritis and mercuric chloride tubulonephritis in the rat. Z Klin Chem Klin Biochem 12:543–550
60. Stroo WE, Hook JB (1977) Enzymes of renal origin in urine as indicators of nephrotoxicity. Toxicol Appl Pharmacol 39:423–434
61. Stroo WE, Hook JB (1977) Renal functional correlates of methyl mercury intoxication: interaction with acute mercuric chloride toxicity. Toxicol Appl Pharmacol 42:399–410
62. Sugihira N, Sagai M, Suzuki KT (1987) Renal damage induced by cadmium-metallothionein: effects on biochemical indicators. Toxicology 44:1–11
63. Suzuki CAM, Cherian MG (1987) Renal toxicity of cadmium-metallothionein and enzymuria in rats. J Pharmacol Exp Ther 240:314–319
64. Tandon SK, Flora SJS, Behari JR, Ashquin M (1984) Vitamin B complex in treatment of cadmium intoxication. Ann Clin Lab Sci 14:487–492
65. Tandon SK, Magos L, Carbal JRP (1980) Protection against mercuric chloride by nephrotoxic agents which do not induce thionein. Toxicol Appl Pharmacol 52:227–236
66. Tohyama C, Sugihira N, Saito H (1987) Critical concentration of cadmium for renal toxicity in rats. J Toxicol Environ Health 22:255–259
67. Vyskocil A, Pancl J, Tusl M, Ettlerova E, Semecky V, Kasparová L, Lauwerys R, Bernard R (1989) Dose-related proximal tubular dysfunction in male rats chronically exposed to lead. J Appl Toxicol 9:395–399

68. Watanabe Y, Nonomura F, Tanahashi N, Sugitani A, Yamada F (1985) Preventive effect of pretreatment with zinc on cadmium nephrotoxicity in rats—evaluation by changes of urinary enzyme activities. Ind Health 23:25–36
69. Yamamoto A, Yoshida K, Wada O (1981) Origin of urinary alkaline phosphatase excreted by cadmium treatment of rabbits. J Toxicol Sci 6:325–333

18.5 Model Substances

E. Bomhard

The rationale for the selection of the following compounds is based on their ability to act at distinct parts of the nephron, the collecting duct or the papilla. Availability of data on enzymuria was another selection criterion. Some very extensively used model compounds, e.g. $HgCl_2$ and hexavalent chromium compounds, are dealt with in the chapter on heavy metals.

18.5.1 Bovine Serum Albumin

High parenteral doses of bovine serum albumin are used to produce overload proteinuria. Glomerular changes consist of swelling of the epithelial cells with blunting and loss of their food processes and accumulation of hyaline droplets in the moderately swollen proximal tubular epithelia [9, 13]. After twice daily 2.5 g/kg i.p. injections into rats on 3 consecutive days a transiently and moderately increased urinary output of lactate dehydrogenase (LDH), *N*-acetyl-*β*-D-glucosaminidase (NAG), gamma-glutamyltransferase (GGT) and glutathione S-transferase (GST) was measured and attributed to the tubular alterations [20].

18.5.2 Puromycin Aminonucleoside

Puromycin aminonucleoside is a polycation neutralizing the polyanions of the glomerular filter, resulting predominantly in epithelial side lesions. Tubules and interstitial parts are also involved. It is used as a model compound for the nephrotic syndrome. Concomitant with the massive proteinuria beginning 5–8 days after parenteral injection there is an increased output of enzymes originating in lysosomes [*β*-galactosidase (GAL), arylsulfatase A (ASA), NAG, acid phosphatase (ACP)], cytosol [phosphohexose isomerase (PHI), GST, LDH], brush border [leucine arylamidase (LAS), GGT, alkaline phosphatase (AP)] and mitochondria (glutamate dehydrogenase, GLDH), while *β*-glucosidase (GLU) excretion was unaffected [1, 3, 27, 31].

18.5.3 Anthracyclines

Rats and rabbits are extremely susceptible to anthracycline (e.g. adriamycin, daunomycin)-induced nephrotic syndrome. Functional and morphological characteristics seem to differ from those associated with puromycin aminonucleoside administration. Morphological changes (visceral epithelial cells with microvilli and vacuoles, lipid inclusion bodies and focal loss of foot processes) are restricted to epithelial cells of the glomerulus. Secondary tubular (atrophy of tubules with extensive filling with protein and cellular casts) as well as interstitial (patchy areas of fibrosis, increased numbers of mononuclear cells) changes have been reported [16, 23]. Free oxygen radical formation with consequent lipid peroxidation of cell membranes and loss of ATPase activity has been suggested as the mechanism [21]. Single i.p. injections of 14 mg adriamycin/kg into female Sprague-Dawley rats increased urinary AP and GGT excretion only very slightly (less than three fold), while NAG was sometimes lower and sometimes higher than in controls [15].

18.5.4 Sodium Maleate

Sodium maleate has been used to create and study renal tubular dysfunction comparable with Fanconi's syndrome, displaying glucosuria, aminoaciduria and phosphaturia. Functional impairment is accompanied by morphological changes of the proximal (S2 and S3) and distal nephron segments [26]. Major changes are confined to increases in endocytic vacuoles and in the mitochondrial inner membrane surface area. A single i.p. injection of 200 mg/kg to male Sprague-Dawley rats significantly increased the urinary output of fructose-1,6-biphosphatase (FBP), LDH, NAG and isocitrate dehydrogenase (ICDH), but not of AP, GGT and pyruvate kinase (PK). The highest increase was found for NAG, followed by FBP and ICDH [26].

18.5.5 Halogenated Alkenes

A number of haloalkenes (e.g. hexachloro-1,3-butadiene, tetrafluoroethylene, chlorotrifluoroethylene, dichloroacetylene, trichloroethylene) are metabolized by the mercapturic acid pathway to nephrotoxic glutathione conjugates [17]. Some of the nephrotoxic cysteine conjugates formed are changed to reactive intermediates by the cysteine conjugate β-lyase pathway or may be direct-acting nephrotoxins. Nephrotoxic lesions appear to be focussed in the S3 segment. After long-term treatment renal tumours (adenomas and adenocarcinomas) have been observed in rats given nephrotoxic doses. Increased urinary excretion of AP, NAG, alanine aminopeptidase (AAP) and GGT has been reported after acute oral administration of hexachloro-1,3-butadiene 50–200 mg/kg to rats [18, 22, 31].

18.5.6 Folic Acid

Parenteral administration of folic acid in doses of more than 35 mg/kg causes acute renal failure in the rat. Accumulation and precipitation of folate is the first step, followed by a proliferative phase leading to a remarkable hypertrophy of the kidneys. Most prominent are the morphological changes in the distal tubules: they contain numerous casts and show severe cellular changes and degeneration. Degeneration is also seen in proximal tubular epithelial cells [5, 14, 28, 30]. Many urinary enzymes have been measured after parenteral administration of doses in the 125–400 mg/kg range (Fig. 1) [11, 20, 24, 25, 29, 32]. These data reveal a pattern which deviates considerably from that after proximal tubular toxins: while LDH excretion is markedly increased, several other enzymes are not substantially changed [ACP, AP, maltase, leucine aminopeptidase (LAP)] or are even reduced (GGT, LAS, NAG, urokinase and kallikrein).

18.5.7 Ethyleneimine, Propyleneimine

Both compounds cause a dose-related renal papillary necrosis first affecting the interstitial cells of the papilla tip and then other anatomical elements of the medulla (for review, see [2]). Single injections of ethyleneimine at doses as low as 0.5 μl/kg were sufficient for induction of this lesion in several species (rat, rabbit, dog) within days. Increased excretion of brush border (AP), lysosomal (ACP, GAL, GLU, NAG) and cytosolic enzymes (GST, LDH) has been reported after treatment of rats or dogs with ethyleneimine [4, 7, 8, 19]. While in some studies NAG proved to be the most sensitive enzyme both in rats and dogs [7, 8], LDH showed a higher sensitivity in other rat studies (Fig. 2) [4, 20]. After a single i.p. injection of propyleneimine into rats, NAG excretion increased immediately, GAL and GLU increased 9 days after administration, while brush border marker enzymes LAP, AAP or AP were not consistently changed [12]. Again, LDH excretion was much more elevated than that of NAG in a study with male and female Wistar rats, even on the 1st day after injection (E. Bomhard et al., unpublished data).

18.5.8 2-Bromoethanamine

Bromoethanamine has been extensively used as a model compound for acutely induced renal papillary necrosis in mice, rats, hamsters and pigs [2, 6, 10]. Degenerative changes begin in the renal medullary interstitial cells, but early

Fig. 1. Excretion (mU/24 h) of 10 urinary enzymes in percentage of controls after single i.p. administration of 125, 250 or 375 mg/kg folic acid to remale Sprague-Dawley rats (D. Maruhn and E. Bomhard, unpublished data). *ALP*, alkaline phosphatase; *ASA*, arylsulfatase A; *GAL*, β-galactosidase; *GGT*, gamma-glutamyltransferase; *GLDH*, glutamate dehydrogenase; *GST*, glutathione S-transferase; *LAS*, leucine arylamidase; *LDH*, lactate dehydrogenase; *NAG*, *N*-acetyl-β-D-glucosaminidase; *PHI*, phosphohexoseisomerase

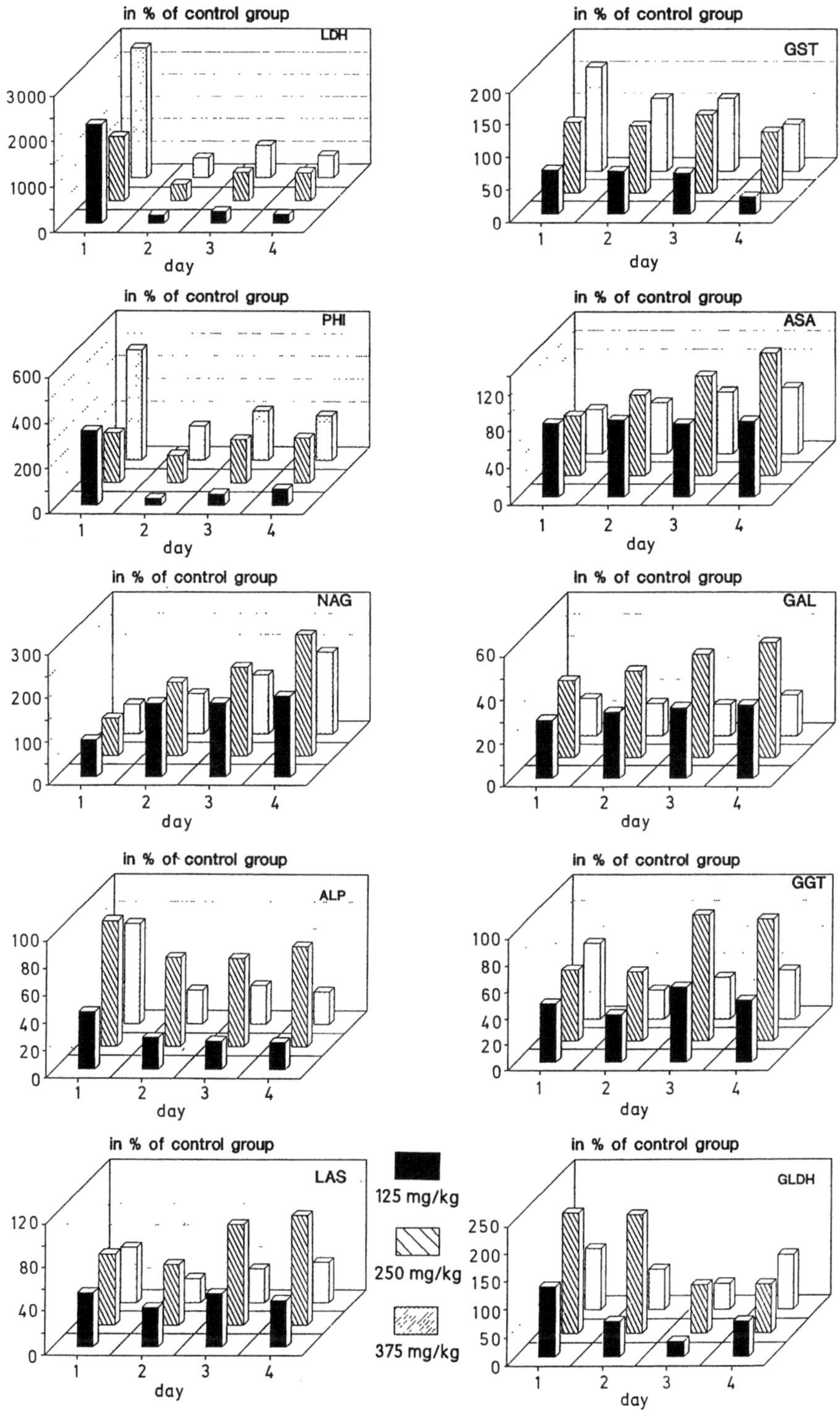
in % of control group
LDH
3000
2000
1000
0
1
2
3
4
day
in % of control group
GST
200
150
100
50
0
1
2
3
4
day
in % of control group
PHI
600
400
200
0
1
2
3
4
day
in % of control group
ASA
120
80
40
0
1
2
3
4
day
in % of control group
NAG
300
200
100
0
1
2
3
4
day
in % of control group
GAL
60
40
20
0
1
2
3
4
day
in % of control group
ALP
100
80
60
40
20
0
1
2
3
4
day
in % of control group
GGT
100
80
60
40
20
0
1
2
3
4
day
in % of control group
LAS
120
80
40
0
1
2
3
4
day
125 mg/kg
250 mg/kg
375 mg/kg
in % of control group
GLDH
250
200
150
100
50
0
1
2
3
4
day

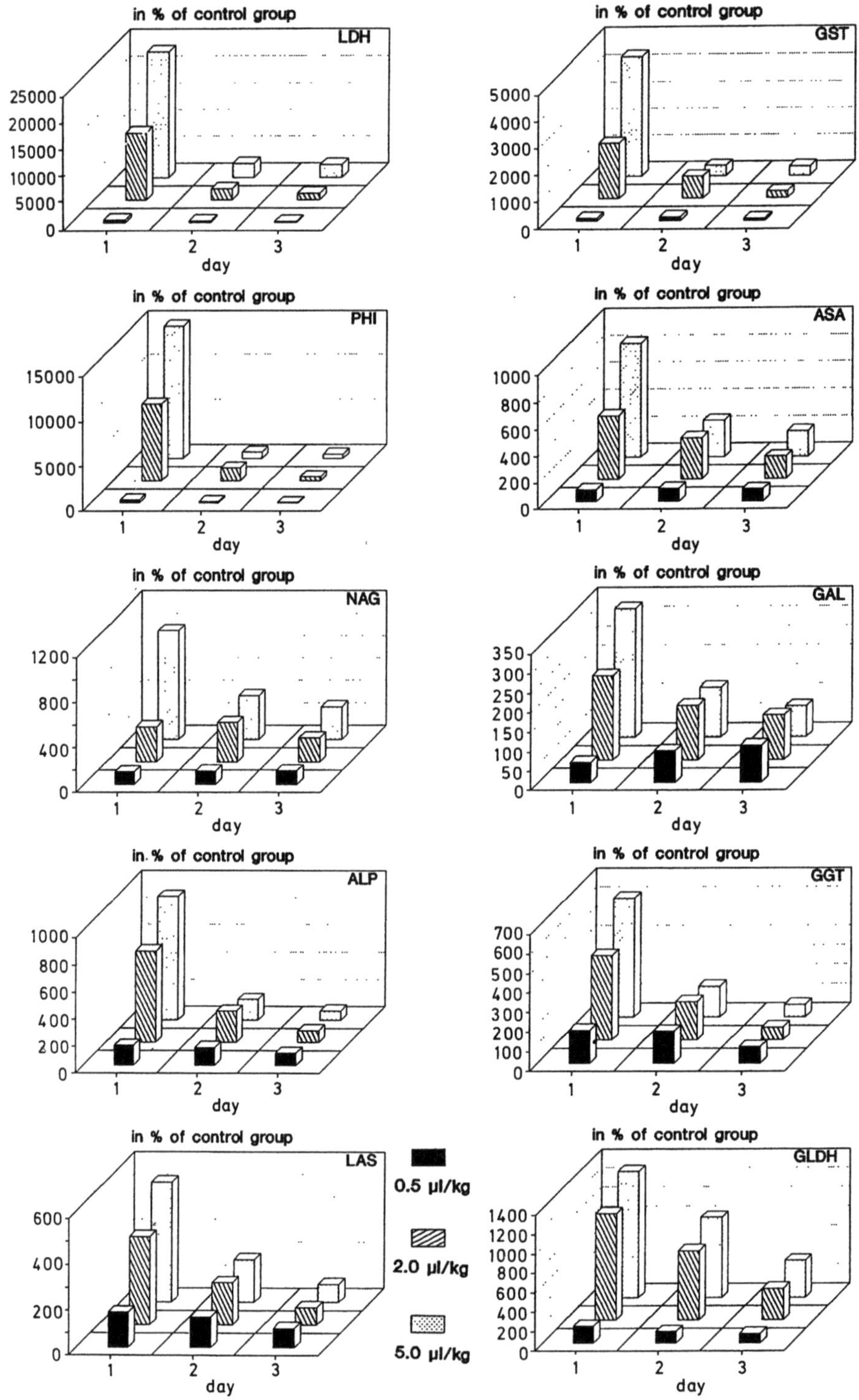

in % of control group
LDH
25000
20000
15000
10000
5000
0
1
2
3
day
in % of control group
GST
5000
4000
3000
2000
1000
0
1
2
3
day
in % of control group
PHI
15000
10000
5000
0
1
2
3
day
in % of control group
ASA
1000
800
600
400
200
0
1
2
3
day
in % of control group
NAG
1200
800
400
0
1
2
3
day
in % of control group
GAL
350
300
250
200
150
100
50
0
1
2
3
day
in % of control group
ALP
1000
800
600
400
200
0
1
2
3
day
in % of control group
GGT
700
600
500
400
300
200
100
0
1
2
3
day
in % of control group
LAS
600
400
200
0
1
2
3
day
0.5 µl/kg
2.0 µl/kg
5.0 µl/kg
in % of control group
GLDH
1400
1200
1000
800
600
400
200
0
1
2
3
day

proximal tubular changes have also been described [2]. An increased urinary excretion of AP and NAG was measured starting 2 days after a single i.p. injection of 50 or 200 mg/kg to male rats [31]. While AP excretion returned to the normal level after 8 days, NAG activity was still elevated after 15 days. In another study with i.p. injection of 50 or 125 mg/kg into male and female Wistar rats, LDH excretion was much more increased than NAG over the subsequent 5 days (E. Bomhard et al., unpublished data).

18.5.9 Conclusions

Excretion of urinary enzymes has not yet been comprehensively investigated with respect to many model compounds, especially those acting primarily outside the proximal tubule. The available information nevertheless indicates that urinary enzyme pattern measurement may provide not only information on the degree of damage but also on the localization of the insult within the nephron. Primary damage even to rather specific functions or structures within certain nephron segments and the papilla, however, is often and sometimes rapidly accompanied by the secondary involvement of other parts. Enzyme release from the latter structures may make it difficult to locate those primarily affected. Therefore, a high frequency of measurements in the early phases of nephrotoxicity may be indicated.

References

1. Abe H, Shibuya T, Odashima S, Arichi S, Nagase S (1988) Alterations in the glomerulus in aminonucleoside nephrosis in analbuminemic rats. Nephron 50:351–355
2. Bach PH, Gregg NJ (1988) Experimentally induced renal papillary necrosis and upper urothelial carcinoma. Int Rev Exp Pathol 30:1–54
3. Bomhard E, Maruhn D (1988) Effects of puromycin aminonucleoside on ten urinary enzymes in rats. Arch Toxicol 12:158–161
4. Bomhard E, Maruhn D, Vogel O, Mager H (1990) Determination of urinary glutathione S-transferase and lactate dehydrogenase for differentiation between proximal and distal nephron damage. Arch Toxicol (Supple) 64:269–278
5. Byrnes KA, John MS, Ghidoni JJ, Suzuki M, Thomas H, Mayfield ED (1972) Response of the rat kidney to folic acid administration. Lab Invest 26:191–200
6. Carlton WW, Engelhardt JA (1989) Experimental renal papillary necrosis in the Syrian hamster. Food Chem Toxicol 27:331–340
7. Ellis BG, Price RG, Topham JC (1973) The effect of papillary damage by ethyleneimine on kidney function and some urinary enzymes in the dog. Chem Biol Interact 7:131–142
8. Ellis BG, Price RG (1975) Urinary enzyme excretion during renal papillary necrosis induced in rats with ethyleneimine. Chem Biol Interact 11:473–482
9. Fisher ER, Hellstrom MD (1962) Mechanism of proteinuria: functional and ultrastructural

Fig. 2. Excretion (mU/24 h) of 10 urinary enzymes in percentage of controls after single i.p. administration of 0.5, 2.0 or 5.0 μl/kg ethyleneimine to female Sprague-Dawley rats (D. Maruhn and E. Bomhard, unpublished data). For abbreviations see Fig. 1

correlation of effects of infusion of homologous and heterologous protein (bovine serum albumin) in the rat. Lab Invest 11:617–637
10. Gregg NJ, Robbin MEC, Hopewell JW, Bach PH (1989) Morphological changes in the pig kidney associated with an acutely induced renal papillary necrosis. In: Bach PH, Lock EA (eds) Nephrotoxicity: in vitro to in vivo—animals to man. Plenum, New York, pp 407–410
11. Halman J, Fowler YSL, Price RG (1985) Urinary enzymes, proteinuria and renal function tests in the assessment of nephrotoxicity in the rat. In: Bach PH, Lock EA (eds) Renal heterogeneity and target cell toxicity. Wiley, Chichester, pp 295–298
12. Halman J, Miller J, Fowler JSL, Price RG (1986) Renal toxicity of propyleneimine: assessment by non-invasive techniques in the rat. Toxicology 41:43–59
13. Howie AJ, Kizaki T, Beaman M, Morland CM, Birtwistle RJ, Adu D, Michael J, Williams AJ, Walls J, Matsuyama M, Shimizu F (1989) Different types of segmental sclerosing glomerular lesions in six experimental models of proteinuria. J Pathol 157:141–151
14. Klingler EL, Evan AP, Anderson RE (1980) Folic acid-induced renal injury and repair. Arch Pathol Lab Med 104:87–93
15. Litterst CL, Copley MP (1985) Cisplatin and doxorubicin: a comparative evaluation of nephrotoxicity using serum chemistry, urinalysis, and histology. In: Bach PH, Lock EA (eds) Renal heterogeneity and target cell toxicity. Wiley, Chichester, pp 393–396
16. Litterst CL, Weiss RB (1987) Clinical and experimental nephrotoxicity of cancer chemotherapeutic agents. In: Bach PH, Lock EA (eds) Nephrotoxicity in the experimental and clinical situation. Nijhoff, Dordrecht, pp 771–804
17. Lock EA (1988) Studies on the mechanism of nephrotoxicity and nephrocarcinogenicity of halogenated alkenes. CRC Crit Rev Toxicol 19:23–42
18. Lock EA, Ishmael J (1979) The acute toxic effects of hexachlorobutadiene on the rat kidney. Arch Toxicol 43:47–57
19. Maruhn D, Paar D, Hartmann HG, Bock KD, Bomhard E, Lorke D (1981) Enzyme patterns of rat urine in folate-induced acute renal failure. In: Brown SS, Davies DS (eds) Organ-directed toxicity chemical indices and mechanisms (IUPAC). Pergamon, Oxford, pp 69–73
20. Maruhn D, Paar D, Bomhard E (1983) Diagnostic sensitivity of urinary enzymes in experimental kidney damage in the rat. In: Galteau MM, Siest G, Heung J (eds) Biologie Prospective—5[e] Colloque International de Pont-a-Mousson. Masson, Paris, pp 943–945
21. Mimnaugh EG, Trush MA, Gram TE (1986) A possible role for membrane lipid peroxidation in anthracycline nephrotoxicity. Biochem Pharmacol 35:4327–4335
22. Nash JA, King LJ, Lock EA, Green T (1984) The metabolism and disposition of hexachloro-1:3-butadiene in the rat and its relevance to nephrotoxicity. Toxicol Appl Pharmacol 73:124–137
23. O'Donnell MP, Michels L, Kasiske B, Raij L, Keane WF (1985) Adriamycin-induced chronic proteinuria: a structural and functional study. J Lab Clin Med 106:62–67
24. Paar D, Maruhn D, Lögering HJ, Bomhard E (1982) Kallikrein- und Urokinaseausscheidung im Harn der Ratte nach experimenteller Nierenschädigung durch Folsäure. In: Loo van de J, Asbeck F (eds) Haemostase, Thrombophilie und Arteriosklerose. Schattauer, Stuttgart, pp 794–797
25. Paar D, Maruhn D, Bock KD, Bomhard E, Lorke D (1982) Urokinase excretion after experimentally induced renal failure in the rat by folic acid. In: Mannucci PM, D'Angelo A (eds) Urokinase: basic and clinical aspects. Academic, London, pp 85–89
26. Pfaller W, Joannidis M, Gstraunthaler G, Kotanko P (1989) Quantitative morphologic changes of nephron structures and urinary enzyme activity pattern in sodium-maleate-induced renal injury. Renal Physiol Biochem 12:56–64
27. Price RG, Ellis BG (1976) Urinary enzyme excretion in aminonucleoside nephrosis in rats. Chem Biol Interact 13:353–358
28. Schmidt U, Dubach UC (1976) Acute renal failure in the folate-treated rat: early metabolic changes in various structures of the nephron. Kidney Int 10:39–45
29. Schmidt U, Schlumpf V, Joesch W, Dubach UC (1974) Acute renal failure in the rat after folate intoxication: diagnostic value of lactate dehydrogenase and alkaline phosphatase measurements in serum and urine. Clin Nephrol 2:106–112
30. Schubert GE (1976) Folic acid-induced acute renal failure in the rat: morphological studies. Kidney Int 10:46–50
31. Stonard MD, Gore CW, Oliver GJA, Smith IK (1987) Urinary enzymes and protein patterns as indicators of injury to different regions of the kidney. Fundam Appl Toxicol 9:339–351
32. Stroo WE, Hook JB (1977) Enzymes of renal origin in urine as indicators of nephrotoxicity. Toxicol Appl Pharmacol 39:423–434

Part 5

Nephrotoxicity

19 Renal Enzymes as the Basis for Alternative Methods of Safety Screening In Vitro and Understanding the Mechanisms of Nephrotoxicity*

Martin F. Wilks and Peter H. Bach

19.1 Introduction

The kidney is an organ of marked structural heterogeneity, in which over 20 different cell types work in concert to fulfill many complex functions. It is adversely affected by a wide variety of exogenous substances (many of which are used therapeutically) that target for discrete cell types whose biochemical characteristics make them susceptible to chemical insult. In addition to the need to understand the toxic effects of therapeutic agents, there are also many chemically induced nephropathies that are excellent models of renal diseases (having other causes) in humans, such as the nephrotic syndrome induced in rats using adriamycin [25]. Thus, a full understanding of the molecular processes that underlie the initial lesion and its progression can lead to the safer use of potentially nephrotoxic therapeutic agents and also contribute to our understanding of other types of nephropathies.

Part of the renal response to injury is an attempt to return function and morphology to normal and to maintain homeostasis. The mechanisms by which this is achieved are highly developed but rather poorly understood. A significant factor is that the kidney is endowed with extra functional capacity over and above that of its daily use, which allows it to call on a reserve when appropriate. Thus, a damaged kidney will often appear to be normal because of the marked functional reserve (both renal and extra-renal), which makes it difficult to diagnose renal injury. As a result, nephrotoxicity is often only recognized at an advanced stage, when progressive damage has led to a severe and irreversible impairment of renal function.

Attempts to develop safer drugs and to undertake risk assessment of chemical exposure depend on a clear identification of which cells are primarily affected, an understanding of the mechanistic basis of the injury and the ability to predict how this may affect renal integrity.

* The research described in this review was supported by the John Hopkins Center for Alternatives to Animals in Testing, Wellcome Trust, the Commission of the European Communities, the Dr Hadwen Trust for Humane Research, the Humane Research Trust and, in part, FRAME, the British Council, the Commonwealth Universities Foundation, the Cancer Research Campaign, the International Agency for Cancer Research and the National Kidney Research Fund. MFW gratefully acknowledges a fellowship from the Deutsche Forschungsgemeinschaft.

Urinary Enzymes
K. Jung/H. Mattenheimer/U. Burchardt (Eds.)

The morphological and functional heterogeneity of the kidney complicates the task of establishing the cellular targets of toxins in the kidney using standard investigation methods of this organ in vivo [14]. Concomitant factors such as age, remaining renal functional reserve capacity, underlying or pre-existing renal and/or extra-renal disease as well as multi-chemical exposure further confuse the situation [8]. Thus the quest to identify the mechanistic basis of nephrotoxicity has moved more towards those morphological techniques that give biochemical information of particular cells, such as enzyme and immuno-histochemistry. Knowledge about the heterogeneous distribution of enzyme in different regions of the nephron has played a major role in developing the concept that target toxicity relates to the biochemical characteristics of cells. In parallel, the use of the isolated perfused kidney, renal slices, glomeruli and tubular fragments and cultured cells has become increasingly popular and useful as a means to promote understanding and as a rapid and often inexpensive way of screening new drug candidates for their potential adverse effects.

19.2 In Vitro Methods for the Assessment of Nephrotoxicity

The study of nephrotoxicity in vitro has greatly advanced our knowledge of its mechanisms and their consequences [12]. More importantly, it has contributed to the development and application of novel markers for toxicity [15]. This overview aims to highlight the important contributions brought about by our understanding of the nephron distribution of enzymes. It will also counterpoint the merits and inherent problems of using measurements of some of these enzymes as markers of cytotoxicity compared with other methods in in vitro systems for toxicity screening and mechanistic studies.

A number of tests have been adapted from standard tissue culture procedures to measure cytotoxicity in isolated renal cells and fragments (Table 1). These include the assessment of cellular functions by monitoring cell growth, membrane integrity (dye uptake, enzyme and potassium release), incorporation of radiolabelled precursors into protein, DNA and RNA and morphological assessment using phase contrast as well as conventional light and electron microscopy.

The systems commonly employed to study nephrotoxicity in vitro are, in order of decreasing morphological integrity, the isolated perfused kidney, renal slices, isolated nephron segments, isolated and cultured cells, subcellular fragments and enzyme systems per se. These in vitro systems have made use of the advances in understanding and description of the physiological processes in renal cells. Methods frequently used for mechanistic investigations (Table 1) include organic ion and sugar transport, measurement of specific enzymes, cellular metabolism such as oxidation of substrates, glucose and lipid metabolism, lipid peroxidation, measurement of intracellular anti-oxidants such as glutathione and other molecules, e.g. ATP or cytochrome P-450.

Table 1. Methods commonly used to assess nephrotoxicity in vitro

Test	Technique	System
Cell growth	Microscopy	CC
Morphology	Microscopy	IPK, slices, CC
Membrane integrity	Lactate dehydrogenase leakage Dye uptake/exclusion Potassium leakage	IC, CC, slices
Transport studies	Organic anions Alpha-methylglucose	Slices, CC
Enzyme measurements	Alkaline phosphatase Na^+/K^+ ATPase Succinate dehydrogenase Maltase *N*-Acetyl-β-glucosaminidase Gamma-glutamyltransferase Leucine aminopeptidase 5′-Nucleotidase Alanine aminopeptidase Phospholipase	All, except slices
Macromolecular synthesis	Radiolabelled precursors	INS, CC
Oxidative metabolism	O_2 consumption CO_2 release	INS, CC
Lipid metabolism	Radiolabelled substrates	INS, CC
Lipid peroxidation	Malondialdehyde production	Slices, INS
Specific molecules	Glutathione Cytochrome P-450 ATP	CC
Ion deregulation	Fluorescent probes	CC

CC, cell culture; IC, isolated cells; INS, isolated nephron segments; IPK, isolated perfused kidney

19.2.1 Isolated Perfused Kidney

The isolated perfused kidney is the only in vitro system which maintains the anatomical integrity of the whole kidney and has therefore been used frequently to evaluate physiological mechanisms of renal function [63, 71, 77, 84]. Routinely measured parameters include glomerular filtration rate, perfusate flow and pressure, electrolyte excretion and reabsorption and water reabsorption [63, 71]. Apart from studying the direct toxicity of chemicals such as cadmium [1] or mercury [37], the main application of the isolated perfused kidney in nephrotoxicity studies lies in the ability to investigate the renal handling and biotransformation of xenobiotics [92] at concentrations of drugs or metabolic inhibitors which would not be possible in vivo. In addition, there are no extra-renal effects, and known concentrations of specific chemicals, including metabolites, can be presented to the cells. For example, the model of the non-filtering kidney has provided evidence for the luminal uptake of aminoglycosides [33] and has shown that filtration of cisplatin is not a prerequisite for the induction of toxicity [66]. The renal metabolism of several chemicals including acetaminophen (paracetamol) and the subsequent formation of reactive metabolites have also been evaluated in this system [35]. The main limitations preventing its more widespread application are that it requires experience, considerable technical skill and an elaborate technical setup and that its function is crucially dependent on the composition of the perfusion medium [63, 86]. Not all of the systems that have been reported provide physiologically or morphologically sound kidneys, and the application is limited to rodents. Although glomerular and tubular functions may be well preserved, there are defects in the distal tubule leading to an impaired electrolyte reabsorption and urinary concentration capacity [63]. Anoxic damage to the thick ascending limb of Henle during cell-free perfusion has been described [3, 85]. Such unavoidable artefacts could cause brush border loss and membrane fragility.

Surprisingly, the measurement of enzymes has been virtually absent from studies with the isolated perfused kidney although, in theory, it should be possible to perform enzyme analysis as easily as in vivo. Furthermore, this would guarantee the absence of interference by post-renal enzymes. This advantage has been used in the characterisation of *N*-acetyl-*β*-glucosaminidase, where the enzyme release from the isolated perfused kidney at similar rates as in situ helped to confirm its renal origin [62].

The isolated perfused kidney could therefore be a valuable tool in the characterization of novel enzymes and other markers for the study of nephrotoxicity, provided the technical limitations are taken into account.

19.2.2 Renal Slices

Renal slices have been frequently used for the study of nephrotoxicity. They are relatively straightforward to prepare and do not require major technical apparatus [23], and the technique can be applied to a number of different species. The anatomical integrity of adjacent nephron segments is at least partially maintained, and it is possible to prepare slices with a particular orientation from major regions of the kidney [81]. The main use of slices in toxicological studies is related to the transport of organic ions such as para-aminohippurate (PAH) and tetraethylammonium (TEA) in the proximal tubule. Effects of toxins such as mercuric chloride [58], potassium dichromate or cephaloridine [67] on PAH and TEA transport are among the most sensitive indicators of tubular damage. Other measurements include oxygen consumption, gluconeogenesis and morphological assessment following exposure to toxins such as heavy metals [54, 80], gentamicin [56], cisplatin [47] or cephalosporins [32]. There is, however, a paucity of data on enzyme leakage.

The cellular heterogeneity of renal slices makes it difficult to discriminate between toxicity to the proximal tubules and other nephron segments such as the glomerulus or distal tubule [13, 92]. In addition, the importance of contraluminal transport may be overemphasized in the light of the limited access of chemicals to the luminal membrane due to collapsed tubules [7]. Assessment of the release of specific marker enzymes in addition to transport measurement might be a way to define cellular targets in slices more clearly but has so far apparently not been undertaken.

19.2.3 Isolated Nephron Segments

Due to the heterogeneity of the kidney, methods have been developed to isolate relatively pure populations of nephron fragments in order to study target-selective toxicity. These include mechanical dispersion of renal tissue followed by enzymatic dispersion [39] or differential sieving [26] and gradient centrifugation [72]. Microdissection allows access to all parts of single nephrons but is time-consuming and renders little material for study [36]. The majority of toxicological studies have therefore been performed on proximal tubular fragments (which are easy to isolate) and, to a lesser extent, on glomeruli. The major limitation of isolated nephron fragments is their limited viability in vitro, which makes it difficult to study effects for longer than 4–6 h [16].

19.2.3.1 *Proximal Tubular Fragments*

The use of proximal tubular fragments offers most of the advantages of renal slices with the exception of nephron integrity and consists of a virtually pure cellular preparation if suitably prepared. Methods for further division into proximal straight and convoluted segments have recently been described [79]. Apart from transport studies similar to those in slices [83], a variety of metabolic processes have been followed for toxicity studies. Nephrotoxicity assessment by measuring cellular ATP content showed that mercuric chloride acts predominantly on the ATP content in the S2 segment [53] and that ochratoxin A tubulotoxicity affected cellular ATP in both S2 and S3 segments [52]. Tubular protein synthesis, as measured by the uptake of radiolabelled amino acids such as proline, was inhibited by various aminoglycosides [57]. Although this process was a sensitive indicator of cytotoxicity in vitro, ranking according to the inhibitory potential showed an inverse relationship to the known clinical nephrotoxicity of aminoglycosides, with streptomycin being the most potent inhibitor in the in vitro system while kanamycin had the least effect. On the other hand, fatty acid oxidation proved to be a more specific, albeit less sensitive parameter of aminoglycoside toxicity, with neomycin being a more potent inhibitor than streptomycin [57]. Similar results were obtained in studies with heavy metals, in which protein synthesis was again a more sensitive but less specific parameter of toxicity [109, 110]. These data underline the importance of the choice of suitable parameters not only for proximal tubular fragments but for all in vitro toxicity studies.

The measurement of marker enzymes is mainly used for characterization of the tubular preparation. For example, the ratio between alkaline phosphatase, a proximal brush border enzyme, and hexokinase, which is located predominantly in distal tubular segments, is a good indicator of the purity of a tubular suspension [43, 79]. In microdissected tubular segments, the activities of enzymes (gamma-glutamyltransferase, alkaline phosphatase and leucine aminopeptidase) were decreased in both S2 and S3 preparations following in vivo treatment with mercuric chloride [36]. In addition, the pattern of urinary enzyme release after mercuric chloride correlated with the intracellular enzyme activities in those two segments, but not with the one found in S1. By contrast, the pattern of urinary enzyme release following gentamicin treatment resembled better the profile found in the S1 segment, although no significant change in intracellular enzyme activity was found in the microdissected segments following gentamicin treatment. It seems, therefore, that urinary enzyme levels are a more sensitive indicator of tubular damage than the determination of the (remaining) tissue enzyme activity [36]. Release of alanine aminopeptidase, leucine aminopeptidase and alkaline phosphatase in tubule suspensions after in vitro exposure to the fungal toxin ocharatoxin A was similar to effects seen in microdissected proximal straight tubules, but the measured enzyme activities were 30%–50% lower in the tubular suspension, indicating a loss of enzyme activity following the isolation procedure [36]. The significance of these findings for studies with other nephrotoxins as well as the question of whether or not the use of different isolation procedures has an influence on the enzyme loss remain to be systematically assessed.

19.2.3.2 *Glomeruli*

Glomeruli can be isolated by the same procedure as proximal tubular fragments. They have been widely used in physiological studies on basement membrane synthesis [30], prostaglandin metabolism [95] or the action of hormones such as angiotensin II [87] and atrial natriuretic peptide [19]. Only recently have isolated glomeruli been evaluated as a tool to study the mechanisms of nephrotoxicity. Glomerular protein synthesis and fatty acid oxidation were found to be particularly sensitive to those heavy metals with the greatest potential to cause acute renal failure in the in vivo situation, such as mercuric chloride and potassium dichromate [109, 110]. These data indicate that direct glomerular toxicity may contribute to the development of acute renal failure. Indeed, parallel in vivo experiments have shown that the distribution of mercury in the early phases of insult is in fact also predominant in the juxtamedullary glomerular population (M.F. Wilks et al. 1988, unpublished observation), but the full functional significance of this observation is not yet certain. There is also evidence that the incubation of isolated glomeruli with several heavy metals leads to an impairment of [^{35}S]sulphate incorporation into proteoglycans in a manner similar to the inhibition of total protein synthesis [101]. "Softer" metal ions such as Cd^{2+} and Hg^{2+} caused a preferential decrease in the production of the more highly charged dermatan sulphate, indicating a possible effect on mesangial cells.

Isolated rat and pig glomeruli are sensitive to the anti-cancer drug adriamycin (L. Delacruz et al. 1987, unpublished data) as assessed by inhibition of protein synthesis. The highly coloured nature of this molecule precludes the assessment of many chromogen-based enzyme marker assays. In recent studies, glomeruli isolated separately from superficial and juxtamedullary nephrons were used to study the role of nephron heterogeneity in the development of nephrotoxicity. Following in vivo treatment with adriamycin, glucose oxidation was stimulated in juxtamedullary but not superficial glomeruli [55], which may be a metabolic expression of the increased protein excretion from juxtamedullary nephrons in vivo [94]. Isolated glomeruli are able to contract [24], an observation which is thought to be related to the contraction of the mesangial cells [4] and which may affect the glomerular microcirculation in vivo [114]. Cyclosporin A, which has profound effects on the renal vascular resistance [21], led to a reduction of the glomerular surface area of superficial glomeruli within 5 min of the start of incubation, whereas isolated juxtamedullary glomeruli were not affected [41]. The underlying differences in structure and/or function between juxtamedullary and superficial glomeruli which may cause the variations in response to toxins remain to be elucidated.

19.2.4 Isolated Renal Cells and Cell Culture

The isolation procedure for glomeruli and tubular fragments can be extended, for example by further collagenase digestion [42], to yield single cells in suspension. These have been used in studies to elucidate the transport of nephrotoxic halogenated hydrocarbons into tubular epithelial cells [59, 60, 61]. Isolated tubular epithelial cells can be characterized using marker enzymes such as gamma-glutamyltransferase [28], transport characteristics [20] or specific molecules such as glutathione or cytochrome P-450 [42]. Recently, a method has been described to separate cells from the convoluted and straight segments of the proximal tubule by free-flow electrophoresis [102], in which knowledge of the enzyme distribution in renal cells was used as the basis for choosing fractions that represented specific regions of the nephron. In common with isolated fragments, the use of isolated cells is limited by their relatively short-lived viability in suspension [16].

This problem can be overcome by primary culture of isolated epithelial cells, a system which is being increasingly employed in toxicity studies. Primary culture of epithelial cells is possible over several weeks, and cells can be passaged a number of times, thus yielding much more cellular material. However, the continued growth is paralleled by dedifferentiation of specific cellular functions, which makes it more difficult to extrapolate data to the in vivo situation [46]. The medium- and time-related dedifferentiation of cells has also been demonstrated by changes in several marker enzymes in primary rat proximal tubule cultures [22], data that emphasize how labile the total concentration of enzyme markers can be in cells. Morin et al. [69] have recently described the change in expression of characteristic marker enzymes in rabbit tubular epithelial cells with increasing time in culture. They showed that the activity of brush border enzymes decreased over the first 6 days in culture by 98% (alkaline phosphatase), 88% (alanine aminopeptidase) and 66% (gamma-glutamyltransferase). By contrast, glycolytic enzymes underwent a 17-fold (hexokinase) and 48-fold (lactate dehydrogenase) increase in activity compared with freshly isolated proximal tubules, indicating a shift in energy metabolism towards glucose utilisation. Similar changes have also been reported in cell lines grown in media containing different carbohydrates [38]. Cytotoxicity tests routinely employed in the study of the effects of drugs and chemicals on cultured epithelial cells include tests for membrane integrity such as dye uptake [105], lactate dehydrogenase activity and potassium leakage [90] as well as assays for characteristic properties of proximal tubular cells such as glutathione or cytochrome P-450 content [91]. The measurement of marker enzymes has also been performed but is not without problems. Acosta et al. [2] compared the effect of cephaloridine on cultured epithelial cells and found that early indicators of toxicity were the mitochondrial succinate dehydrogenase activity and the plasma membrane Na^+/K^+-ATPase activity, whereas effects on the cytoplasmic lactate dehydrogenase and the brush border enzyme alkaline phosphatase were not seen until later. In studies using cisplatin, protein, DNA and RNA synthesis (measured using incorporation of radiolabelled precursors) were affected as early as 6 h after the beginning of incubation, whereas enzyme activity (Na^+/K^+-ATPase, alkaline phosphatase, succinic dehydrogenase) was altered only after 24 h [99]. The different sensitivity of these changes may represent the primary role of cisplatin in affecting cellular genetic expression.

The use of continuous cell lines has become increasingly popular because of the virtually unlimited material available. The most commonly used renal cell lines are the LLC-PK_1 and the MDCK cells, epithelial cells derived from the pig and dog kidney, respectively. The problem of dedifferentiation of normal cellular functions is more evident with cell lines, and much work has been done to characterize these cells and define their likely origin within the nephron (for a review see [44]). The enzymatic profile of these two cell lines suggests a proximal tubular origin for the LLC-PK_1 and a distal tubular origin for the MDCK cells, although the expression of some marker enzymes notably differs from in vivo data [45]. Examples of recent toxicological studies using LLC-PK_1 cells include studies on lipid alterations after exposure to mercuric chloride [103], cellular viability upon aminoglycoside exposure [111], uptake studies with radiolabelled cisplatin [31], effects of hexachlorobutadiene conjugates on organic ion transport [65] and measurement of unscheduled DNA synthesis after exposure to halogenated hydrocarbons [104]. Despite detailed information on the enzymatic profile of LLC-PK_1 cells [45], this knowledge has so far not been employed for toxicity studies.

19.2.5 Subcellular Fragments and Purified Enzymes

The knowledge of compartmentalization of specific functions to discrete cell types and within specific organelles in renal cells has led to the development of sensitive tests for the assessment of nephrotoxicity in particular fragments of renal cells isolated by differential centrifugation. Enzyme measurement is an important parameter frequently used in studies with subcellular fragments. Aminoglycoside antibiotics have a membrane stabilizing effect on isolated lysosomes at low concentrations as assessed by the release of *N*-acetyl-β-D-glucosaminidase, while higher concentrations led to a dose-dependent release of this enzyme, paralleling the clinically observed potential for producing nephrotoxicity [76]. Comparable results were obtained when measuring the inhibitory effect of aminoglycosides on lysosomal phospholipase activity in vitro [29]. Marker enzymes have been used to characterize brush border membrane vesicles [40] employed in the study of the effect of aminoglycosides on calcium binding and transport [40]. The comparison of effects of brush border membrane and basolateral vesicles has been useful in the characterization of the tubular transport of amninoglycosides [5]. The potential of cyclosporin A to cause lipid peroxidation has been evaluated in studies with renal microsomes [50].

The main potential for the use of subcellular fragments lies in the study of known in vivo effects under well-defined experimental conditions and has helped to clarify some of the effects of proximal tubular toxins. The appropriate choice of enzymes can be used to screen for chemicals that are likely to undergo parallel metabolic activation in vitro.

Metabolic activation of chemicals is an important mechanism of nephrotoxicity, as the enzymes responsible are localized to those regions that are affected [11, 49, 82]. These systems have been studied in vitro using renal slices or cells/fragments and progressively cytosolic, mitochondrial or microsomal enzymes (as outlined below). They highlight the potential to use appropriate enzymes in a test tube. Thus, for example, isolated renal microsomes (as a source of the mixed functional oxidase) cause metabolic activation and protein binding of radiolabelled chemicals, and the inhibition of such binding may be manipulated by substances known to block these enzymes. There is a sex-related sensitivity of male mice to chloroform nephrotoxicity [75], which is based on the mixed functional oxidase (cytochrome P-450) mediated metabolic activation of this compound. Both cortical slices and microsomes from male mice [74] generate the reactive intermediate phosgene (which binds to macromolecules) in the presence of NADPH (reduced form of nicotinamide-adenine dinucleotide phosphate) and O_2, a process that can be inhibited by carbon monoxide [88, 89] and such effects can be shown to be induced by phenobarbital in microsomes from the rabbit [18]. Similarly, the target-selective toxicity of several haloalkenes for the proximal tubule is dependent on the formation of the GSH (reduced glutathione) conjugate taking place in the liver. During its renal excretion it is degraded to a cysteine conjugate that accumulates in renal cells through an anion transport mechanism and is activated by the renal enzyme cysteine conjugate β-lyase (glutamine transaminase K) which cleaves the C-S bond, generating 1 mole of pyruvate and 1 mole of ammonia plus a reactive mercaptan moiety that binds to biological macromolecules. These processes have all been studied in vitro [34, 51, 96, 112], where the use of inhibitors such as amino-oxyacetic acid (pyridoxal phosphate-dependent inhibitor of β-lyase), phenylalanylglycine (competitively blocks cysteinglycine dipeptidase) and AT-125 (irreversible inhibitor of gamma-glutamyl transferase) have all served to help elucidate the mechanism of haloalkene toxicity and in vitro test systems.

Prostaglandin synthase consists of two enzyme activities: cyclo-oxygenase and hydroperoxidase. The hydroperoxidase (and other peroxidative enzymes) has been shown to convert a number of chemicals to reactive intermediates, thus providing an explanation for toxicological processes that cannot be cytochrome P-450 mediated [64]. The prostaglandin synthase (and therefore peroxidase) localization to the inner medulla has been strongly implicated in analgesic nephropathy [10]. The metabolic activation of acetaminophen is arachidonic acid-dependent in the papilla and leads to the formation of reactive intermediates that covalently bind to medullary microsomes. Metabolic activation is reduced by inhibitors of prostaglandin synthase and by antioxidants and is thought to be mediated by a reactive intermediate of the quinone-imine type [68, 113] that can also covalently bind nucleic acid. One flaw in this argument is that acetaminophen does not cause analgesic nephropathy in animals [10, 11].

Prostaglandin-mediated peroxidation is closely paralleled by horseradish peroxidase [70] which, because of its commercial availability, has become the simplest model system for studying such processes. *N*-Phenylanthranilic acid, a model compound that causes renal papillary necrosis [48], is in the presence of horseradish peroxidase mediated peroxidation converted to about 20 metabolites [107, 108] via a quinone-imine reactive intermediate. Peroxidatively activated *N*-phenylanthranilic acid binds to purified protein and DNA in vitro (M. Feldman 1986, unpublished observation). Microsomes from pig renal cortex and medulla show the preferential arachidonate-mediated activation of this compound to an intermediate that also binds to protein (M. Feldman 1986, unpublished observation).

Thus, the use of an appropriate enzyme can give information in vitro on the metabolic activation of a chemical, and it is likely that the sophistication of such systems and the parallel use of the appropriate cell type in culture could help screen for chemicals with the potential to cause papillotoxicity.

19.3 New Developments: The Use of Fluorescent Probes

The application of a variety of vital fluorescent probes to in vitro investigations has opened up an exciting new vista for studying cells and their response to chemicals (see 6, 27, 73, 93, 100] for review). While many of these investigations have focussed on intracellular calcium and pH changes, it is important to stress that several probes do not permeate cell membranes unless their esters are used. The active probe is then released inside the cell due to the presence of non-specific esterases. Interestingly, the use of fluorescein diacetate, a probe that is often employed to assess cell viability, on freshly isolated glomeruli shows two distinct populations: those that fluoresce and those that do not (P. H. Bach 1987, unpublished observation). At present there are no morphological data to indicate the basis for these differences, but we have recently shown that the isolated juxtamedullary glomeruli are more sensitive to adriamycin treatment than those from the superficial region [55], indicating a metabolic heterogeneity of glomerular populations which may be related to the fluorescence pattern described above.

There are also a number of fluorescent probes measuring intracellular enzyme activities [9] that have been applied to freshly isolated renal glomeruli and tubular fragments, freshly isolated cells and primary and continuous lines derived from different regions of the kidney. 2′,7′-Dichlorofluorescein is oxidized to fluorescein and can be used to help identify the biochemical characteristics (such as the presence of hydrogen peroxide, organic peroxides and peroxidative enzymes) that may predispose medullary interstitial cells to chemical injury like that associated with analgesics and other papillotoxic substances [9]. Similarly, there are probes such as the resorufins that undergo microsomal dealkylation through the mixed functional oxidase system which can be used to show the localization of cytochrome P-450. A variety of substrates are available that could be used to assess cellular peptidase, β-glucuronidase and β-galactosidase fluorescent activities, although there appears to be no data on their in vitro use for studying renal cells. In our hands the methylumbelliferyl substrates for measuring sulphatase, β-glucuronidase and β-galactosidase all showed fluorescence when incubated with proximal tubular fragments and cells, but the rapid diffusion of the fluorophore away from the cell largely negates the use of these umbelliferyl derivatives (P.H. Bach 1987, unpublished observation). It does, however, highlight that other substrates may be more useful in identifying the cell type and assessing the degree of dedifferentiation or adaptation in vitro. There is a very active programme of fluorescent probe design and development, and it is likely that other probes of interest will become available for studying the enzymic processes that occur in live renal cells in vitro.

19.4 Rational Basis for the Selection of In Vitro Tests for Toxicity Studies

Nephrotoxicity is determined in vitro usually by a combination of several of the test systems detailed above. However, effects seen in vitro may not necessarily reflect the in vivo situation, and the extrapolation of data to humans in particular is, at best, difficult. The rational choice of methods for the study of target-selective nephrotoxicity has therefore to be based on the individual strengths and weaknesses of the various systems available. The major limitations in interpretation of data obtained using the above-mentioned methods include: species and strain differences in the renal response to chemicals, disruption of anatomical integrity during the isolation procedure, absence of normal renal and extra-renal xenobiotic metabolism (which may form a proximate toxin), dedifferentiation of cellular functions in vitro, an inappropriately high or low concentration of toxins in the incubation media, an abnormal route of administration of chemicals and arbitrary choice of exposure time and vehicle [8, 12]. These limitations are valid for all parameters measured in vitro, and, to a varying degree, for the various systems in use.

Bearing these potential problems in mind, the use of in vitro techniques offers unique advantages in the study of the mechanistic basis of nephrotoxicity. Thus, direct effects of chemicals on renal tissue can be evaluated, the experimental conditions can be precisely controlled and manipulated, samples may be obtained from individual animals, the interaction of chemicals can easily be evaluated and effects can be tested in well-defined areas of the kidney [12, 92].

19.5 Problem Areas and Potential Solutions

The pattern of enzyme distribution along the nephron segments has been well documented by microdissection [78] and enzyme histochemistry (see [17, 106]). Despite the very wide range of enzymes that have been reported, relatively few have been used for in vitro investigations. It is apparent from the data presented above that the potential of using renal enzyme assessment for the estimation of cytotoxic effects and the mechanistic basis of injury in vitro has not been fully realized. Despite the extensive data on renal enzyme distribution, there are still some areas of ignorance on the distribution of enzymes in individual cell types. For example, Takahashi et al. [98] have shown very high levels of adenosine deaminase and purine nucleoside phosphorylase in glomeruli. Glomeruli, however, contain at least five morphologically and biochemically distinct cell types [97], each of which will express a different profile of enzyme activities. Thus, there is a need to define in which cell type the particular enzyme activities are expressed. This will aid isolation of pure cells of each type for use as markers of target-selective cell injury in whole glomeruli, and such data can help us to understand the molecular basis of injury.

It is not clear why estimations of the function, presence, absence, leakage, induction or inhibition of renal enzymes have not been more widely used in monitoring in vitro systems and studying mechanisms of toxicity. Some of the possible explanations include the fact that most of the short-term methods of investigating renal tissues induce some cellular injury, which may be reflected as increased membrane fragility, leakage of markers and altered genetic control of enzyme systems, to mention the obvious. Some of the failure to realize the fuller potentials of this approach revolves around technical difficulties. Enzyme leakage is assessed in terms of the total activity that is available from the tissue at zero time, but this normally requires the use of sonication, detergents, etc., and there does not appear to be a clear idea of how to deal with the anomalies that are encountered but rarely reported. This includes leakage of more enzyme than that which appears to be totally available, denaturation of pro-enzymes, etc. In addition, there is the same problem that pervades the use of urinary enzymes as markers of injury in vivo i.e. the presence of proteolytic enzymes and temperature-related instability, which may under in vitro incubation conditions promote the loss of enzyme activity. Some of the compounds studied in nephrotoxicity assessment, especially the heavy metals and reactive intermediates, may also inactivate enzymes, and many chemicals interfere with spectrophotometric assays. This could explain the high background counts and variance of data that are so often encountered, and why most researchers appear to have opted for technically simple methods (such as lactate dehydrogenase and *N*-acetyl-*β*-D-glucosaminidase) to process large numbers of specimens.

There are examples of enzyme leakage data that are at variance with other types of cytotoxicity measurements. This is quite common and reflects aspects of the inherent strengths and weaknesses of each technique (as shown above): some are very sensitive, others selective, but rarely do they have both attributes, and often tests of cytotoxicity possess neither. This highlights the need to use a number of different techniques that fundamentally depend on widely different biological processes for screening. Mechanistic investigations can often be rationalized on a specific aspect of cell biophysiology, but the inclusion of other types of assessment of cellular function can be useful.

19.6 Conclusions

In vitro methods have become an integral part of research in nephrotoxicity. They offer the unique possibility of gaining insight into the primary mechanisms of renal injury, thus providing the necessary information on the formation of chemically induced lesions and the basis for the development of new diagnostic and therapeutic strategies. Knowledge of the heterogeneous distribution of enzymes along the nephron has been a key factor in the development of many of these methods. The measurement of enzymes has been frequently employed in the characterization of in vitro models. Although individual enzymes have been used as parameters in cytotoxicity assays, there are more sensitive and specific tests available to study mechanisms of nephrotoxicity. The unique value of enzyme measurement lies in the ability of enzymes to act as markers for specific nephron segments and individual cell types, thereby ensuring the quality and specificity of in vitro systems and individual preparations. It is in this respect that they will continue to play an important role in the assessment of nephrotoxicity in vitro.

The increasing availability of mono- and polyclonal antibodies will also play a key role in the identification of the cell type and its phenotypic expressions in the future. A large number of test systems and methods is available, and new developments in cellular and molecular biology will further add to the list. Individual tests offer little information per se and are open to over- and misinterpretation. Only the combination of suitable methods with constant comparison with in vivo models can provide the information necessary to extrapolate data to the human situation.

Acknowledgements. We are indebted to Gail Sutherland and Monique Rivet for help with the manuscript and to Dr Neill J Gregg for critical comments.

References

1. Abel J, Höhr D, Schurek HJ (1987) Renal handling of cadmium and cadmium-metallothionein: studies on the isolated perfused rate kidney. Arch Toxicol 60:370–375
2. Acosta D, Sorensen EMB, Anuforo DC, Mitchell DB, Ramos K, Santone KS, Smith MA (1985) An in vitro approach to the study of target organ toxicity of drugs and chemicals. In Vitro Cell Dev Biol 21:495–504
3. Alcorn D, Emslie KR, Ross BD, Ryan GB, Tange JD (1981) Selective distal nephron damage during isolated kidney perfusion. Kidney Int 19:638–647
4. Andrews PM, Coffey AK (1983) Cytoplasmic contractile elements in glomerular cells. Fed Proc 42:3046–3051
5. Aramaki Y, Takahashi M, Inaba A, Ishii Y, Tsuchiya S (1986) Uptake of aminoglycoside antibiotics into brush-border membrane vesicles and inhibition of (Na^+–K^+)-ATPase activity of basolateral membrane. Biochim Biophys Acta 862:111–118
6. Arndt-Jovin DJ, Robert-Nicoud M, Kaufman SJ, Jovin TM (1985) Fluorescence digital imaging microscopy in cell biology. Science 230:247–256
7. Arthus MF, Bergeron M, Scriver CR (1982) Topology of membrane exposure in the renal cortex slice. Studies of glutathione and maltose cleavage. Biochim Biophys Acta 692:371–376
8. Bach PH (1989) Detection of chemically induced renal injury: the cascade of degenerative morphological and functional changes that follow the primary nephrotoxic insult and evaluation of these changes by in-vitro methods. Toxicol Lett 46:237–249
9. Bach PH (1990) Fluorescent probes for understanding the mechanisms of chemically induced cellular injuries. Mar Environ Res 28:351–356
10. Bach PH, Bridges JW (1984) The role of prostaglandin synthase mediated metabolic activation of analgesics and non-steroidal anti-inflammatory drugs in the development of renal papillary necrosis and upper urothelial carcinoma. Prostaglandins Leukotrienes Med 15:251–274
11. Bach PH, Bridges JW (1985) Chemically induced renal papillary necrosis and upper urothelial carcinoma. CRC Crit Rev Toxicol 15:217–439
12. Bach PH, Kwizera EN (1988) Nephrotoxicity: a rational approach to target cell injury in vitro in the kidney. Xenobiotica 18:685–698
13. Bach PH, Lock EA, (1982) The use of renal tissue slices, perfusion and infusion techniques to assess renal function and malfunction. In: Bach PH, Bonner FW, Bridges JW, Lock EA (eds) Nephrotoxicity, assessment and pathogenesis. Wiley, Chichester, pp 128–143
14. Bach PH, Lock EA, (eds) (1985) Renal heterogeneity and target cell toxicity. Wiley, Chichester
15. Bach PH, Lock EA, (eds) (1989) Nephrotoxicity: in vitro to in vivo, animals to man. Plenum, New York
16. Bach PH, Ketley CP, Dixit M, Ahmed I (1985) The use of isolated and cultured renal cells in

nephrotoxicity—practice, potential and problems. In: Bach PH, Lock EA (eds) Renal heterogeneity and target cell toxicity. Wiley, Chichester, pp 505–518
17. Bach PH, Gregg NJ, Wachsmuth ED (1987) The application of histochemistry at the light microscopic level to the study of nephrotoxicity. In: Bach PH, Lock EA (eds) Nephrotoxicity in the experimental and the clinical situation. Nijhoff, Dordrecht, pp 19–84
18. Bailie MB, Smith JH, Newton JF, Hook JB (1984) Mechanism of chloroform nephrotoxicity. IV. Phenobarbital potentiation of in vitro chloroform metabolism and toxicity in rabbit kidneys. Toxicol Appl Pharmacol 74:285–292
19. Barrio V, De Arriba G, Lopez-Novoa JM, Rodriguez-Puyol D (1987 Atrial natriuretic peptide inhibits glomerular contraction induced by angiotensin II and platelet activating factor. Eur J Pharmacol 135:93–96
20. Barron ET, Pratt I, Ryan MP (1991) Development of a primary rat cell culture model for investigations of renal proximal tubular damage. In: Bach PH, Delacruz L, Gregg NJ, Wilks MF (eds) Nephrotoxicity: the basis for early diagnosis, modulation of effects and improved therapeutic management. Dekker, New York pp 439–441
21. Barros EJG, Boim MA, Ajzen H, Ramos OL, Schor N (1987) Glomerular hemodynamics and hormonal participation on cyclosporine nephrotoxicity. Kidney Int 32:19–25
22. Bellemann P (1980) Primary monolayer culture of liver parenchymal cell and kidney cortical tubules as a useful new model for biochemical pharmacology and experimental toxicology. Arch Toxicol 44:63–84
23. Berndt WO (1976) Use of the tissue slice technique for evaluation of renal transport processes. Environ Health Perspect 15:73–88
24. Bernik MB (1969) Contractile activity of human glomeruli in culture. Nephron 6:1–10
25. Bertani T, Poggi A, Pozzoni R, Delaini F, Sacchi G, Thoua Y, Mecca G, Remuzzi G, Donati MB (1982) Adriamycin induced nephrotic syndrome in rats. Sequence of pathological events. Lab Invest 46:16–23
26. Brendel K, Meezan E (1975) Isolation and properties of proximal kidney tubules obtained without collagenase treatment. Fed Proc 34:308
27. Bright GR, Rogowska J, Fisher GW, Taylor DL (1987) Fluorescence ratio imaging microscopy temporal and spatial measurements in single living cells. BioTechniques 5:556–563
28. Cal JC, Merlet D, Cambar J (1989) In vitro and in vivo $HgCl_2$-induced nephrotoxicity assessed by tubular enzymes release. In: Bach PH, Lock EA (eds) Nephrotoxicity: in vitro to in vivo, animals to man. Plenum, New York, pp 99–106
29. Carlier MB, Laurent G, Tulkens P (1984) In vitro inhibition of lysosomal phospholipases by aminoglycoside antibiotics: a comparative study. Arch Toxicol [Suppl] 7:282–285
30. Carlson EC (1987) Morphology of isolated renal tubular and glomerular basement membranes. In: Price RG, Hudson BG (eds) Renal basement membranes in health and disease. Academic, London, pp 99–113
31. Casey B, McGuinness S, Pratt I, Ryan MP, McAuliffe CA, Sharma HL, Tinker ND (1989) Uptake of cisplatin (^{195m}Pt) into $LLCPK_1$ cells in the presence of diethyldithiocarbamate (DDTC), mercaptoethanesulphate (MESNA) and amiloride. In: Bach PH, Lock EA (eds) Nephrotoxicity: in vitro to in vivo, animals to man. Plenum, New York, pp 353–356
32. Cojocel C, Gottsche U, Tollt K, Baumann K (1988) Nephrotoxic potential of first-, second- and third-generation cephalosporins. Arch Toxicol 62:458–464
33. Collier VU, Lietman PS, Mitch WE (1979) Evidence for luminal uptake of gentamicin in the perfused rat kidney. J Pharmacol Exp Ther 210:247–251
34. Dohn DR, Anders MW (1982) Assay of cysteine conjugate β-lyase activity with S-(2-benzothiazolyl) cysteine as the substrate. Anal Biochem 120:379–386
35. Emslie KR, Calder IC, Hart SJ, Tange JD (1981) Induction of paracetamol metabolism in the isolated perfused kidney. Xenobiotica 11:579–587
36. Endou H, Koseki C, Yamada H, Obara T (1986) Evaluation of nephrotoxicity using isolated nephron segments. In: Tanabe T, Hook IB, Endou H (eds) Nephrotoxicity of antibiotics and immunosuppressants. Elsevier, New York, pp 207–216
37. Endre ZH, Nicholls LG, Ratcliffe PJ, Ledingham JGG (1989) Prevention and reversal and mercuric chloride-induced increases in renal vascular resistance by captopril. In: Bach PH, Lock EA (eds) Nephrotoxicity: in vitro to in vivo, animals to man. Plenum, New York, pp 103–106
38. Frick H, Gstraunthaler G, Pfaller W (1989) Modification of membrane protein expression and protein secretion in $LLC\text{-}PK_1$ grown on different carbohydrates. Renal Physiol Biochem 12:393–399

39. Fry JR, Wiebkin P, Kao J, Jones CA, Gwynn J, Bridges JW (1978) A comparison of drug-metabolizing capability in isolated viable rat hepatocytes and renal tubule fragments. Xenobiotica 8:113–120
40. Godson C, Ryan MP (1989) Investigation of gentamicin nephrotoxicity using renal brush border membrane vesicles. In: Bach PH, Lock EA (eds) Nephrotoxicity: in vitro to in vivo, animals to man. Plenum, New York, pp 247–252
41. Gonzales R, Lakhdar B, Redon P, Potaux L, Cambar J, Aparicio M (1989) Differential effects of cyclosporin on response of superficial and juxtamedullary isolated human glomeruli. Nephrol Dial Transplant 4:513–514 (abstr)
42. Gordon EM, Whiting PH, Simpson JG, Hawksworth GM (1989) Isolated rat renal proximal tubular cells: a model to investigate drug induced nephrotoxicity. In: Bach PH, Lock EA (eds) Nephrotoxicity: in vitro to in vivo, animals to man. Plenum, New York, pp 725–730
43. Green CE, Dabbs JE, Allen KL, Tyson CA, Rauckman EJ (1989) Characterization of isolated renal proximal tubules for nephrotoxicity studies. In: Bach PH, Lock EA (eds) Nephrotoxicity: in vitro to in vivo, animals to man. Plenum, New York, pp 719–723
44. Gstraunthaler GJA (1988) Epithelial cells in tissue culture. Renal Physiol Biochem 11:1–42
45. Gstraunthaler GJA, Pfaller W, Kotanko P (1985) Biochemical characterization of renal epithelial cell cultures (LLC-PK_1 and MDCK). Am J Physiol 248:F536–F544
46. Handler JS (1986) Studies of kidney cells in culture. Kidney Int 30:208–215
47. Hannemann J, Baumann K (1988) Cisplatin-induced lipid peroxidation and decrease of gluconeogenesis in rat kidney cortex: different effects of antioxidants and radical scavengers. Toxicology 51:119–132
48. Hardy TL, Bach PH (1984) The effects of *N*-phenylanthranilic acid-induced renal papillary necrosis on urinary acidification and renal electrolyte handling. Toxicol Appl Pharmacol 75:265–277
49. Hook JB, McCormack KM, Kluwe WM (1979) Biochemical mechanisms of nephrotoxicity. Rev Biochem Toxicol 1:53–78
50. Inselmann G, Blank M, Baumann K (1989) Cyclosporin A-induced lipid peroxidation in rat renal microsomes and effect on glucose uptake by renal brush border membrane vesicles. In: Bach PH, Lock EA (eds) Nephrotoxicity: in vitro to in vivo, animals to man. Plenum, New York, pp 303–308
51. Jaffe DR, Hassall CD, Brendel K, Gandolfi AJ (1983) In vivo and in vitro nephrotoxicity of the cysteine conjugate of hexachlorobutadiene. J Toxicol Environ Health 11:857–867
52. Jung KY, Endou H (1989) Nephrotoxicity assessment by measuring cellular ATP content. II. Intranephron site of ochratoxin A nephrotoxicity. Toxicol Appl Pharmacol 100:383–390
53. Jung KY, Uchida S, Endou H (1989) Nephrotoxicity assessment by measuring cellular ATP content. I. Substrate specificities in the maintenance of ATP content in isolated rat nephron segments. Toxicol Appl Pharmacol 100:369–382
54. Kacew S, Hirsch GH (1981) Evaluation of nephrotoxicity of various compounds by means of in vitro techniques and comparison to in vivo methods. In: Hook JB (ed) Toxicology of the kidney. Raven, New York, pp 77–98
55. Kastner S, Wilks MF, Soose M, Bach PH, Stolte H (1991) Metabolic studies on isolated rat glomeruli; a valuable tool to investigate glomerular damage. In: Bach PH, Gregg NJ, Wilks MF Delacruz L (eds) Nephrotoxicity: the basis for early diagnosis, modulation of effects and improved therapeutic management. Dekker, New York pp 467–473
56. Kluwe WM, Hook JB (1978) Functional nephrotoxicity of gentamicin in the rat. Toxicol Appl Pharmacol 35:355–364
57. Kwizera EN, Wilks MF, Bach PH (1990) Effects of aminoglycosides on the incorporation of amino acids and on fatty acid oxidation in freshly isolated rat renal proximal tubules. In Vitro Toxicol 3:243–253
58. Kyle GM, Luthra R, Bruckner JV, Mackenzie WF, Acosta D (1983) Assessment of functional, morphological and enzymatic tests for acute nephrotoxicity induced by mercuric chloride. J Toxicol Environ Health 12:99–117
59. Lash LH, Anders MW (1986) Bioactivation and cytotoxicity of nephrotoxic amino acid and glutathione S-conjugates. Comments Toxicol 1:87–107
60. Lash LH, Anders MW (1989) Uptake of nephrotoxic S-conjugates by isolated rat renal proximal tubular cells. J Pharmacol Exp Ther 248:531–537
61. Lash LH, Jones DP (1985) Uptake of the glutathione conjugate *S*-(1,2-dichlorovinyl)glutathione by renal basal-lateral membrane vesicles and isolated kidney cells. Mol Pharmacol 28:278–282

62. Lockwood TD, Bosmann HB (1979) The use of urinary *N*-acetyl-*β*-glucosaminidase in human renal toxicology. I. Partial biochemical characterization and excretion in humans and release from the isolated perfused rat kidney. Toxicol Appl Pharmacol 49:323–336
63. Maack T (1980) Physiological evaluation of the isolated perfused rat kidney. Am J Physiol 238:F71–78
64. Mernett LJ, Eling TE (1983) Cooxidation during prostaglandin biosynthesis: a pathway for the metabolic activation of xenobiotics. In: Hodgson E, Bend JR, Philpot RM (eds) Reviews in biochemical toxicology, vol 5. Elsevier, New York, pp 135–172
65. Mertens JJWM, Weijnen JGJ, van Doorn WJ, Spenkelink B, Ternink JHM, van Bladeren PJ (1989) Use of LLC-PK_1 monolayers as an in vitro model for nephrotoxicity. In: Bach PH, Lock EA (eds) Nephrotoxicity: in vitro to in vivo, animals to man. Plenum, New York, pp 591–594
66. Miura K, Goldstein RS, Pasino DA, Hook JB (1987) Cisplatin nephrotoxicity: role of filtration and tubular transport of cisplatin in isolated perfused kidneys. Toxicology 44:147–158
67. Miyajima H, Hewitt WR, Cote MG, Plaa GL (1983) Relationships between histological and functional indices of acute chemically induced nephrotoxicity. Fund am Appl Toxicol 3:543–551
68. Mohandas J, Duggin GG, Horvath JS, Tiller DJ (1981) Metabolic oxidation of acetaminophen (paracetamol) mediated by cytochrome P-450 mixed function oxidase and prostaglandin endoperoxide synthase in rabbit kidney. Toxicol Appl Pharmacol 61:252–259
69. Morin JP, Toutain H, Vouclin-Jacques N, Fillastre JP (1991) Biochemical and functional characterisation of a primary culture of rabbit proximal tubule cells: a time-course study. In: Bach PH, Delacruz L, Gregg NJ, Wilks MF (eds) Nephrotoxicity: the basis for early diagnosis, modulation of effects and improved therapeutic management. Dekker, New York pp 433–438
70. Nelson SD, Dahlin DC, Rauckman EJ, Rosen GM (1981) Peroxidase-mediated formation of reactive metabolites of acetaminophen. Mol Pharmacol 20:195–199
71. Newton JF, Hook JB (1981) The isolated perfused kidney. Methods Ensymol 77:94–105
72. Norgaard JOR (1976) A new method for the isolation of ultrastructurally preserved glomeruli. Kidney Int 9:278–285
73. Phelps PC, Smith MW, Trump BF (1989) Cytosolic ionized calcium and bleb formation after acute cell injury of cultured rabbit renal tubule cells. Lab Invest 60:630–642
74. Pohl LR (1979) Biochemical toxicology of chloroform. Rev Biochem Toxicol 1:79–107
75. Pohl LR, George JW, Satoh H (1984) Strain and sex differences in chloroform induced nephrotoxicity. Different rates of metabolism of chloroform to phosgene by the mouse kidney. Drug Metab Dispos 12:304–308
76. Powell JH, Reidenberg MM (1982) In vitro response of rat and human kidney lysosomes to aminoglycosides. Biochem Pharmacol 31:3447–3453
77. Ross BD (1978) The isolated perfused rat kidney. Clin Sci Mol Med 55:513–521
78. Ross BD, Guder WG (1982) Heterogeneity and compartmentation in the kidney. In: Sies H (ed) Metabolic compartmentation. Academic, New York, pp 363–409
79. Ruegg CE, Mandel LJ (1991) Bulk isolation of straight and convoluted proximal tubules from rabbit kidneys: differential glucose-dependent protection during anoxia. In: Bach PH, Gregg NJ, Wilks MF, Delacruz L, (eds) Nephrotoxicity: the basis for early diagnosis, modulation of effects and improved therapeutic management. Dekker, New York pp 459–465
80. Ruegg CE, Gandolfi AJ, Nagle RB, Brendel K (1987) Differential patterns of injury to the proximal tubule of renal cortical slices following in vitro exposure to mercuric chloride, potassium dichromate or hypoxic conditions. Toxicol Appl Pharmacol 90:261–273
81. Ruegg CE, Gandolfi AJ, Nagle RB, Krumdieck CL, Brendel K (1987) Preparation of positional renal slices for study of cell-specific toxicity. J Pharmacol Methods 17:111–123
82. Rush GF, Smith JH, Newton JF, Hook JB (1984) Chemically induced nephrotoxicity: role of metabolic activation. CRC Crit Rev Toxicol 13:99–160
83. Rylander LA, Phelps JS, Gandolfi AJ, Brendel K (1987) In vitro nephrotoxicity: response of isolated renal tubules to cadmium chloride and dichlorovinyl cysteine. In Vitro Toxicol 1:111–127
84. Schurek HJ (1980) Application of the isolated perfused rat kidney in nephrology. Contrib Nephrol 19:176–190
85. Schurek HJ (1988) Die Nierenmarkhypoxie: ein Schlüssel zum Verständnis des akuten Nierenversagens? Klin Wochenschr 66:828–835
86. Schurek HJ, Alt JM (1981) Effect of albumin on the function of perfused rat kidney. Am J Physiol 240:F569–F576

87. Skorecki KL, Ballermann BJ, Rennke HG, Brenner BM (1983) Angiotensin II receptor regulation in isolated renal glomeruli. Fed Proc 42:3064–3071
88. Smith JH, Hook JB (1983) Mechanism of chloroform nephrotoxicity. II. In vitro evidence for renal metabolism of chloroform in mice. Toxicol Appl Pharmacol 70:480–485
89. Smith JH, Hook JB (1984) Mechanism of chloroform nephrotoxicity. III. Renal and hepatic microsomal metabolism of chloroform in mice. Toxicol Appl Pharmacol 73:511–524
90. Smith MA, Acosta D (1987) Cephaloridine toxicity in primary cultures of rat renal cortical epithelial cells. Toxic in Vitro 1:23–29
91. Smith MA, Acosta D, Brucknen JV (1986) Development of a primary culture system of rat kidney cortical cells to evaluate the nephrotoxicity of xenobiotics. Food Chem Toxicol 24:551–556
92. Smith MA, Hewitt WR, Hook JB (1988) In vitro methods in renal toxicology. In: Atterwill CK, Steele CE (eds) In vitro methods in toxicology. Cambridge University Press, pp 13–35
93. Smith MW, Ambudkar IS, Phelps PC, Regec AL, Trump BF (1987) $HgCl_2$-induced changes in cytosolic Ca^{2+} of cultured rabbit renal tubular cells. Biochim Biophys Acta 931:130–142
94. Soose M, Haberstroh U, Rovira-Halbach G, Brunkhorst R, Stolte H (1988) Heterogeneity of glomerular barrier function in early adriamycin nephrosis of MWF rats. Clin Physiol Biochem 6:310–315
95. Sraer JD, Baud L, Sraer J, Delarue F, Ardaillou R (1982) Stimulation of PGE_2 synthesis by mercuric chloride in rat glomeruli and glomerular cells in vitro. Kidney Int 21:S63–S68
96. Stevens JL (1985) Cysteine conjugate β-lyase activities in rat kidney cortex: subcellular localisation and relationship to the hepatic enzyme. Biochem Biophys Res Commun 129:499–504
97. Striker GE, Lange MA, MacKay K, Bernstein K, Striker LJ (1987) Glomerular cells in vitro. Adv Nephrol 16:169–186
98. Takahashi T, Kakuno K, Yamada H, Endou H (1989) Intranephron distribution of purine-metabolizing enzymes in rats. Renal Physiol Biochem 12:287–294
99. Tay LK, Bregman CL, Masters BA, Williams PD (1988) Effects of *cis*-diamminedichloroplatinum (II) on rabbit kidney in vivo and on rabbit renal proximal tubule cells in culture. Cancer Res 48:2538–2543
100. Taylor DL, Salmon ED, (1989) Fluorescence microscopy of living cells in culture. Methods Cell Biol 29(A):20–23
101. Templeton DM, Chaitu N (1990) Effects of divalent metals on the isolated rat glomerulus. Toxicology 61: 119–133
102. Toutain H, Morin JP (1990) Separation of cells from the convoluted and straight portions of the proximal tubule from the rabbit kidney by free-flow electrophoresis. In: Bach PH, Gregg NJ, Wilks ME, Delacruz L, (eds) Nephro-toxicity: the basis for early diagnosis, modulation of effects and improved therapeutic management. Dekker, New York 451–457
103. Troyer DA, Kreisberg JI, Venkatachalam MA (1986) Lipid alterations in LLC-PK_1 cells exposed to mercuric chloride. Kidney Int 29:530–538
104. Vamvakas S, Dekant W, Schiffmann D, Henschler D (1989) Characterization of an unscheduled DNA synthesis assay with a cultured line of porcine kidney cells (LLC-PK_1). In: Bach PH, Lock EA (eds) Nephrotoxicity: in vitro to in vivo, from animals to man. Plenum, New York, pp 749–754
105. Viano I, Eandi M, Santiano M (1983) Toxic effects of some antibiotics on rabbit kidney cells. Int J Tissue React V:181–186
106. Wachsmuth ED (1985) Renal heterogeneity at a light microscopic level. In: Bach PH, Lock EA (eds) Renal heterogeneity and target cell toxicity. Wiley, Chichester, pp 13–30
107. Whittingham A (1987) Metabolism of *N*-phenylanthranilic acid and peroxidation of arylamines. Master's thesis, University of Surrey, Guildford
108. Whittingham A, Delacruz L, Gregg NJ, Albores A, Ijomah P, Bach PH (1989) The kinetics of papillotoxic doses of ^3H-*N*-phenylanthranilic acid in rats. Renal Physiol Biochem 12:406–412
109. Wilks MF, Kwizera EN, Bach PH (1990) Assessment of heavy metal nephrotoxicity in vitro using isolated rat glomeruli and proximal tubular fragments. Renal Physiol Biochem 13:275–284
110. Wilks MF, Kwizera EN, Bach PH (1990) Effects of heavy metals on metabolic functions in isolated glomeruli and proximal tubular fragments. In: Bach PH, Gregg NJ, Wilks MF, Delacruz L, (eds) (1991) Nephrotoxicity: the basis for early diagnosis, modulation of effects and improved therapeutic management. Dekker, New York pp 363–366

111. Williams PD, Laska DA, Hottendorf GH (1986) Comparative toxicity of aminoglycoside antibiotics in cell cultures derived from human and pig kidney. In Vitro Toxicol 1:23–32
112. Wolf CR, Berry PN, Nash JA, Green T, Lock EA (1984) Role of microsomal and cytosolic glutathione-S-transferases in the conjugation of hexachloro-1,3-butadiene and its possible relevance to toxicity. J Pharmacol Exp Ther 228:202–208
113. Zenser TV, Mattamal MB, Brown WW, Davis BB (1979) Cooxygenation by prostaglandin cyclooxygenase from rabbit inner medulla. Kidney Int 16:688–694
114. Zimmerhackl B, Parekh N, Kücherer H, Steinhausen M (1985) Influence of systemically applied angiotensin II on the microcirculation of glomerular capillaries in the rat. Kidney Int 27:17–24

Subject Index

MIX
Papier aus verantwortungsvollen Quellen
Paper from responsible sources
FSC® C105338

FSC
www.fsc.org

If you have any concerns about our products,
you can contact us on
ProductSafety@springernature.com

In case Publisher is established outside the EU,
the EU authorized representative is:
Springer Nature Customer Service Center GmbH
Europaplatz 3, 69115 Heidelberg, Germany

Printed by Libri Plureos GmbH
in Hamburg, Germany